Biology

... or GCE Biology A

WITHDRAWN

8 .12.16

Askham Bryan College

072249

ASK...

LEA...... ...JURCES

Heinemann is an imprint of Pearson Education Limited, a company incorporated in England and Wales, having its registered office at Edinburgh Gate, Harlow, Essex CM20 2JE. Registered company number: 872828

www.heinemann.co.uk

Heinemann is a registered trademark of Pearson Education Limited

Text © Pearson Education Ltd 2008

First published 2008

12 11 10 09 08
10 9 8 7 6 5 4 3 2 1

British Library Cataloguing in Publication Data
A catalogue record for this book is available from the British Library

ISBN 978 0 435691 90 5

Copyright notice
All rights reserved. No part of this publication may be reproduced in any form or by any means (including photocopying or storing it in any medium by electronic means and whether or not transiently or incidentally to some other use of this publication) without the written permission of the copyright owner, except in accordance with the provisions of the Copyright, Designs and Patents Act 1988 or under the terms of a licence issued by the Copyright Licensing Agency, Saffron House, 6–10 Kirby Street, London EC1N 8TS (www.cla.co.uk). Applications for the copyright owner's written permission should be addressed to the publisher.

Edited by Liz Jones
Index compiled by Wendy Simpson
Designed by Kamae Design
Project managed and typeset by Wearset Ltd, Boldon, Tyne and Wear
Original illustrations © Pearson Education Limited 2008
Illustrated by Wearset Ltd, Boldon, Tyne and Wear
Picture research by Wearset Ltd, Boldon, Tyne and Wear
Cover photo showing a cross-sectional MRI scan of a human brain © Neil Borden/Science Photo Library
Printed in China (SWTC/01)

Acknowledgements
We would like to thank the following for their invaluable help in the development and trialling of this course: Alan Cadogan, Rob Duncan, Richard Fosbery, Amanda Hawkins, Dave Keble, Maggie Perry, Maggie Sindall and Jenny Wakefield-Warren.

The authors and publisher would like to thank the following for permission to reproduce photographs:

p3 Medi-Mation/Science Photo Library; p4 M John Beatty/Science Photo Library; p4 MR W.K. Fletcher/Science Photo Library; p9 Mickey McLean; p24 Biophoto Associates/Science Photo Library; p29 Zephyr/Science Photo Library; p32 Conge, ISM/Science Photo Library; p35 Medi-Mation/Science Photo Library; p38 B M.I. Walker/Science Photo Library; p38 BR Biophoto Associates/Science Photo Library; p42 T SIU/Science Photo Library; p42 B Manfred Kage/Science Photo Library; p43 CNRI/Science Photo Library; p57 Eric Grave/Science Photo Library; p58 T Eye Of Science/Science Photo Library; p58 B Jeff Kubina; p59 M M.I. Walker/Science Photo Library; p59 TR Michael Abbey/Science Photo Library; p60 T Dr Kari Lounatmaa/Science Photo Library; p60 M Francis Leroy, Biocosmos/Science Photo Library; p67 Ludek Pesek/Science Photo Library; p79 Georgette Douwma/Science Photo Library; p86 ISM/Science Photo Library; p93 Tim Vickers; p94 S. Purdy Matthews/Getty Images; p95 Andrew Syred/Science Photo Library; p103 Alfred Pasieka/Science Photo Library; p104 Alfred Pasieka/Science Photo Library; p105 Professor Oscar L. Miller/Science Photo Library; p106 Hybrid Medical Animation/Science Photo Library; p107 M Dr Elena Kiseleva/Science Photo Library; p107 MR Dr Elena Kiseleva/Science Photo Library; p108 TEK Image/Science Photo Library; p110 Janice Fee; p111 Louise Murray/Science Photo Library; p115 Eye of Science/Science Photo Library; p117 PSChemp; p118 Adrian T. Sumner/Science Photo Library; p122 Ron Buskirk/Alamy; p123 Georgette Doumwa/Science Photo Library; p126 M Eye Of Science/Science Photo Library; p126 MR Sue Ford/Science Photo Library; p127 Robert Scarth; p128 Michael P. Gadomski/Science Photo Library; p129 T Kenneth H. Thomas/Science Photo Library; p129 TR Johan Openheimer; p129 M Mike Spence/Alamy; p130 Jeanot; p132 T Oak Ridge National Laboratory/US Dept of Energy/Science Photo Library; p132 T Blickwinkel/Alamy; p140 Ysangsok; p141 Photofrenetic/Alamy; p144 T C. Michael Hogan; p144 M James King-Holmes/Science Photo Library; p145 James King-Holmes/Science Photo Library; p145 David Monninaux; p151 Sinclair Stammers/Science Photo Library; p152 Pete Kennedy; p154 Paul Connell; p155 Science Photo Library; p158 US Federal Govt; p159 T Peggy Greb/US Department of Agriculture/Science Photo Library; p159 M Michael Abbey/Science Photo Library; p160 Wildscape/Alamy; p162 Pete Kennedy; p163 Pete Kennedy; p164 T Victor De Schwanberg/Science Photo Library; p164 M PHOTOTAKE Inc./Alamy; p164 BR Colin Underhill/Alamy; p164 B United States Department of Agriculture; p167 TL Dr E. Walker/Science Photo Library; p167 TR Paul Asman and Jill Lenoble; p167 BL Dr Gary Guegler/Science Photo Library; p167 BR Jeremy Walker/Science Photo Library; p180 The Golden Rice Humanitarian Board; p191 David Hay Jones/Science Photo Library; p192 Mitch Reardon/Science Photo Library; p196 M Artic_Images/CORBIS; p196 B Rosenfeld Images Ltd/Science Photo Library; p197 T Bill Barksdale/AGStockUSA/Science Photo Library; p197 M R. Maisonneuve, Publiphoto Diffusion/Science Photo Library; p198 M Pierre Vauthey/Corbis SYGMA; p198 B Robert Gill PAPILIO/Corbis; p200 Pete Kennedy; p201 M Pete Kennedy; p201 B Martyn F. Chillmaid/Science Photo Library; p203 Dr Jermey Burgess/Science Photo Library; p206 M.I. Walker/Science Photo Library; p208 T Bob Gibbons/Science Photo Library; p208 B Leslie J. Borg/Science Photo Library; p209 TM David Aubrey/Science Photo Library; p209 TR Mike Read/rspb-images; p210 M Nigel Cattlin/ALAMY; p210 B Geoff Kidd/Science Photo Library; p211 John and Karen Hollingsworth; p212 T Brandon Cole/Visuals Unlimited, Inc.; p212 M Sue Ford/Science Photo Library; p212 B Peter Arnold, Inc./Alamy; p219 Martin Shields/Science Photo Library; p221 B J.C. Revy/Science Photo Library; p221 BM J.C. Revy/Science Photo Library; p222 M.I. Walker/Science Photo Library; p224 Nigel Cattlin/Visuals Unlimited, Inc.; p225 Sylvan Wittwer/Visuals Unlimited, Inc.; p226 T Martyn F. Chillmaid/Science Photo Library; p226 B Ria Novosti/Science Photo Library; p226 BM Alvis Upitis/AGSTOCKUSA/Science Photo Library; p227 P.G. Adam, Publiphoto Diffusion/Science Photo Library; p235 Dr Gladden Willis; p239 Josh Plueger; p241 Richard Becker/Alamy; p243 Walter Dawn/Science Photo Library; p243 UCL Institute of Cognitive Neuroscience; p244 Paul Souders/Corbis

The authors and publisher would like to thank the following for permission to use copyright material:

WoltersKluwer (Figure 3, p169), Henry Jakubowski (Figure 1, p175), W.H. Freeman and Company (Figure 2 , p177), GCS Research Society (Figure 2, p181), Danny Nicholson (Figure 1, p232), Rwandan Gorilla Project (Figure 1, p244)

Every effort has been made to contact copyright holders of material reproduced in this book. Any omissions will be rectified in subsequent printings if notice is given to the publisher.

Websites
There are links to websites relevant to this book. In order to ensure that the links are up-to-date, that the links work, and that links are not inadvertently made to sites that could be considered offensive, we have made the links available on the Heinemann website at www.heinemann.co.uk/hotlinks. When you access the site, the express code is 1905P.

Exam Café student CD-ROM
© Pearson Education Limited 2008
The material in this publication is copyright. It may be edited and printed for one-time use as instructional material in a classroom by a teacher, but may not be copied in unlimited quantities, kept on behalf of others, passed on or sold to third parties, or stored for future use in a retrieval system. If you wish to use the material in any way other than that specified you must apply in writing to the publisher.

Original illustrations, screen designs and animation by Michael Heald
Photographs © iStock Ltd
Developed by Elektra Media Ltd

Technical problems
If you encounter technical problems whilst running this software, please contact the Customer Support team on 01865 888108 or email software.enquiries@heinemann.co.uk

OCR Biology

A2

Exclusively endorsed by OCR for GCE Biology A

Sue Hocking
Pete Kennedy
Frank Sochacki
Mark Winterbottom
Series editor: Sue Hocking

www.heinemann.co.uk
✓ Free online support
✓ Useful weblinks
✓ 24 hour online ordering

01865 888080

In Exclusive Partnership

Contents

Contents

Introduction

How to use this book

In this book you will find a number of features planned to help you.

- **Module opener pages** – these carry an introductory paragraph to set the context for the topics covered in the module. They also have a short set of questions that you should already be able to answer from your previous science courses or from your general knowledge.
- **Double-page spreads** filled with information about each topic together with some questions you should be able to answer when you have worked through the spread. The final question may be more challenging.
- **End-of-module summary pages** to help you link all the topics within each module together.
- **End-of-module examination questions** selected to show you the types of question that may appear in your examination.

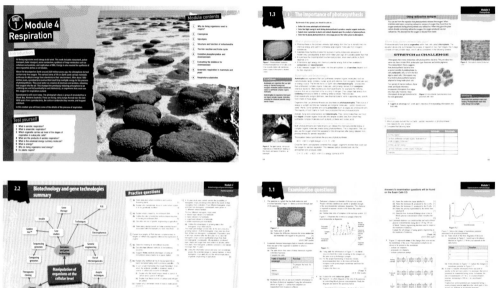

Within each double-page spread you will find other features to highlight important points.

Learning objectives

How Science Works

Term in bold

Stretch and Challenge

Questions

- **Learning objectives** – these are taken from the Biology A2 specification to highlight what you need to know and to understand.
- **Key definitions** – these terms appear in the specification. You must know the definitions and how to use them.
- **Terms in bold** – these draw attention to terms that you are expected to know. These are important terms with specific meanings used by biologists. Each term in bold is listed in the glossary at the end of the book.
- **Examiner tips** – these will help you avoid making common errors in the examination.
- **Worked examples** – these show you how calculations should be set out.
- **How Science Works** – this book has been written to reflect the way that scientists work. Certain sections have been highlighted as good examples of how science works.
- **Stretch and Challenge** – these boxes will help you to develop the skills you need to tackle stretch and challenge questions on your final examination. You do not have to learn extra information, but you will need to make links across work you've met at AS and A2 and to be able to apply your existing knowledge in unfamiliar contexts.
- **Questions** – there are a few questions at the end of each spread that you should be able to answer after studying the spread.

The examination

It is useful to know some of the language used by examiners. Often this means just looking closely at the wording used in each question on the paper. When you first read a question, do so carefully. Once you have read something incorrectly, it is very difficult to get the incorrect wording out of your head.

- Look at the number of **marks allocated** to a part question – ensure you write enough points to gain these marks. Do not repeat yourself. Different marks are for different ideas. The number of marks is a guide to the depth of treatment required for the answer.
- Look for words in **bold**. These are meant to draw your attention.
- **Diagrams and tables** often help communicate an idea. In longer answers you may gain marks by making an *annotated* diagram.
- If you are given **data** in tables or graphs, read the queston carefully and, if it asks you to describe the data, make data quotes to illustrate your answer. If it asks you to explain the data, give reasons for the trends you see.
- Write legibly. You cannot be given any marks for something that is illegible.

Eash question has an **action word**. Some action words are explained below.

- *Define:* only a formal statement of a definition is required.
- *Explain:* a supporting argument is required using your knowledge of biology. The depth of the answer should be judged from the mark allocated.
- *State:* a concise answer is expected, with little or no supporting argument.
- *List:* give a number of points with no elaboration. If you are asked for *two* points then only give two!
- *Describe:* state in words, using data from the question or diagrams where appropriate, the main points of the topic.
- *Discuss:* give a detailed account of the points involved in the topic.
- *Deduce/Predict:* draw conclusions from pieces of information given.
- *Outline:* restrict the answer to essential detail only.
- *Suggest:* apply your knowledge and understanding to a 'novel' situation that you may not have covered in the specification.
- *Calculate:* generate a numerical answer, with working shown.
- *Determine:* the quantity cannot be obtained directly. A sequence of calculations may be required.
- *Sketch:* a diagram is required. A graph need only be qualitatively correct, but important points on the graph may require numerical values.

Checking your work is essential. Check that an answer is realistic; check that it answers the question. If you do have time at the end of the examination, read through your descriptive answers to ensure that what you wrote is what you intended to write.

NewScientist

Reinforce your learning and keep up to date with recent developments in science by taking advantage of Heinemann's unique partnership with New Scientist. Visit www.heinemann.co.uk/newscientistmagazine for guidance and discounts on subscriptions.

Communication and homeostasis

Introduction

Living things rely on the action of enzymes. Enzymes ensure that the chemical reactions of life occur at a suitable rate. They require a fairly narrow range of environmental conditions in which to work well. When enzymes make certain reactions occur the reactions may manufacture products that alter the conditions. They may also use energy or release energy as heat.

Many living things consist of a large number of cells. Each of these cells may be carrying out a range of reactions that is manufacturing a range of products. In addition to this the external environment may keep changing. It is therefore essential to ensure that the environment inside the body remains constant.

Homeostasis is the maintenance of the internal environment in a constant state despite changes in the environment. Among the conditions homeostasis must keep constant in the body are:

- temperature
- pH
- water potential
- concentrations of salts
- blood glucose concentration.

Most of the organs in the body are involved in homeostasis. This includes the:

- brain
- heart
- lungs
- liver
- pancreas
- kidneys.

Communication between these organs is key to the success of homeostasis. It is achieved by the nervous system and the endocrine system. These two systems monitor changes and coordinate responses to those changes through precise cell signalling. Signals sent from one organ to another can enable these organs to work together to ensure that homeostasis is effective.

Test yourself

1 What effect does temperature change have on enzyme action?
2 What other environmental factors inhibit the action of enzymes?
3 List three changes to the external environment to which we might need to respond.
4 What is the main role of: **(a)** the heart   **(b)** the lungs   **(c)** the kidneys?
5 What is meant by cell signalling?
6 In what other process in the body is cell signalling particularly important?
7 Explain the role of cell surface receptors in cell signalling.

Module contents

By the end of this spread, you should be able to ...

* Outline the need for communication systems within multicellular organisms, with reference to the need to respond to changes in the internal and external environment and to coordinate the activities of different organs.
* State that cells need to communicate with each other by a process called cell signalling.
* State that neuronal and hormonal systems are examples of cell signalling.

Keeping cells active

All living things need to maintain a certain limited set of conditions inside their cells. This is because the cellular activities rely on the action of enzymes. You will recall from your AS work that enzymes need a specific set of conditions in which to work efficiently. These include:
* a suitable temperature
* a suitable pH
* an aqueous environment that keeps the substrates and products in solution
* freedom from toxins and excess inhibitors.

Stimulus and response

External environments

All living organisms have an external environment that consists of the air, water or soil around them. This external environment will change. As it changes it may place stress on the living organism. For instance, a cooler environment will cause greater heat loss. If the organism is to remain active and survive, the changes in the environment must be monitored and the organism must change its behaviour or physiology to reduce the stress. The environmental change is a **stimulus** and the way in which the organism changes its behaviour or physiology is its **response**.

Key definitions

A **stimulus** is any change in the environment that causes a response.

A **response** is a change in behaviour or physiology as a result of a change in the environment.

Figure 1   Arctic fox in **(a)** winter coat and **(b)** summer coat

The environment may change slowly, for example as the seasons pass or because of global warming. This change will elicit a gradual response. For example the arctic fox (*Alopex lagopus*) has a much thicker white coat in winter and a thinner grey/brown coat in summer. The different coats adapt it to the changing conditions.

However, the environment may change much more quickly. The changes from day to night or walking from bright sunlight into an unlit room are rapid changes. Again, the change (a stimulus) must be monitored and the organism must respond to the change.

Internal environments

Most multicellular organisms have a range of tissues and organs. Many of the cells and tissues are not exposed to the external environment, they are protected by epithelial tissues and organs such as skin or bark. In animals the internal cells and tissues are bathed in tissue fluid. This is the environment of the cells.

Module 1
Communication and homeostasis
The need for communication

As cells undergo their various metabolic activities they use up substrates and produce products. Some of these products may be unwanted or may even be toxic. These substances diffuse out of the cells into the tissue fluid. Therefore, the activities of the cells alter their own environment. One waste product is carbon dioxide. If this is allowed to build up in the tissue fluid outside the cells it could disrupt the action of enzymes by changing the pH of the environment around the cell.

It is clear that accumulation of excess waste or toxins in this internal environment must act as a stimulus to cause removal of these wastes so that the cells can survive. This stimulus may act directly on the cells which respond by reducing their activities so that less waste is produced. However, this response may not be good for the whole organism.

Maintaining the internal environment of the cells

The composition of the tissue fluid is maintained by the blood. Blood flows throughout the body and transports substances to and from the cells. Any wastes or toxins accumulating in the tissue fluid are likely to enter the blood and be carried away. In order to prevent their accumulation in the blood they must be removed from the body, that is, **excreted**.

It is important that the concentrations of wastes and all substances in the blood are monitored closely. This ensures that the body does not excrete too much of any useful substance but removes enough of the wastes to maintain good health. It also ensures that all the cells of the body are supplied with the substrates they need.

Coordination

A multicellular organism is more efficient than a single-celled organism, as its cells can be differentiated. This means that its cells can be specialised to perform particular functions. Groups of cells specialised in this way form tissues and organs. As a result the cells that monitor the blood may be in a different part of the body and well away from the cells that release a substance into the blood or well away from the organ that removes that substance from the blood.

Therefore a good communication system is required to ensure that these different parts of the body work together effectively.

A good communication system will:
- cover the whole body
- enable cells to communicate with each other
- enable specific communication
- enable rapid communication
- enable both short-term and long-term responses.

Cell signalling

Cells communicate with each other by the process of cell signalling. This is a process in which one cell will release a chemical that is detected by another cell. The second cell will respond to the signal released by the first cell.

There are two major systems of communication that work by cell signalling.
- The neuronal system is an interconnected network of neurones that signal to each other across synapse junctions. The neurones can conduct a signal very quickly and enable rapid responses to stimuli that may be changing quickly.
- The hormonal system uses the blood to transport its signals. Cells in an endocrine organ release the signal (a hormone) directly into the blood. It is carried all over the body but is only recognised by specific target cells. The hormonal system enables longer-term responses to be coordinated.

Questions

1 (a) State three examples of a stimulus and a corresponding response.
   (b) For each of your examples, suggest whether it uses the neuronal system or the hormonal system for communication.

2 Explain the advantages to the arctic fox of having a thicker, white coat in winter.

3 What are the requirements of a good communication system?

By the end of this spread, you should be able to . . .

✳ Define the terms *negative feedback*, *positive feedback* and *homeostasis*.
✳ Explain the principles of homeostasis in terms of receptors, effectors and negative feedback.

Homeostasis

Homeostasis can be defined as keeping the internal environment constant despite external changes. Many living organisms have to keep a great number of conditions constant inside the body. These may include:

- body temperature
- blood glucose concentration
- blood salt concentration
- water potential of the blood
- blood pressure
- carbon dioxide concentration.

Negative feedback

In order to maintain a constant internal environment a number of processes must occur.
- Any change to the internal environment must be detected.
- The change must be signalled to other cells.
- There must be a response that reverses the change.

The reversal of a change in the internal environment to return to a steady state or optimum position is known as **negative feedback**.

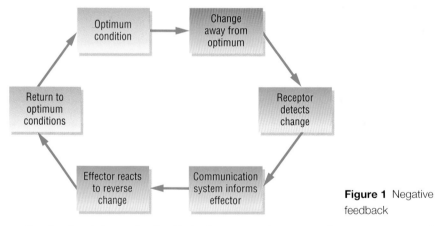

Figure 1 Negative feedback

For negative feedback to work effectively there must be a complex arrangement of structures that are all coordinated through cell signalling. These structures are part of a standard pathway that is used to produce a suitable response to a stimulus.

stimulus → receptor → communication pathway (cell signalling) → effector → response

There are a number of structures required for this pathway to work.
- Sensory receptors, such as temperature receptors or glucose concentration receptors. These receptors are internal and monitor conditions inside the body. If they detect a change they will be stimulated to send a message.
- A communication system such as the nervous system or the hormonal system. This acts by signalling between cells. It is used to transmit a message from the receptor cells to the effector cells. The message may or may not pass through a coordination centre such as the brain.
- Effector cells, such as liver cells or muscle cells. These cells will bring about a response that reverses the change detected by the receptor cells.

Key definition

Homeostasis is the maintenance of the internal environment in a constant state despite external changes.

Examiner tip

Students are often asked to define homeostasis. There are three main points to remember:
- maintenance of internal environment
- constant
- despite changes

Key definition

Negative feedback is a process that brings about a reversal of any change in conditions. It ensures that an optimum steady state can be maintained, as the internal environment is returned to its original set of conditions after any change. It is essential for homeostasis.

Module 1
Communication and homeostasis
Homeostasis and negative feedback

Positive feedback

Positive feedback is less common than negative feedback. When positive feedback occurs the response is to increase the original change. This destabilises the system and is usually harmful. An example is when the body gets too cold. Below a certain core body temperature the enzymes become less active. If they are less active the exergonic reactions that release heat are slower and release less heat. This allows the body to cool further and slows down the enzyme-controlled reactions even more, so that the body temperature spirals downwards.

STRETCH and CHALLENGE

Enzyme action and temperature regulation

As core body temperature rises the increase will affect the activity of enzymes. This can lead to heat exhaustion and even death.

Questions

A Describe the effect of increasing temperature on enzyme action.
B Suggest what actually causes death as the body temperature rises.

There are, however, some occasions when positive feedback can be beneficial. Positive feedback is used to stimulate an increase in a change.

An example is seen at the end of pregnancy to bring about dilation of the cervix. As the cervix begins to stretch the change is signalled to the anterior pituitary gland, stimulating it to secrete the hormone **oxytocin**. Oxytocin increases the uterine contractions, which stretch the cervix more, which causes secretion of more oxytocin. Once the cervix is fully dilated, the baby can be born.

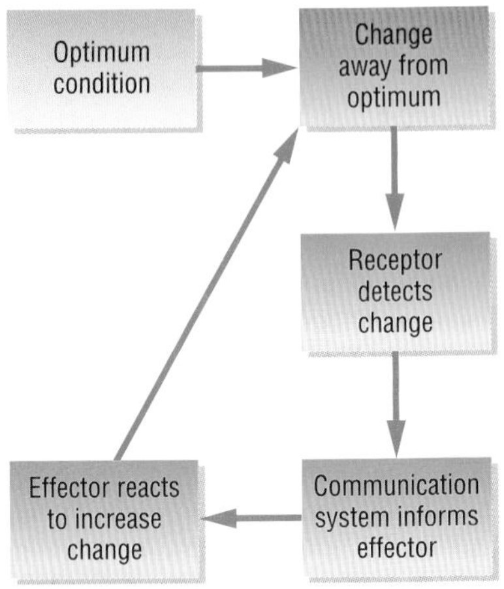

STRETCH and CHALLENGE

The stress response

The usual response to stress is to release the hormone adrenaline. This hormone has a wide range of target cells and prepares the body for activity. The activity may be to stay and fight or it may be to run away. The hormone is known as the 'flight or fight' hormone. When under stress women also release the hormone oxytocin. This results in a tendency to pacify or protect. It has been called the 'tend and befriend' hormone. Oxytocin prompts a mother to protect her children.

Question

C Suggest how these responses to adrenaline and oxytocin may have evolved.

The meaning of constant

A negative feedback system can maintain a reasonably constant set of conditions. However, the conditions will never remain perfectly constant. There will be some variation about the mean or optimum condition. As long as this variation is not too great then the conditions will remain acceptable. For example, a heated room will never get too cold or too hot. When this is applied to living systems it means that the conditions inside a living organism will remain within a relatively narrow range, 'warm' enough to allow enzymes to continue functioning efficiently but 'cool' enough to not damage the body's many other proteins.

Key definition

Positive feedback is a process that increases any change detected by the receptors. It tends to be harmful and does not lead to homeostasis.

Figure 2 Positive feedback

Questions

1 Explain the difference between negative feedback and positive feedback.
2 In the lists below match each component of a heating system to its role in negative feedback.

thermostat	effector
boiler	communication
electric cables	receptor

By the end of this spread, you should be able to . . .

* Describe the physiological and behavioural responses that maintain a constant core body temperature in ectotherms.

The need to maintain body temperature

Changes in body temperature can have a dramatic effect upon the structure of proteins, including enzymes. Enzymes are globular proteins, and their structure is very specific to their function. The activity of enzymes is affected if they are not kept at, or close to, their optimum temperature, so temperature affects their ability to function inside cells. If enzymes do not function properly, the level of activity that can be achieved by the organism will be dramatically altered.

The core temperature is the important factor as all the vital organs are found within the main part of the body. Peripheral parts of the body may be allowed to increase or decrease in temperature without affecting the survival of the individual.

Ectotherm or endotherm

Endotherms

Endotherms can maintain the temperature of their body within fairly strict limits. They are largely independent of the external temperature.

Ectotherms

The body temperature of an ectotherm tends to fluctuate with the external temperature. Ectotherms are not able to increase respiration rates to generate heat internally. They rely on external sources of heat to keep warm. However, despite this, many ectotherms can successfully regulate their body temperature under all but the most extreme conditions.

There are some advantages to being an ectotherm.
* Ectotherms use less of their food in respiration.
* They need to find less food and may be able to survive for long periods without eating. For example, a snake can last several weeks between meals.
* A greater proportion of the energy obtained from food can be used for growth.

However, there are disadvantages.
* They are less active in cooler temperatures, and may need to warm up in the morning before they can be active. This puts them at greater risk of predation. Lizards can often be seen basking in the sun during the early morning.
* They may not be capable of activity during the winter as they never warm up sufficiently. This means that they must have sufficient stores of energy to survive over the winter without eating.

Temperature regulation in ectotherms

Ectotherms do not use internal energy sources to maintain their body temperature when cold. However, once they are active their muscle contractions will generate some heat from increased respiration.

Temperature regulation relies upon increasing the exchange of heat with their environment.
* When an ectotherm is cold it will change its behaviour or physiology to increase absorption of heat from its environment.

Key definition

An **ectotherm** is an organism that relies on external sources of heat to regulate its body temperature.

Examiner tip

Do not use the terms 'warm-blooded' and 'cold-blooded'.

Endothermic organisms used to be described as warm-blooded. Ectothermic organisms used to be described as cold-blooded. However, this terminology has fallen out of use as research has shown it to be inaccurate. Many ectotherms can maintain their body temperature at 37 °C or even higher on warm days. This cannot be considered to be cold.

- When an ectotherm is hot it will change its behaviour or physiology to decrease absorption of heat and increase loss of heat to its environment.

In order to warm up they will bask in the sun or lie on a warm surface. If they are too hot they will stay underground or in the shade. Many ectotherms have developed this to a fine art. Locusts (*Schistocerca* spp.) will orientate themselves to be side-on to the sun when they need to warm up. However, if they are too hot they will move higher up a plant to get away from the hot ground and also face into the sun – this exposes a smaller surface area to the sun.

Some ectotherms possess physiological or anatomical adaptations to help exchange heat with their environment. For example, the horned lizard (*Phrynosoma* spp.) can alter its surface area by expanding or contracting its rib cage, and the frilled lizard (*Chlamydosaurus kingii*) may use its frill to help absorb heat from the sun. Locusts have also been shown to increase their abdominal breathing movements when hot. This will increase evaporation of water and aid cooling.

Figure 1 A horned lizard basking in the sun with its ribcage expanded

Adaptation	How it helps regulate temperature	Example
Expose body to sun	Enables more heat to be absorbed	Snakes
Orientate body to sun	Exposes larger surface area for more heat absorption	Locusts
Orientate body away from sun	Exposes lower surface area so that less heat is absorbed	Locusts
Hide in burrow	Reduces heat absorption by keeping out of the sun	Lizards
Alter body shape	Exposes more or less surface area to sun	Horned lizards
Increase breathing movements	Evaporates more water	Locusts

Table 1 How ectotherms can regulate their temperature

STRETCH and CHALLENGE

Temperature regulation in bee swarms

Bees are ectothermic. However, it has been shown that the temperature of a bee swarm can be maintained accurately to within one degree of 35 °C. This is achieved by bees moving to different parts of the swarm and by allowing passages for air flow through the swarm. Also, adult worker bees in a hive often display forms of behaviour that help to maintain the internal temperature of that hive. Such behaviours include extra movement within the hive to generate heat from muscular contraction and flapping of wings at the hive entrance to create air movement.

Question

A Suggest how movement of bees within a swarm and air movement through the swarm can help to maintain the temperature of the swarm.

Questions

1 Explain why it is important to maintain a constant body temperature.
2 Make a list of five ectotherms.
3 Explain how basking on a hot rock in the sun can help an ectotherm to regulate its body temperature.

④ Maintaining body temperature – endotherms

By the end of this spread, you should be able to...

* Describe the physiological and behavioural responses that maintain a constant core body temperature in endotherms, with reference to peripheral temperature receptors, the hypothalamus and effectors in skin and muscles.

Endotherms

Endotherms use internal sources of heat to help maintain their body temperature. Many chemical reactions in the body are exergonic – they release energy in the form of heat. Endotherms can increase the rate of respiration in the liver (an exergonic reaction) simply to release heat; they are using some of their energy intake to stay warm.

Endotherms, like ectotherms, can use behavioural mechanisms as well to help maintain body temperature (basking in the sun, for example). They also have useful physiological mechanisms, such as redirecting blood to or away from the skin.

There are many advantages of endothermy:
- a fairly constant body temperature whatever the temperature is externally
- activity possible when external temperatures are cool – such as at night, early in the morning or during the winter
- ability to inhabit colder parts of the planet.

However, there are disadvantages:
- a significant part of the energy intake used to maintain body temperature in the cold
- more food required (a shrew has to eat almost its own body mass of food each day to avoid starving to death)
- less of the energy from food is used for growth, or more food is needed in order to grow.

Temperature regulation in endotherms

Several behavioural and physiological responses are used by endothermic organisms to maintain their constant body temperature (Table 1).

Physiological mechanisms to maintain body temperature		
Component of body involved	**Response if core body temperature is too high**	**Response if core body temperature is too low**
Sweat glands in skin	Secrete more sweat onto skin; water in sweat evaporates, using heat from blood to supply latent heat of vaporisation	Less sweat is secreted; less evaporation of water, so less loss of latent heat
Lungs, mouth and nose	Panting increases evaporation of water from lungs, tongue and other moist surfaces, using latent heat as above	The animal does not pant, so less water evaporates
Hairs on skin	Hairs lie flat, providing little insulation, and thus more heat can be lost by convection and radiation	Hairs are raised to trap a layer of insulating air, reducing the loss of heat from the skin
Arterioles leading to capillaries in skin	Vasodilation allows more blood into capillaries near the skin surface; more heat can be radiated from the skin, which, in pale-skinned people, may look red	Vasoconstriction reduces the flow of blood through capillaries near the surface of skin; less heat is radiated

Key definition

An **endotherm** is an organism that can use internal sources of heat, such as heat generated from metabolism in the liver, to maintain its body temperature.

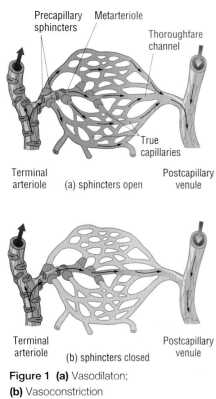

Figure 1 (a) Vasodilaton; **(b)** Vasoconstriction

Liver cells	Rate of metabolism is reduced; less heat is generated from exergonic reactions such as respiration	Rate of metabolism is increased, therefore respiration generates more heat, which is transferred to the blood
Skeletal muscles	No spontaneous contractions	Spontaneous contractions (shivering) generate heat as muscle cells respire more

Behavioural mechanisms to maintain body temperature	
Behaviour if too hot	**Behaviour if too cold**
Move into shade or hide in burrow	Move into sunlight
Orientate body to decrease surface area exposed to sun	Orientate body to increase surface area exposed to sun
Remain inactive and spread out the limbs to increase surface area	Move about to generate heat in muscles (except in extreme cold when it is better to keep still and roll into a ball to reduce surface area)

Table 1 Physiological and behavioural mechanisms for temperature regulation in endotherms

Control of temperature regulation

The maintenance of core body temperature is important. Endotherms monitor the temperature of their blood in the hypothalamus of the brain. If the core temperature drops below optimum the hypothalamus sends signals to reverse the change. This will involve several changes:

- increased rate of metabolism in order to release more heat from exergonic reactions
- release of heat through extra muscular contraction
- decreased loss of heat to the environment.

If core temperature rises above the optimum the hypothalamus sends signals that bring about the opposite changes. This is an example of negative feedback.

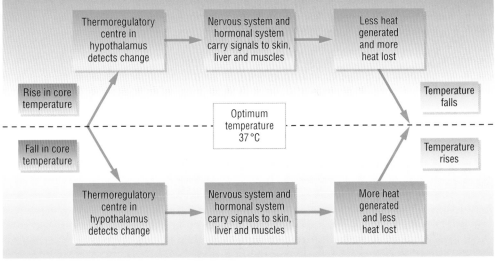

Figure 2 Controlling body temperature by negative feedback

The role of peripheral temperature receptors

The thermoregulatory centre in the hypothalamus monitors blood temperature and detects any changes in the core body temperature. However, an early warning that the body temperature may change could help the hypothalamus to respond more quickly and avoid too much variation in core body temperature.

If the extremities start to cool down or warm up this may eventually affect the core body temperature. The peripheral temperature receptors in the skin monitor the temperature in the extremities. This information is fed to the thermoregulatory centre in the hypothalamus. If it signals to the brain that the external environment is very cold or very hot, the brain can initiate behavioural mechanisms for maintaining body temperature, such as moving into shade.

STRETCH and CHALLENGE

Should mountain rescue dogs carry brandy?

In the early part of the twentieth century St Bernard dogs were used for mountain rescues. Traditionally they carried a small container of brandy for the lost or injured climber to drink.

Alcohol causes vasodilation.

Question

A Explain why drinking brandy is not a good idea for someone who is lost or injured and exposed to cold weather.

Questions

1 Explain why a shrew has to eat almost its own body mass each day, but an elephant eats less than one percent of its body mass each day.

2 Suggest why:
   (a) the fairy penguin of Australia grows to about 25 cm in height while the emperor penguin of Antarctica grows to a metre in height;
   (b) penguins huddle together.

3 Explain how vasoconstriction helps to reduce heat loss.

11

By the end of this spread, you should be able to ...

* Outline the roles of sensory receptors in mammals in converting different forms of energy into nerve impulses.
* Describe, with the aid of diagrams, the structure and functions of sensory and motor neurones.

Receptors	Energy changes detected
Light-sensitive cells (rods and cones) in the retina of the eye	Light intensity and range of wavelengths (colour)
Olfactory cells lining the inner surface in the nasal cavity	Presence of volatile chemicals
Taste buds in the tongue, hard palate, epiglottis and the first part of the oesophagus	Presence of soluble chemicals
Pressure receptors (Pacinian corpuscles) in the skin	Pressure on skin
Sound receptors in the inner ear (cochlea)	Vibrations in air
Muscle spindles (proprioceptors)	Length of muscle fibres

Table 1 Sensory receptors

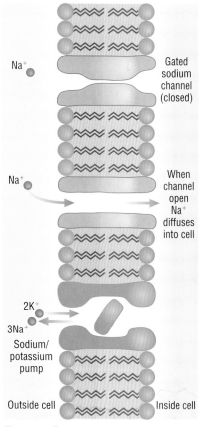

Figure 1 Gated sodium ion channels and a sodium/potassium pump in the cell membrane

Na⁺

Gated sodium channel (closed)

Na⁺

When channel open Na⁺ diffuses into cell

2K⁺
3Na⁺
Sodium/ potassium pump

Outside cell · Inside cell

Sensory receptors

Sensory receptors are specialised cells that can detect changes in our surroundings. They are energy *transducers* that convert one form of energy to another. Each type of transducer is adapted to detect changes in a particular form of energy. This may be a change in light levels, a change in pressure on the skin or one of many other energy changes. Each change in energy levels in the environment is called a *stimulus*. Whatever the stimulus the sensory receptors convert the energy into a form of electrical energy called a nerve impulse. Table 1 shows a number of receptors and the energy changes that they detect.

Generating nerve impulses

Changing membrane permeability

As you will recall from your AS Biology work, all cell surface membranes contain proteins. Some proteins are channels that allow the movement of ions (charged particles) across the membrane. If the channel is permanently open then the ions can diffuse across the membrane and will do so until concentrations of that particular ion are equal on both sides of the membrane.

Neurones (nerve cells) have more specialised channel proteins that are specific either to sodium or potassium ions. They also possess a gate that can open or close the channel. When open, the permeability of the membrane to that particular ion is increased; when closed the permeability is reduced. The channels are usually kept closed.

Nerve cell membranes also contain carrier proteins that actively transport sodium ions (Na⁺) out of the cell and potassium ions (K⁺) into the cell. These are called sodium/ potassium ion pumps. More sodium ions are actively transported out of the cell than potassium ions are actively transported into the cell. The inside of the cell is negatively charged with respect to the outside. The cell membrane is said to be **polarised**.

A nerve impulse is created by altering the permeability of the nerve cell membrane to sodium ions. As the sodium ion channels open the membrane permeability is increased and sodium ions can move across the membrane down their concentration gradient into the cell. The movement of ions across the membrane creates a change in the potential difference (charge) across the membrane. The inside of the cell becomes less negative (compared to the outside) than usual. This is called **depolarisation**.

Generator potentials

Receptor cells respond to changes in the environment. The gated sodium ion channels open, allowing sodium ions to diffuse across the membrane into the cell. A small

Key definitions

A **polarised** membrane is one that has a potential difference across it. This is the resting potential.

Depolarisation is the loss of polarisation across the membrane. It refers to the period when sodium ions are entering the cell making the inside less negative with respect to the outside.

change in potential caused by one or two sodium ion channels opening is called a **generator potential**. The larger the stimulus (the change in energy levels in the environment) the more gated channels will open. If enough sodium ions enter the cell the potential difference changes significantly and will initiate an impulse or **action potential**.

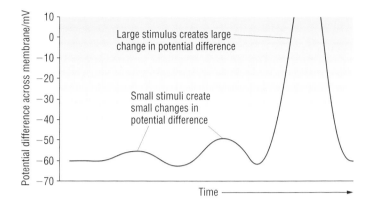

Figure 2 Changing the potential difference across a cell membrane

Key definitions

A **generator potential** is a small depolarisation caused by sodium ions entering the cell.

An **action potential** is achieved when the membrane is depolarised to a value of about +40 mV. It is an all-or-nothing response. In the events leading up to an action potential, the membrane depolarises and reaches a threshold level, then lots of sodium ions enter the axon and an action potential is reached.

Sensory and motor neurones

Once a stimulus has been detected and its energy has been converted to a depolarisation of the receptor cell membrane, the impulse must be transmitted to other parts of the body. The impulse is transmitted along neurones as an action potential.

There are a number of different types of neurone. These include:
- sensory neurones that carry the action potential from a sensory receptor to the central nervous system (CNS)
- motor neurones that carry an action potential from the CNS to an effector such as a muscle or gland
- relay neurones that connect sensory and motor neurones.

The function of the neurone is to transmit the action potential from one part of the body to another. Most neurones have a very similar basic structure that enables them to carry out this function. They are specialised cells with the following features:
- Many are very long so that they can transmit the action potential over a long distance.
- The cell surface (plasma) membrane has many gated ion channels that control the entry or exit of sodium, potassium or calcium ions.
- They have sodium/potassium ion pumps that use ATP to actively transport sodium ions out of the cell and potassium ions into the cell.
- They maintain a potential difference across their cell surface (plasma) membrane.
- They are surrounded by a fatty sheath called the myelin sheath (actually a series of cells called Schwann cells) that insulates the neurone from the electrical activity in nearby cells. There are gaps where the Schwann cells meet called nodes of Ranvier.
- They have a cell body that contains the nucleus, many mitochondria and ribosomes.
- Motor neurones have their cell body in the central nervous system and have a long axon that carries the action potential out to the effector.
- Sensory neurones have a long dendron carrying the action potential from a sensory receptor to the cell body, which is positioned just outside the central nervous system. They then have a short axon carrying the action potential into the central nervous system.
- Both sensory and motor neurones have numerous dendrites connected to other neurones.

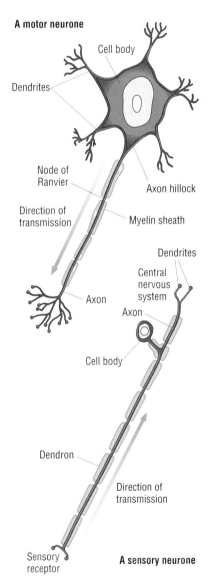

Figure 3 A sensory neurone and a motor neurone

Questions

1 Use a table to compare and contrast sensory neurones and motor neurones.
2 Suggest why neurones need to contain a large number of mitochondria.
3 Why do membranes need special channels for the diffusion of charged ions?

By the end of this spread, you should be able to ...

* Describe and explain how the resting potential is established and maintained.
* Describe and explain how an action potential is generated.
* Interpret graphs of the voltage changes taking place during the generation and transmission of an action potential.

A resting neurone

When a neurone is not transmitting an action potential it is said to be at rest. In fact it is actively transporting ions across its cell surface (plasma) membrane. Sodium/potassium ion pumps use ATP to pump three sodium ions out of the cell for every two potassium ions that are pumped in. The plasma membrane is more permeable to potassium ions than to sodium ions and many diffuse out again. The cell cytoplasm also contains large organic anions (negatively charged ions). Hence, the interior of the cell is maintained at a negative potential compared to the outside. The cell membrane is said to be **polarised**. The potential difference across the cell membrane is about –60 mV. This is called the **resting potential**.

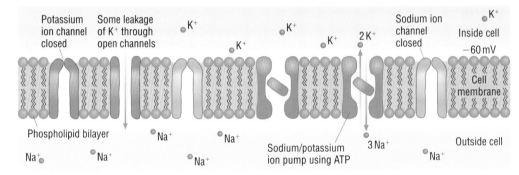

Figure 1 Cell surface membrane of a neurone at rest

An action potential

At rest the gated sodium ion channels are kept closed. The sodium/potassium ion pump uses ATP to actively transport three sodium ions out for every two potassium ions brought into the axon. A few of these potassium ions diffuse back out as some potassium channels are open. If some of the sodium ion channels are opened then sodium ions will quickly diffuse down their concentration gradient into the cell from the surrounding tissue fluid. This causes a depolarisation of the membrane. In the generator region of receptor cells the gated channels are opened by energy changes in the environment. For example, the gates in a Pacinian corpuscle, which detects pressure change, are opened by deformation. The gates further along the neurone are opened by changes in the potential difference (voltage) across the membrane. They are called **voltage-gated channels**. These channels respond to depolarisations of the membrane.

All or nothing

Generator potentials in the sensory receptor are depolarisations of the cell membrane. A small depolarisation will have no effect on the voltage-gated channels. However, if the depolarisation is large enough to reach **threshold potential**, it will open some nearby voltage-gated channels. This causes a large influx of sodium ions and the depolarisation reaches +40 mV, which is an action potential. Once this value is reached the neurone will transmit an **action potential** because many voltage-gated sodium ion channels open. The action potential is self-perpetuating – once it starts at one point in the neurone, it will continue along to the end of the neurone.

Key definition

The **resting potential** is the potential difference or voltage across the neurone cell membrane while the neurone is at rest. It is about –60 mV inside the cell compared with the outside. Other cells may also maintain a resting potential that might change under certain circumstances.

Key definitions

Voltage-gated channels are channels in the cell membrane that allow the passage of charged particles or ions. They have a mechanism called a gate which can open and close the channel. In these channels the gates respond to changes in the potential difference across the membrane.

The **threshold potential** is a potential difference across the membrane of about –50 mV. If the depolarisation of the membrane does not reach the threshold potential then no action potential is created. If the depolarisation reaches the threshold potential then an action potential is created.

The **action potential** is a depolarisation of the cell membrane so that the inside is more positive than the outside, with a potential difference across the membrane of +40 mV. This can be transmitted along the axon or dendron plasma membrane.

Ionic movements

An action potential consists of a set of ionic movements. The ions move across the cell membrane when the correct channels are opened.

An action potential consists of the following stages (the numbers refer to Figure 2):

1 The membrane starts in its resting state – **polarised** with the inside of the cell being –60 mV compared to the outside.
2 Sodium ion channels open and some sodium ions diffuse into the cell.
3 The membrane **depolarises** – it becomes less negative with respect to the outside and reaches the threshold value of –50 mV.
4 Voltage-gated sodium ion channels open and many sodium ions flood in. As more sodium ions enter, the cell becomes positively charged inside compared with outside.
5 The potential difference across the plasma membrane reaches +40 mV. The inside of the cell is positive compared with the outside.
6 The sodium ion channels close and potassium channels open.
7 Potassium ions diffuse out of the cell bringing the potential difference back to negative inside compared with outside – this is called **repolarisation**.
8 The potential difference overshoots slightly, making the cell **hyperpolarised**.
9 The original potential difference is restored so that the cell returns to its resting state.

After an action potential the sodium and potassium ions are in the wrong places. The concentrations of these ions inside and outside the cell must be restored by the action of the sodium/potassium ion pumps.

For a short time after each action potential it is impossible to stimulate the cell membrane to reach another action potential. This is known as the **refractory period** and allows the cell to recover after an action potential. It also ensures that action potentials are transmitted in only one direction.

Questions

1 Explain why a neurone is active while it is said to be resting.
2 What is the role of the organic anions inside the neurone?
3 What is the difference between the gated channels in the generator region and those further along the neurone?
4 Explain why it is impossible to stimulate another action potential in a cell membrane that is hyperpolarised.

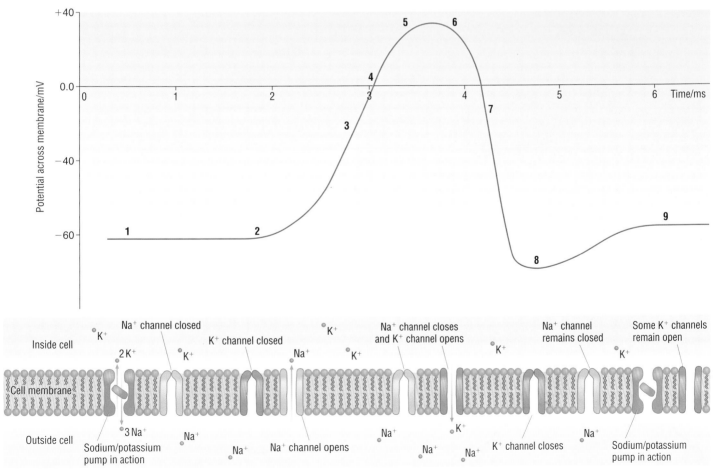

Figure 2 An action potential

By the end of this spread, you should be able to ...

✳ Describe and explain how an action potential is transmitted in a myelinated neurone, with reference to the roles of voltage-gated sodium ion and potassium ion channels.

Transmission of action potentials

In spread 1.1.6 we saw how an action potential can be generated. We also saw how the action potential consists of a series of movements of ions across the membrane of the neurone. The role of the neurone is to transmit information in the form of action potentials to other parts of the body. So how does that action potential travel along the neurone?

Local currents

The opening of sodium ion channels at one particular point of the neurone upsets the balance of sodium and potassium ions (the resting potential) created by the sodium/potassium ion pumps. This creates **local currents** in the cytoplasm of the neurone. These local currents cause sodium ion channels further along the membrane to open. This can be explained in more detail:

- When an action potential occurs the sodium ion channels open at a particular point along the neurone.
- This allows sodium ions to diffuse across the membrane from the region of higher concentration outside the neurone into the neurone.
- The movement of sodium ions into the neurone upsets the balance of ionic concentrations created by the sodium/potassium ion pumps.
- The concentration of sodium ions inside the neurone rises at the point where the sodium ion channels are open.
- This causes the sodium ions to diffuse sideways, away from this region of increased concentration.
- The movement of charged particles is a current called a local current.

Voltage-gated sodium ion channels

Further along the membrane are more gated sodium ion channels. The gates on these channels are operated by changes in the voltage across the membrane. At rest the voltage across the membrane will be at resting potential (–60 mV inside the neurone).

Examiner tip

Remember that the movements of ions associated with an action potential are all by diffusion. Diffusion requires a concentration gradient (diffusion gradient). This is created by the action of the sodium/potassium pumps, using ATP, while the neurone is at rest.

Key definition

Local currents are the movements of ions along the neurone. The flow of ions is caused by an increase in concentration at one point, which causes diffusion away from the region of higher concentration.

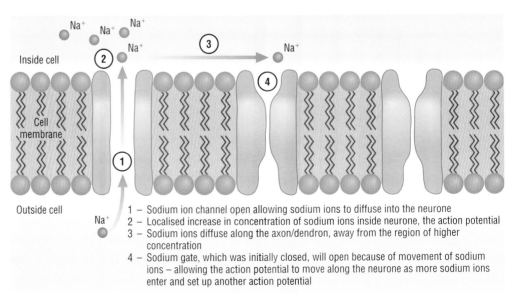

1 – Sodium ion channel open allowing sodium ions to diffuse into the neurone
2 – Localised increase in concentration of sodium ions inside neurone, the action potential
3 – Sodium ions diffuse along the axon/dendron, away from the region of higher concentration
4 – Sodium gate, which was initially closed, will open because of movement of sodium ions – allowing the action potential to move along the neurone as more sodium ions enter and set up another action potential

Figure 1 The creation of local currents

Module 1
Communication and homeostasis
Transmission of action potentials

The movement of sodium ions along the neurone alters the potential difference across the membrane. When the potential difference across the membrane is reduced, the gates open. This allows sodium ions to enter the neurone at a point further along the membrane. The action potential has moved along the neurone.

STRETCH and CHALLENGE

When sodium ions diffuse along the neurone they reduce the potential difference or voltage across the membrane. This is the start of depolarisation of the adjacent area of the axon membrane. The decrease in potential difference causes the sodium ion gated channels to open, which allows more sodium ions to diffuse into the neurone.

Question

A What sort of feedback is operating in this example?

The myelin sheath

The myelin sheath is an insulating layer of fatty material. Sodium and potassium ions cannot diffuse through this fatty layer. Therefore, the ionic movements that create an action potential cannot occur over much of the length of the neurone. The gaps in the myelin sheath are gaps between the Schwann cells that make up the myelin sheath. These gaps are called nodes of Ranvier. The ionic exchanges that cause an action potential occur only at the nodes of Ranvier. In myelinated neurones the local currents are elongated and sodium ions diffuse along the neurone from one node of Ranvier to the next. This means that the action potential appears to jump from one node to the next. This is called **saltatory** (Latin, meaning to jump) conduction.

Key definition

Saltatory conduction means 'jumping conduction'. It refers to the way that the action potential appears to jump from one node of Ranvier to the next.

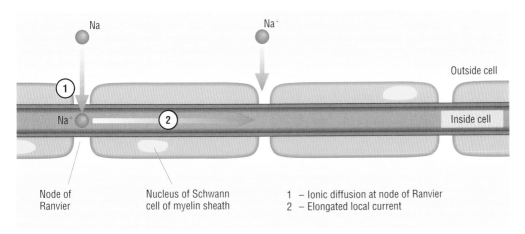

Figure 2 Saltatory conduction

The advantages of saltatory conduction

The myelin sheath means that action potentials can only occur at the gaps between the Schwann cells that make up the myelin sheath. This speeds up the transmission of the action potential. Myelinated neurones conduct action potentials more quickly than non-myelinated neurones. A myelinated neurone can conduct an action potential at up to 120 m s^{-1}.

Questions

1 How are the sodium ion channels opened?
2 Explain how diffusion gradients are created so that sodium ions can diffuse into the neurone.
3 Explain how the myelin sheath causes saltatory conduction.

By the end of this spread, you should be able to ...

* Describe, with the aid of diagrams, the structure of a cholinergic synapse.
* Outline the role of neurotransmitters in the transmission of action potentials.

The structure of a cholinergic synapse

A synapse is a junction between two or more neurones. It is where one neurone can communicate with, or signal to, another neurone. Between the two neurones is a small gap called the synaptic cleft, which is approximately 20 nm wide. An action potential is produced by the movements of ions across the neurone membrane. The action potential cannot bridge the gap between two neurones. Instead, the presynaptic action potential causes the release of a chemical (the transmitter substance) that diffuses across the gap and generates a new action potential in the postsynaptic neurone. Synapses that use acetylcholine as the **neurotransmitter** are called **cholinergic synapses**.

The synaptic knob

The presynaptic neurone ends in a swelling called the **synaptic knob**. This knob contains a number of specialised features:

* many mitochondria – indicating that an active process, needing ATP, is involved
* a large amount of smooth endoplasmic reticulum
* vesicles of a chemical called acetylcholine, the transmitter substance that will diffuse across the synaptic cleft
* voltage-gated calcium ion channels in the membrane.

Key definitions

A **neurotransmitter** (transmitter substance) is a chemical that diffuses across the cleft of the synapse to transmit a signal to the postsynaptic neurone.

Cholinergic synapses are those that use acetylcholine as their transmitter substance.

The **synaptic knob** is a swelling at the end of the presynaptic neurone.

(a)

Five protein subunits make up the sodium ion channel

Acetylcholine receptor site

Postsynaptic membrane

(b)

Acetylcholine   Na⁺

Channel closed   Channel open

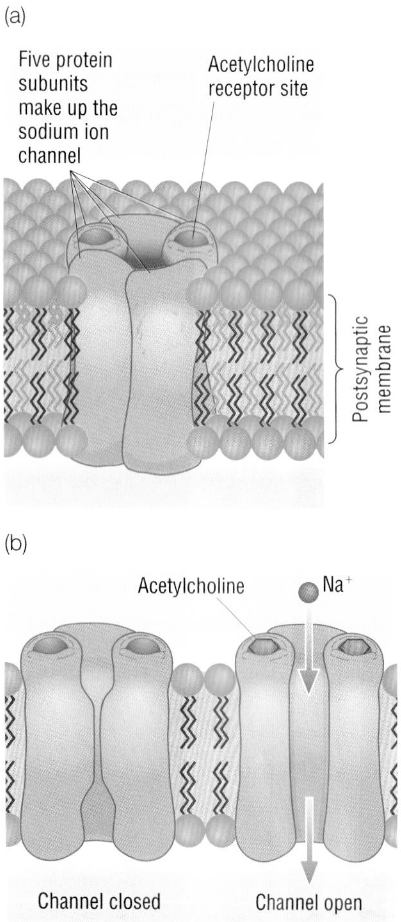

Figure 2 (a) The postsynaptic membrane with a sodium ion channel; **(b)** sodium ion channel opened by acetylcholine

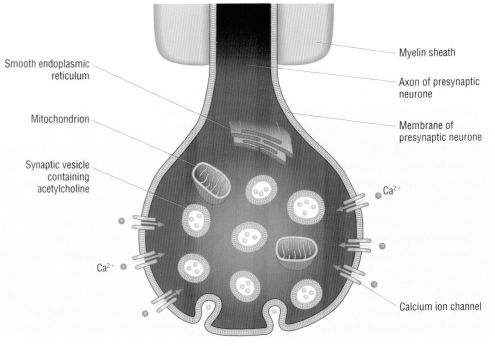

Smooth endoplasmic reticulum

Mitochondrion

Synaptic vesicle containing acetylcholine

Myelin sheath

Axon of presynaptic neurone

Membrane of presynaptic neurone

Ca²⁺

Ca²⁺

Calcium ion channel

Figure 1 The synaptic knob

The postsynaptic membrane

The postsynaptic membrane contains specialised sodium ion channels that can respond to the transmitter substance. These channels consist of five polypeptide molecules. Two of these polypeptides have a special receptor site that is specific to acetylcholine. The receptors have a complementary shape to the acetylcholine molecule and acetylcholine will fit and bind into the site. When acetylcholine binds to the two receptors the sodium ion channel opens.

Transmission across the synapse

The transmission of a signal across the synaptic cleft is a straightforward sequence of events.

- An action potential arrives at the synaptic knob.
- The voltage-gated calcium ion channels open.
- Calcium ions diffuse into the synaptic knob.
- The calcium ions cause the synaptic vesicles to move to and fuse with the presynaptic membrane.
- Acetylcholine is released by exocytosis.
- Acetylcholine molecules diffuse across the cleft.
- Acetylcholine molecules bind to the receptor sites on the sodium ion channels in the postsynaptic membrane.
- The sodium ion channels open.
- Sodium ions diffuse across the postsynaptic membrane into the postsynaptic neurone.
- A generator potential or excitatory postsynaptic potential (EPSP) is created.
- If sufficient generator potentials combine then the potential across the postsynaptic membrane reaches the threshold potential.
- A new action potential is created in the postsynaptic neurone.

Once an action potential is achieved it will pass down the postsynaptic neurone.

The role of acetylcholinesterase

Acetylcholinesterase is an enzyme found in the synaptic cleft. It hydrolyses the acetylcholine to ethanoic acid and choline. This stops the transmission of signals so that the synapse does not continue to produce action potentials in the postsynaptic neurone. The ethanoic acid and choline are recycled. They re-enter the synaptic knob by diffusion and are recombined to acetylcholine using ATP from respiration in the mitochondria. The recycled acetylcholine is stored in synaptic vesicles for future use.

Examiner tip

Remember that the acetylcholine diffuses across the cleft – it is not squirted out of the cells and it does not jump across the cleft.

Key definition

Acetylcholinesterase is an enzyme in the synaptic cleft. It breaks down the transmitter substance acetylcholine.

STRETCH and CHALLENGE

The cytoplasm in the synaptic knob has a high proportion of certain organelles. These include smooth endoplasmic reticulum, mitochondria and vesicles. Each organelle has a specific role to play in the functioning of the cell.

Question

A Describe the role of each of these organelles and explain why they are found in relatively large numbers in the synaptic knob.

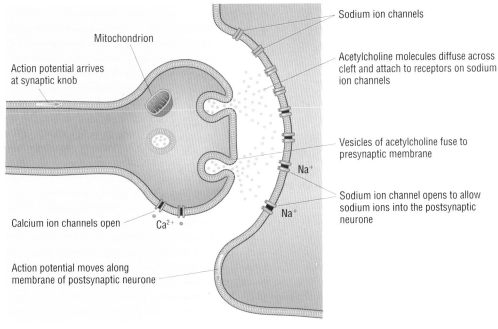

Figure 3 Transmission across the synaptic cleft

Questions

1 Explain why the calcium ion channels need to be voltage-gated.
2 Why is it important that the synaptic cleft contains the enzyme acetylcholinesterase?
3 Suggest why the presynaptic neurone ends in a synaptic knob.

By the end of this spread, you should be able to ...

* Outline the roles of synapses in the nervous system.
* Outline the significance of the frequency of impulse transmission.
* Compare and contrast the structure and function of myelinated and non-myelinated neurones.

Key definitions

All or nothing refers to the fact that a neurone either conducts an action potential or it does not. All action potentials are of the same magnitude, +40 mV.

Summation is a term that refers to the way that several small potential changes can combine to produce one larger change in potential difference across the membrane.

(a)

One action potential in the presynaptic neurone does not produce an action potential in the postsynaptic neurone – it requires a series of action potentials in the presynaptic neurone

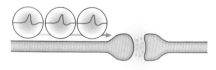

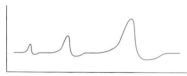

Small EPSPs (excitatory postsynaptic potentials) in postsynaptic neurone do not create an action potential until they act together

(b)

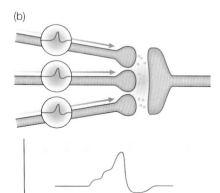

Several presynaptic neurones may each contribute to producing an action potential in the postsynaptic neurone

Figure 1 (a) Temporal summation;
(b) spatial summation

Action potentials and cell signalling

An action potential is an **all or nothing** response. Once the action potential starts, a neurone will conduct it all along its length. The action potential does not vary in size or intensity. At the end of the neurone the presynaptic knob releases transmitter substance into the synaptic cleft. The signal sent to the next neurone is a molecule of acetylcholine. These processes are the same in all neurones and cholinergic synapses.

The roles of synapses in the nervous system

The main role of synapses is to connect two neurones together so that a signal can be passed from one to the other. However, they are also able to perform a number of other important functions.

* Several presynaptic neurones might converge to one postsynaptic neurone. This would allow signals from different parts of the nervous system to create the same response. This could be useful where several different stimuli are warning us of danger.
* One presynaptic neurone might diverge to several postsynaptic neurones. This would allow one signal to be transmitted to several parts of the nervous system. This is useful in a reflex arc. One postsynaptic neurone elicits the response while another informs the brain.
* Synapses ensure that signals are transmitted in the correct direction – only the presynaptic knob contains vesicles of acetylcholine.
* Synapses can filter out unwanted low-level signals. If a low-level stimulus creates an action potential in the presynaptic neurone it is unlikely to pass across a synapse to the next neurone because several vesicles of acetylcholine must be released to create an action potential in the postsynaptic neurone.
* Low-level signals can be amplified by a process called **summation**. If a low-level stimulus is persistent it will generate several successive action potentials in the presynaptic neurone. The release of many vesicles of acetylcholine over a short period of time will enable the postsynaptic generator potentials to combine together to produce an action potential. Summation can also occur when several presynaptic neurones each release small numbers of vesicles into one synapse.
* Acclimatisation – after repeated stimulation a synapse may run out of vesicles containing the transmitter substance. The synapse is said to be fatigued. This means the nervous system no longer responds to the stimulus. It explains why we soon get used to a smell or a background noise. It may also help to avoid overstimulation of an effector, which could damage it.
* The creation of specific pathways within the nervous system is thought to be the basis of conscious thought and memory.

Above all it is the pathways created by synapses that enable the nervous system to convey a wide range of messages. The brain perceives light when it receives signals from the light receptors in the eyes. It perceives sound when it receives signals from the ears. This is because it 'knows' where the signals are coming from, because the neurones from specific receptors always connect to specific regions of the brain.

Module 1
Communication and homeostasis
Signals and messages

The frequency of transmission

The complex interconnections of the nervous system are not the only way that different messages can be conveyed using the same signals. A signal coming from the light receptors in the eye to the relevant optical centre in the brain will inform the brain that light is falling on the eye. However, it says nothing about the intensity of the light.

When a stimulus is at higher intensity the sensory receptor will produce more generator potentials. This will cause more frequent action potentials in the sensory neurone. When these arrive at a synapse they will cause more vesicles to be released. In turn, this creates a higher frequency of action potentials in the postsynaptic neurone. Our brain can determine the intensity of the stimulus from the frequency of signals arriving. A higher frequency of signals means a more intense stimulus.

Myelinated and non-myelinated neurones

Around one third of the peripheral neurones in vertebrates are myelinated. That is, they are insulated by an individual myelin sheath. The sheath is created by a series of separate cells called Schwann cells. These are wrapped around the neurone so the sheath actually consists of several layers of membrane and thin cytoplasm from the Schwann cell. At intervals of 1–3 mm along the neurone there are gaps in the myelin sheath. These are called the nodes of Ranvier. Each node is very short (about 2–3 μm long).

The remainder of the peripheral neurones and the neurones found in the CNS are not myelinated. Non-myelinated neurones are still associated with Schwann cells, but several neurones may be enshrouded in one loosely wrapped Schwann cell. This means that the action potential moves along the neurone in a wave rather than jumping from node to node as seen in myelinated neurones.

What is the advantage of myelination?

Myelinated neurones can transmit an action potential much more quickly than non-myelinated neurones can. Typical speed of transmission in myelinated neurones is 100–120 m s^{-1}. An non-myelinated neurone may only reach a speed of 2–20 m s^{-1}.

Myelinated neurones carry signals from sensory receptors to the CNS and from the CNS to effectors. They carry signals over long distances – the longest neurone in a human can be about 1 m in length. The increased speed of transmission means that the signal reaches the end of the neurone much more quickly. This enables a more rapid response to a stimulus. Non-myelinated neurones tend to be shorter and carry signals only over a short distance. They are used in coordinating body functions such as breathing, and the action of the digestive system. Therefore the increased speed of transmission is not so important.

(a)

A myelinated neurone

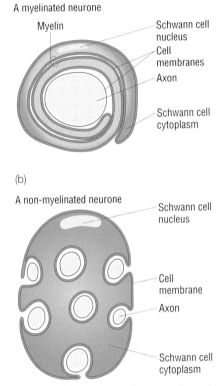

Figure 2 Myelinated and non-myelinated neurones

STRETCH and CHALLENGE

Demyelination is the loss of myelin from neurones that are normally myelinated. Several degenerative diseases are due to the loss of myelin in certain neurones. The loss of muscle coordination faced by people with multiple sclerosis is due to the degeneration of the myelin sheath in neurones involved in the movement of muscles. The disease is suspected to be an autoimmune disorder – the immune system attacks the myelin sheaths. The immune system may cause inflammation due to overproduction of cytokines or other factors such as interferon.

Question

A What role do cytokines play in the immune system? Suggest how overproduction of cytokines could lead to inflammation and cell damage.

Questions

1 Explain what is meant by summation.
2 Draw a table to summarise the differences between the structure and function of myelinated and non-myelinated neurones.

By the end of this spread, you should be able to . . .

* Define the terms *endocrine gland*, *exocrine gland*, *hormone* and *target tissue*.
* Explain the meaning of the terms *first messenger* and *second messenger*, with reference to adrenaline and cyclic AMP (cAMP).
* Describe the functions of the adrenal glands.

Signalling by using hormones

The endocrine system is another communication system in the body. The endocrine system uses the blood circulation to transport its signals. The blood system transports materials all over the body; therefore any signal released into the blood will be transported throughout the body. The signals released by the endocrine system are molecules called **hormones**.

Hormones are released directly into the blood from glands called **endocrine glands**. These are ductless glands – they consist of a group of cells that produce and release the hormone straight into the blood capillaries running through the gland.

Endocrine or exocrine

There are two types of gland in our bodies. As explained above, endocrine glands release or secrete their hormones directly into the blood. The other type of gland is an **exocrine gland**. These glands do not release hormones. Exocrine glands have a small tube or duct that carries their secretion to another place. For example, the salivary glands secrete saliva into a duct. The saliva flows along the duct into the mouth.

Targeting the signal

The cells receiving a hormone signal must possess a specific complementary receptor on their plasma membrane. The hormone binds to this receptor. If all the cells in the body possess such a receptor then all the cells can respond to the signal. Each hormone is different from all the others. This means that a hormone can travel around in the blood without affecting cells that do not possess the correct specific receptor. The cells that possess the specific receptor on their cell surface membranes are called **target cells**. They are usually grouped together to form the **target tissue**.

For this reason the endocrine system can be used to send signals all over the body at the same time. But it can also be used to send very specific signals.

The nature of hormones

There are two types of hormone:
* protein and peptide hormones, and derivatives of amino acids (for example adrenaline, insulin and glucagon)
* steroid hormones (for example the sex hormones).

These two types of hormone work in different ways. Proteins are not soluble in the phospholipid membrane and do not enter the cell. Steroids, however, can pass through the membrane and actually enter the cell to have a direct effect on the DNA in the nucleus.

The action of adrenaline

Adrenaline is an amino acid derivative. It is unable to enter the target cell. Therefore, it must cause an effect inside the cell without entering the cell itself. The adrenaline receptor on the outside of the cell surface membrane has a shape complementary to the shape of the adrenaline molecule. This receptor is associated with an enzyme on the inner surface of the cell surface membrane. The enzyme is called **adenyl cyclase**.

Key definitions

Hormones are molecules that are released by endocrine glands directly into the blood. They act as messengers, carrying a signal from the endocrine gland to a specific target organ or tissue.

An **endocrine gland** is a gland that secretes hormones directly into the blood. Endocrine glands have no ducts.

An **exocrine gland** is a gland that secretes molecules into a duct that carries the molecules to where they are used.

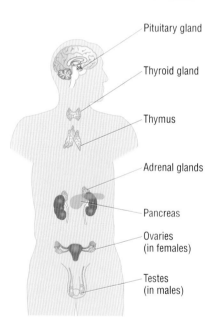

Pituitary gland

Thyroid gland

Thymus

Adrenal glands

Pancreas

Ovaries (in females)

Testes (in males)

Figure 1 The glands in the endocrine system

Key definitions

The **target cells** are those that possess a specific receptor on their plasma (cell surface) membrane. The shape of the receptor is complementary to the shape of the hormone molecule. Many similar cells together form a **tissue**.

Adenyl cyclase is an enzyme associated with the receptor for many hormones, including adrenaline. It is found on the inside of the cell surface membrane.

Module 1
Communication and homeostasis
The endocrine system

Adrenaline in the blood binds to its specific receptor on the cell surface membrane. The adrenaline molecule is called the **first messenger**. When it binds to the receptor it activates the enzyme adenyl cyclase. The adenyl cyclase converts ATP to cyclic AMP (cAMP). The cAMP is a **second messenger** inside the cell. The cAMP can then cause an effect inside the cell by activating enzyme action.

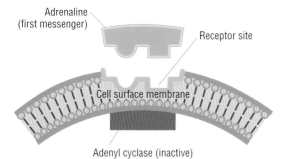

1. Adrenaline receptor site has shape complementary to adrenaline

2. Adrenaline activates the enzyme adenyl cyclase

3. Adenyl cyclase converts ATP to cAMP, which can activate other enzymes inside the cell

Figure 2 The action of adrenaline

> **Key definition**
>
> The **first messenger** is the hormone that transmits a signal around the body. The **second messenger** is cAMP, which transmits a signal inside the cell.

The functions of the adrenal glands

The adrenal glands are found lying anterior to (just above) the kidneys – one on each side of the body. Each gland can be divided into a medulla region and a cortex region.

The adrenal medulla

The medulla is found in the centre of the gland. The cells in the medulla manufacture and release the hormone adrenaline in response to stress such as pain or shock. The effects of adrenaline are widespread and most cells have adrenaline receptors. The effect of adrenaline is to prepare the body for activity. This includes the following effects:

- relax smooth muscle in the bronchioles
- increase stroke volume of the heart
- increase heart rate
- cause general vasoconstriction to raise blood pressure
- stimulate conversion of glycogen to glucose
- dilate the pupils
- increase mental awareness
- inhibit the action of the gut
- cause body hair to erect.

The adrenal cortex

The adrenal cortex uses cholesterol to produce certain steroid hormones. These have a variety of roles in the body.

- The mineralocorticoids (e.g. aldosterone) help to control the concentrations of sodium and potassium in the blood.
- The glucocorticoids (e.g. cortisol) help to control the metabolism of carbohydrates and proteins in the liver.

Questions

1 Explain the difference between exocrine and endocrine glands.
2 Explain why target cells must have specific receptors.
3 Explain how the effects of adrenaline prepare the body for activity.

By the end of this spread, you should be able to ...

* Describe, with the aid of diagrams and photographs, the histology of the pancreas, and outline its role as an endocrine and exocrine gland.
* Explain how blood glucose concentration is regulated, with reference to insulin, glucagon and the liver.

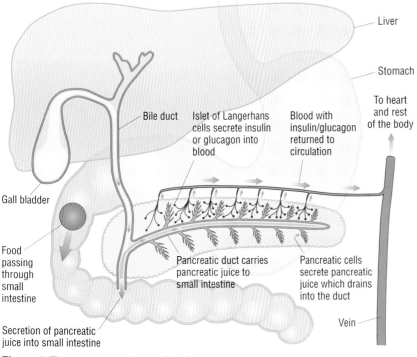

Liver

Stomach

To heart and rest of the body

Bile duct

Islet of Langerhans cells secrete insulin or glucagon into blood

Blood with insulin/glucagon returned to circulation

Gall bladder

Food passing through small intestine

Pancreatic duct carries pancreatic juice to small intestine

Pancreatic cells secrete pancreatic juice which drains into the duct

Vein

Secretion of pancreatic juice into small intestine

Figure 1 The pancreas and associated organs

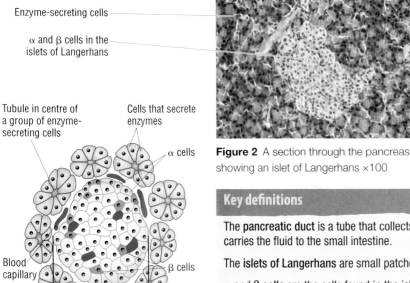

Enzyme-secreting cells

α and β cells in the islets of Langerhans

Tubule in centre of a group of enzyme-secreting cells

Cells that secrete enzymes

α cells

Blood capillary

β cells

Islet of Langerhans

Figure 3 The α and β cells in islet of Langerhans

The pancreas

The **pancreas** is a small organ lying below the stomach. It is an unusual organ in that it has both **exocrine** and **endocrine** functions.

Secretion of enzymes

The majority of cells in the pancreas manufacture and release digestive enzymes. This is the exocrine function of the pancreas. The cells are found in small groups surrounding tiny tubules into which they secrete digestive enzymes. The tubules join to make up the **pancreatic duct** which carries the fluid containing the enzymes into the first part of the small intestine.

The fluid contains the following enzymes:
* amylase – a carbohydrase
* trypsinogen – an inactive protease
* lipase.

The fluid also contains sodium hydrogencarbonate, which makes it alkaline. This helps to neutralise the contents of the digestive system that have just left the acid environment of the stomach.

Secretion of hormones

Certain areas of the pancreas called the **islets of Langerhans** contain different types of cells. There are two types of cell. α **cells** manufacture and secrete the hormone **glucagon**. β **cells** manufacture and secrete the hormone **insulin**. The islets are well supplied with blood capillaries and these hormones are secreted directly into the blood. This is the endocrine function of the pancreas.

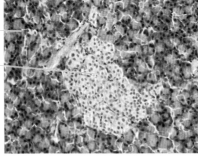

Figure 2 A section through the pancreas showing an islet of Langerhans ×100

Key definitions

The **pancreatic duct** is a tube that collects all the secretions from the exocrine cells in the pancreas and carries the fluid to the small intestine.

The **islets of Langerhans** are small patches of tissue in the pancreas that have an endocrine function.

α and β **cells** are the cells found in the islets of Langerhans.
α cells secrete the hormone glucagon.
β cells secrete the hormone insulin.

Insulin is the hormone, released from the pancreas, that causes blood glucose levels to go down.

Glucagon is the hormone that causes blood glucose levels to rise.

Module 1
Communication and homeostasis
The regulaton of blood glucose

The control of blood glucose

The concentration of blood glucose is carefully regulated. The cells in the islets of Langerhans monitor the concentration of glucose in the blood. The normal blood concentration of glucose is 90 mg 100 cm^{-3} (90 mg in every 100 cm^3 of blood). This can also be expressed as between 4 and 6 mmol dm^{-3}. If the concentration rises or falls away from the acceptable concentration then the α and β cells in the islets of Langerhans detect the change and respond by releasing a hormone.

If blood glucose rises too high

A high blood glucose concentration is detected by the β cells. In response the β cells secrete insulin into the blood. The target cells are the liver cells or **hepatocytes**, muscle cells and some other body cells including those in the brain. These possess the specific membrane-bound receptors for insulin. When the blood passes these cells the insulin binds to the receptors. This activates the adenyl cyclase inside each cell which converts ATP to cAMP (cyclic AMP). The cAMP activates a series of enzyme-controlled reactions in the cell.

Insulin has several effects on the cell:

- More glucose channels are placed into the cell surface membrane.
- More glucose enters the cell.
- Glucose in the cell is converted to glycogen for storage (glycogenesis).
- More glucose is converted to fats.
- More glucose is used in respiration.

The increased entry of glucose, through the specific channels, reduces the blood glucose concentration.

If blood glucose drops too low

A low blood glucose concentration is detected by the α cells. In response the α cells secrete the hormone glucagon. Its target cells are the hepatocytes (liver cells) which possess the specific receptor for glucagon. The effects of glucagon include:

- conversion of glycogen to glucose (glycogenolysis)
- use of more fatty acids in respiration
- the production of glucose by conversion from amino acids and fats (gluconeogenesis).

The overall effect of these changes is to increase the blood glucose concentration.

> ### Key definition
>
> **Hepatocytes** are liver cells. They are specialised to perform a range of metabolic functions.

Examiner tip

It is easy to mix up the spelling of the terms 'glucose', 'glucagon' and 'glycogen'. Add in the terms 'glycogenesis', 'gluconeogenesis' and 'glycogenolysis' and things get very confusing. Write these terms out on a separate page of your notebook along with their meanings. Make sure you can spell them accurately.

STRETCH and CHALLENGE

Most cells are specialised to perform their functions. The two hormones secreted by the pancreas are proteins.

Question

A Describe and explain how the ultrastructure of the α and β cells in the islets of Langerhans is specialised to manufacture and secrete hormones.

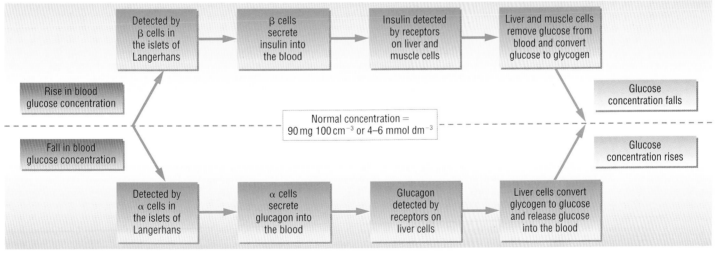

Figure 4 Controlling the concentration of glucose in the blood

Questions

1 Why do hepatocytes have specialised receptors for both insulin and glucagon?
2 State which molecules named on this spread are first messengers and which are second messengers.

By the end of this spread, you should be able to . . .

＊ **Outline how insulin secretion is controlled, with reference to potassium ion channels and calcium ion channels in β-cells in the islets of Langerhans.**

＊ **Compare and contrast the causes of type I (insulin-dependent) and type II (non-insulin-dependent) diabetes mellitus.**

＊ **Discuss the use of insulin produced by genetically modified bacteria, and the potential use of stem cells, to treat diabetes mellitus.**

The importance of regulating insulin levels

Insulin brings about effects that reduce the blood glucose concentration. If the blood glucose concentration is too high then it is important that insulin is released from the β cells. However, if the blood glucose concentration drops too low it is important that insulin secretion stops.

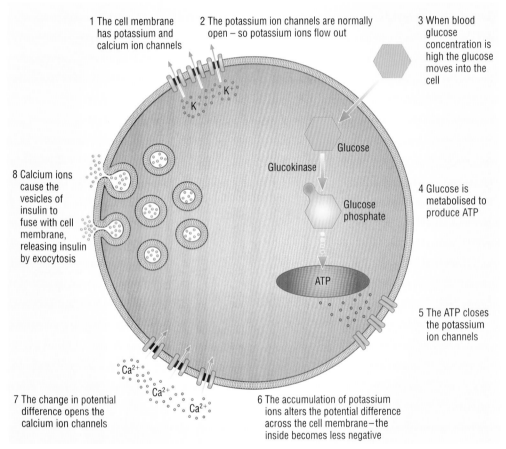

1 The cell membrane has potassium and calcium ion channels

2 The potassium ion channels are normally open – so potassium ions flow out

3 When blood glucose concentration is high the glucose moves into the cell

4 Glucose is metabolised to produce ATP

Glucose

Glucokinase

Glucose phosphate

ATP

5 The ATP closes the potassium ion channels

8 Calcium ions cause the vesicles of insulin to fuse with cell membrane, releasing insulin by exocytosis

7 The change in potential difference opens the calcium ion channels

6 The accumulation of potassium ions alters the potential difference across the cell membrane – the inside becomes less negative

Figure 1 The mechanism of insulin secretion

The control of insulin secretion

1 The cell membranes of the β cells contain both calcium ion channels and potassium ion channels.

2 The potassium ion channels are normally open and the calcium ion channels are normally closed. Potassium ions diffuse out of the cell making the inside of the cell more negative; at rest the potential difference across the cell membrane is about −70 mV.

3 When glucose concentrations outside the cell are high, glucose molecules diffuse into the cell.

4 The glucose is quickly used in metabolism to produce ATP.

5 The extra ATP causes the potassium channels to close.

6 The potassium can no longer diffuse out and this alters the potential difference across the cell membrane – it becomes less negative inside.

7 This change in potential difference opens the calcium ion channels.

8 Calcium ions enter the cell and cause the secretion of insulin by making the vesicles containing insulin move to the cell surface membrane and fuse with it, releasing insulin by **exocytosis**.

Diabetes mellitus

Blood glucose levels never remain absolutely constant. After a meal the concentration will rise and during exercise the concentration will fall. However, using a negative feedback mechanism the body is able to keep the blood concentration fairly well controlled within certain limits.

Diabetes mellitus is a disease in which the body is no longer able to control its blood glucose concentration. This can lead to very high concentrations (**hyperglycaemia**)

Key definitions

Diabetes mellitus is a disease in which blood glucose concentrations cannot be controlled effectively.

Hyperglycaemia is the state in which the blood glucose concentration is too high (hyper = above, glyc = glucose, aemia = blood).

Module 1
Communication and homeostasis
Regulation of insulin levels

of glucose after a meal rich in sugars and other carbohydrates. It can also lead to the concentration dropping too low (**hypoglycaemia**) after exercise or after fasting.

Type I diabetes

Type I diabetes is also known as insulin-dependent diabetes. It is often called juvenile-onset diabetes because it usually starts in childhood. It is thought to be the result of an autoimmune response in which the body's own immune system attacks the β cells and destroys them. It may also result from a viral attack. The body is no longer able to manufacture sufficient insulin and cannot store excess glucose as glycogen.

Type II diabetes

Type II diabetes is also known as non-insulin-dependent diabetes. A person with type II diabetes can still produce insulin. However, as people age, their responsiveness to insulin declines. This is probably because the specific receptors on the surface of the liver and muscle cells decline and the cells lose their ability to respond to the insulin in the blood. The levels of insulin secreted by the β cells may also decline. It is thought that anyone who lives long enough will eventually become diabetic – but this may not be until they are about 120 years old! Certain factors seem to bring an earlier onset of type II diabetes. These include:

* obesity
* a diet high in sugars, particularly refined sugars
* being of Asian or Afro-Caribbean origin
* family history.

Treatment of diabetes

* Type II diabetes is usually treated by careful monitoring and control of the diet. Care is taken to match carbohydrate intake and use. This may eventually be supplemented by insulin injections or use of other drugs which slow down the absorption of glucose from the digestive system.
* Type I diabetes is treated using insulin injections. The blood glucose concentration must be monitored and the correct dose of insulin must be administered to ensure that the glucose concentration remains fairly stable.

The source of insulin

Insulin used to be extracted from the pancreas of animals – usually from pigs as this matches human insulin most closely. However, more recently, insulin can be produced by bacteria that have been **genetically engineered** to manufacture human insulin.

The advantages of using insulin from genetically engineered bacteria include:

* It is an exact copy of human insulin, therefore it is faster acting and more effective.
* There is less chance of developing tolerance to the insulin.
* There is less chance of rejection due to an immune response.
* There is a lower risk of infection.
* It is cheaper to manufacture the insulin than to extract it from animals.
* The manufacturing process is more adaptable to demand.
* Some people are less likely to have moral objections to using the insulin produced from bacteria than to using that extracted from animals.

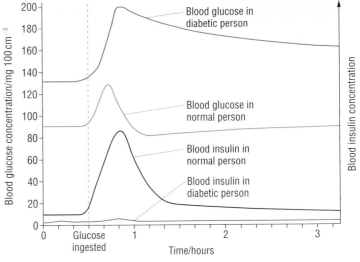

Figure 2 Blood glucose and insulin levels in a diabetic and a healthy person

Key definitions

Hypoglycaemia is the state in which the blood glucose concentration is too low (hypo = under).

Genetically engineered bacteria are those in which the DNA has been altered. In this case a gene coding for human insulin has been inserted into the DNA of the bacteria.

Stem cells are unspecialised cells that have the potential to develop into any type of cell.

A new way to treat diabetes

Recent research has shown that it may be possible to treat type I diabetes using **stem cells**. Stem cells are not yet differentiated and can be induced to develop into a variety of cell types. The most common sources of stem cells are bone marrow and the placenta. However, scientists have recently found precursor cells in the pancreas of adult mice. These cells are capable of developing into a variety of cell types and may be true stem cells. If similar cells can be found in the human pancreas then they could be used to produce new β cells in patients with type I diabetes.

Questions

1 Insulin is a protein. Why must insulin be injected rather than being taken orally?
2 Explain the meaning of the terms *hyperglycaemia* and *hypoglycaemia*.
3 Compare the mechanism of insulin secretion with the secretion of neurotransmitter at a synapse.

By the end of this spread, you should be able to ...

* Outline the hormonal and nervous mechanisms involved in the control of heart rate in humans.

The human heart

The heart pumps blood around the circulatory system. Blood supplies the tissues with oxygen and nutrients such as glucose, fatty acids and amino acids. It also removes waste products, such as carbon dioxide and urea, from the tissues so that they do not accumulate and inhibit **cell metabolism**.

The requirements of the cells vary according to their level of activity. When you are being physically active your muscle cells need more oxygen and glucose and your heart muscle cells need more oxygen and fatty acids. These cells also need to remove more carbon dioxide.

It is, therefore, important that the heart can adapt to meet the requirements of the body.

How the heart adapts to supply more oxygen and glucose

- The most obvious change in the activity of the heart is an increase in the number of beats per minute. This is known as the heart rate.
- The heart can increase the strength of its contractions.
- It can also increase the volume of blood pumped per beat (the stroke volume).

Control of heart rate

The rate at which the heart beats is affected by a number of factors.
- The heart muscle is **myogenic**.
- The heart contains its own **pacemaker** – this is called the **sinoatrial node (SAN)**. The SAN is a region of tissue that can initiate an action potential, which travels as a wave of excitation over the atria walls, through the **AVN (atrioventricular node)** and down the Purkyne (sometimes spelt Purkinje) fibres to the ventricles, causing them to contract.
- The heart is supplied by nerves from the **medulla oblongata** of the brain. These nerves connect to the SAN. These do not initiate a contraction, but they can affect the frequency of the contractions. Action potentials sent down the accelerator nerve increase the heart rate. Action potentials sent down the vagus nerve reduce the heart rate.
- The heart muscle responds to the presence of the hormone adrenaline in the blood.

Interaction between control mechanisms

The various factors that affect heart rate must interact in a coordinated way to ensure that the heart beats at the most appropriate rate.

Under resting conditions the heart rate is controlled by the SAN. This has a set frequency, varying from person to person, at which it initiates waves of excitation. The frequency of waves is typically 60–80 per minute. However, the frequency of these excitation waves can be controlled by the **cardiovascular centre** in the medulla oblongata.

There are many factors that affect the heart rate.
- Movement of the limbs is detected by stretch receptors in the muscles. These send impulses to the cardiovascular centre informing it that extra oxygen may soon be needed. This tends to increase heart rate.
- When we exercise the muscles produce more carbon dioxide. Some of this reacts with the water in the blood plasma and reduces its pH. The change in pH is detected by chemoreceptors in the carotid arteries, the aorta and the brain. The chemoreceptors send impulses to the cardiovascular centre, which increases the heart rate.

Key definition

Cell metabolism is the result of all the chemical reactions taking place in the cytoplasm.

Key definitions

Myogenic muscle tissue can initiate its own contractions.

The **pacemaker** is a region of tissue in the right atrium wall that can generate an impulse and initiates the contraction of the chambers.

The **medulla oblongata** is found at the base of the brain. It is the region of the brain that coordinates the unconscious functions of the body such as breathing rate and heart rate.

The **accelerator nerve** and the **vagus nerve** run from the medulla oblongata to the heart.

The **cardiovascular centre** is a specific region of the medulla oblongata that receives sensory inputs about levels of physical activity, blood carbon dioxide concentration and blood pressure. It sends nerve impulses to the SAN in the heart to alter the frequency of excitation waves.

- When we stop exercising the concentration of carbon dioxide in the blood falls. This reduces the activity of the accelerator pathway. Therefore the heart rate declines.
- Adrenaline is secreted in response to stress, shock, anticipation or excitement. The presence of adrenaline in the blood increases the heart rate. This helps to prepare the body for activity.
- Blood pressure is monitored by stretch receptors in the walls of the carotid sinus. This is a small swelling in the carotid artery. If blood pressure rises too high, perhaps during vigorous exercise, the stretch receptors send signals to the cardiovascular centre. This responds by reducing the heart rate.

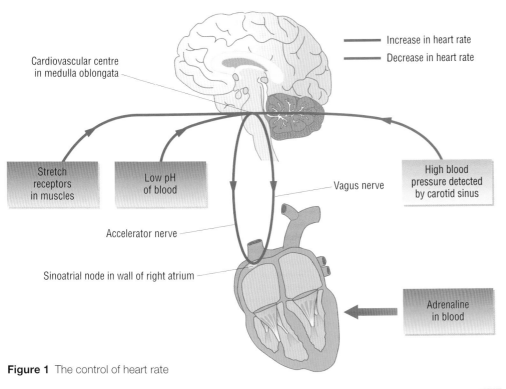

Figure 1 The control of heart rate

Examiner tip

Always refer to the medulla oblongata by its full name. Other organs have a medulla (think of the kidney) – it just means the inside bit. The medulla oblongata is a specific part of the brain.

Artificial pacemakers

If the mechanism controlling the heart rate fails then an artificial pacemaker must be fitted. A pacemaker is a device that delivers an electrical impulse to the heart muscle. The first pacemaker in 1928 was a device that included a needle electrode which was inserted into the heart wall. This was not portable as it needed to be connected to a light fitting!

Further development produced devices that could deliver impulses via an electrode pad on the skin. These still needed mains circuit electricity to function. Also, the method of delivering the impulse to the body was a little similar to the way in which people are executed on the electric chair. Patients, unsurprisingly, complained that it was painful.

Real advances came in the design of pacemakers when miniaturisation became possible during the 1950s. This allowed the patient to wear a small plastic box with wires inserted through the skin to act as electrodes on the heart muscle.

Modern pacemakers are only about 4 cm long. They may be implanted under the skin and fat on the chest (or sometimes within the chest cavity) and are capable of responding to the activity of the patient.

Some pacemakers deliver impulses to the ventricle walls. This deals with conditions where the AVN (atrioventricular node) that normally relays the impulse from atria to ventricles, via the Purkyne fibres, is not functioning but the SAN may be functioning.

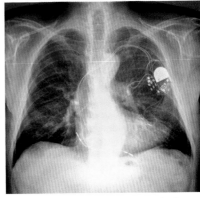

Figure 2 An artificial pacemaker in the chest of a patient

Questions

1 Explain what is meant by the term *myogenic*.
2 Why must the heart be able to respond to increased physical activity?

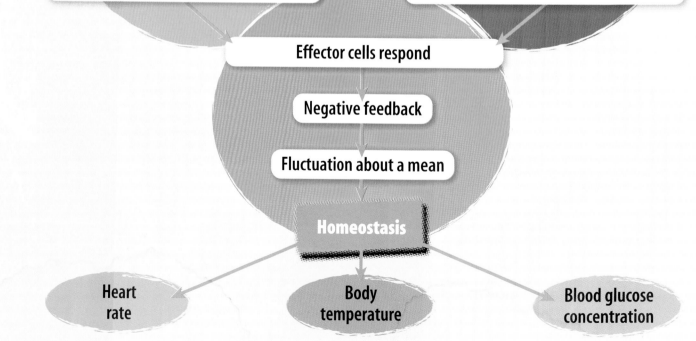

Environmental change
Could be due to external changes or changes brought about by the activities of the cells

Sensory receptors detect a change in the environment

Communication between cells
Sensory receptors that detect the change and signal to cells that respond

Nervous communication

Hormonal communication

Rapid response made direct by complex connections between neurones

Slower, longer-term response made specific by use of complementary cell surface receptors

Action potentials: all or nothing, signal strength made variable by altering frequency of action potentials

Signal strength made variable by altering concentration of hormone

Effector cells respond

Negative feedback

Fluctuation about a mean

Homeostasis

Heart rate

Body temperature

Blood glucose concentration

Practice questions

1. List three environmental changes that may cause an organism stress. [3]

2. Define the terms 'stimulus' and 'response'. [4]

3. What is meant by the term 'cell signalling'? [2]

4. State four conditions inside the body that must be kept constant. [4]

5. Complete the table:

Response to feeling warm	Explanation – how does this response help to cool the body?
The skin releases more sweat.	
The blood vessels in the skin dilate.	
	Less air is trapped close to the skin so that the body is less well insulated. More heat can be lost by convection and radiation.
Sit in the shade.	

[4]

6. Sensory receptors act as transducers. What is meant by the term 'transducer'? Use an example to explain your answer. [4]

7. How are generator potentials created? [2]

8. (a) Describe the sequence of events in the neurone membrane during an action potential. [5]

    (b) How does an action potential move along the neurone? [3]

9. (a) Explain the difference between myelinated neurones and non-myelinated neurones. [4]

    (b) What effect does myelination have on the neurone? [2]

10. (a) Outline the roles of synapses. [5]

    (b) Describe what happens in the synaptic knob as a result of an action potential passing along the neurone. [4]

    (c) How are postsynaptic membranes specialised? [3]

11. List the organs in the endocrine system. [5]

12. Describe how different hormones can produce specific responses. [4]

13. Describe how the concentration of glucose in the blood is monitored and kept constant. [9]

14. What are the differences between type I and type II diabetes? [4]

15. Describe and explain the effect of exercise on the heart rate. [8]

1 The pancreas is a gland that has both endocrine and exocrine functions. Figure 1.1 shows a section through part of the pancreas.

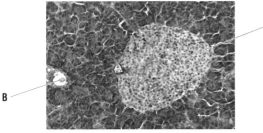

Figure 1.1

(a) Name parts **A** and **B**. [2]
(b) Explain the difference between the terms **endocrine** and **exocrine** with regard to the pancreas. [4]

[Total: 6]
(OCR 2805 Jun06)

2 In animals chemical messengers help to transfer information from one part of the organism to another to achieve coordination.

(a) The table below lists some of these chemicals together with their functions.
Complete the table.

Name of chemical messenger	Function
	To prepare the body for activity
Insulin	
Glucagon	

[3]

(b) Mammals also rely on nerves to transfer information in the form of electrical impulses. Using the information shown in Figure 2.1, outline how impulses are transmitted from receptor to effector.

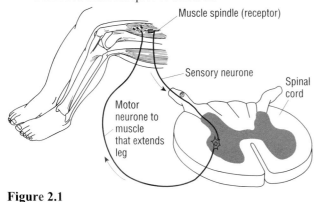

Figure 2.1

In your answer you should use appropriate technical terms spelled correctly. [9]

[Total: 12]
(OCR 2804 Jan06)

3 Parkinson's disease is a disorder of the nervous system. People with this condition are unable to produce enough of the neurotransmitter substance dopamine. This chemical is required in neurone circuits in the brain that control movement.

(a) Outline **two** roles of synapses in the nervous system. [2]

Figure 3.1 illustrates the events at a synapse where the neurotransmitter is dopamine.

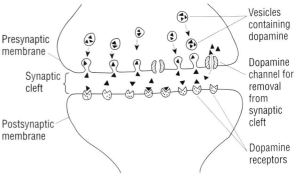

Figure 3.1

(b) Using **only** the information in Figure 3.1, list **three** ways in which the events occurring at this synapse are the same as at a cholinergic synapse. [3]
(c) For the proper functioning of neurone circuits, neurotransmitters have to be removed from the receptors in the postsynaptic membrane and from the synaptic cleft.
Explain why this is so. [2]

[Total: 7]
(OCR 2804 Jun06)

4 (a) Explain the term **endocrine gland**. [2]
(b) Figure 4.1 is a flow diagram of the role of the pancreas in controlling blood glucose concentrations. Study the diagram and answer the questions below.

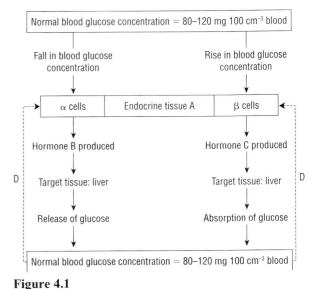

Figure 4.1

Answers to examination questions will be found on the Exam Café CD.

   (i) Name the endocrine tissue labelled **A**. [1]
   (ii) Name the hormone, **B**, produced by the α cells. [1]
   (iii) Name the hormone, **C**, produced by the β cells. [1]
   (iv) Name the process represented by the dotted lines labelled **D**. [1]
   (v) Describe how hormone **B** brings about a rise in blood glucose concentration when it reaches the liver. [5]
(c) Untreated diabetes is a condition that can lead to blood glucose concentrations rising above 120 mg 100 cm^{-3} of blood. Genetic engineering has been used to improve the treatment of diabetes.
Explain the advantages of using genetic engineering in the treatment of diabetics. [3]
[Total: 14]
(OCR 2804 Jun05)

5 Figure 5.1 represents **some** of the changes that occur across the membrane of the axon. Three protein complexes are shown to be present in the membrane:
- sodium ion channels
- potassium ion channels
- sodium–potassium pumps.

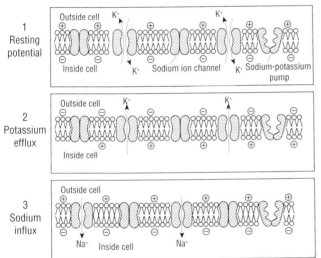

Figure 5.1

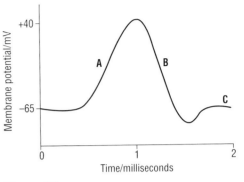

Figure 5.2

Figure 5.2 shows the change of membrane potential associated with an action potential.
(a) (i) State which of the three diagrams of the axon membrane in Figure 5.1 match up to the phases labelled in Figure 5.2. Write your answers in the table below.

Phase	Number
A	
B	
C	

[1]
   (ii) With reference to Figure 5.1, explain the changes in membrane potential in Figure 5.2. [5]
(b) In order to transfer information from one point to another in the nervous system, it is necessary that action potentials be transmitted along axons. In humans, the rate of transmission is 0.5 m s^{-1} in a nonmyelinated neurone, increasing to 100 m s^{-1} in a myelinated neurone.
Explain how action potentials are transmitted along a nonmyelinated neurone **and** describe which parts of this process are different in myelinated neurones.
No credit will be given for reference to events at the synapse.
In your answer you should use appropriate technical terms spelled correctly. [8]
[Total: 14]
(OCR 2804 Jan07)

33

Introduction

All cells perform a variety of metabolic functions – these are essentially chemical reactions taking place inside the cell. These reactions produce substances that the body needs, such as proteins or hormones. However, they may also produce by-products that the body does not need or that may be toxic if allowed to accumulate. It is the removal of these by-products that we call excretion.

The liver is the most metabolically active organ in the body. The liver has many functions including:
- adjusting the concentrations of certain substances in the blood
- removing toxic compounds from the blood
- converting toxic compounds to less toxic compounds.

However, the liver cannot dispose of many of the waste products it creates. It must return them to the blood.

The kidneys are the main organs that can remove wastes from the blood and dispose of them. The kidneys filter blood and produce urine. Urine is used to remove a wide variety of wastes from the blood and therefore from the body. These include:
- excess water
- excess salts
- nitrogenous wastes such as urea
- breakdown products of hormones.

The contents of the urine can tell us a lot about what is happening in the body. For instance, the presence of human pregnancy hormone (chorionic gonadotrophin) can be used to detect pregnancy and the presence of steroid metabolites can be used to detect the use of anabolic steroids by athletes.

Test yourself

1 What is meant by the term excretion?
2 How does excretion differ from egestion?
3 Carbon dioxide is an excretory product. What organ is used to excrete carbon dioxide?
4 What substances are found in urine?
5 Where is urine stored before being released from the body?
6 Why must toxic compounds be removed from the body?
7 What functions do hormones perform?

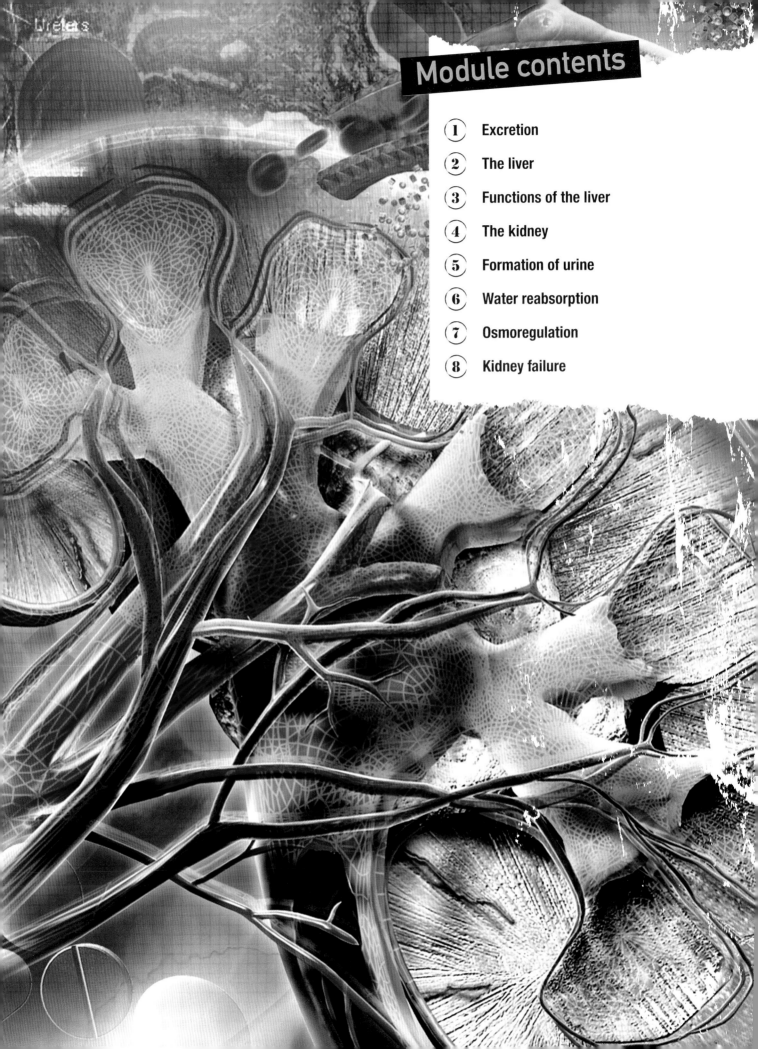

Ureters

Module contents

By the end of this spread, you should be able to ...

* Define the term *excretion*.
* Explain the importance of removing metabolic wastes from the body, including carbon dioxide and nitrogenous waste.

Excretion

Excretion is the removal of **metabolic waste** from the body, that is the removal, from the body, of by-products or unwanted substances from normal cell processes.

The main substances to be excreted

There are many substances that need to be excreted – almost any cell product that is formed in excess by the chemical processes occurring in the cells must be excreted. However, there are two products that are produced in very large amounts:

* carbon dioxide from respiration
* nitrogen-containing compounds such as **urea**.

Where are these substances produced?

* Carbon dioxide is produced by every living cell in the body as a result of respiration.
* Urea is produced in the liver from excess amino acids.

Where are these substances excreted?

Carbon dioxide is passed from the cells of respiring tissues into the bloodstream. It is transported in the blood (mostly in the form of hydrogencarbonate ions) to the lungs. In the lungs the carbon dioxide diffuses into the alveoli to be excreted as we breathe out.

Urea is produced by breaking down excess amino acids in the liver. This process is called **deamination**. The urea is then passed into the bloodstream to be transported to the kidneys. It is transported in solution – dissolved in the plasma. In the kidneys the urea is removed from the blood to become a part of urine. Urine is stored in the bladder before being excreted via the urethra.

Key definitions

Excretion is the removal of metabolic waste from the body.

Metabolic waste consists of waste substances that may be toxic or are produced in excess by the reactions inside cells.

Examiner tip

Many students confuse egestion, or elimination, with excretion. Egestion is the removal of undigested food by the process of defecation. These substances have never been in cells and so cannot be excreted.

Key definition

Deamination is the removal of the amine group from an amino acid to produce ammonia.

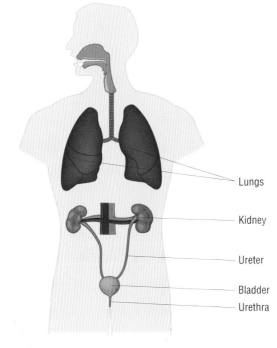

Lungs

Kidney

Ureter

Bladder

Urethra

Figure 1 A diagram showing the positions of the main excretory organs

Why must these substances be removed?

Carbon dioxide

Excess carbon dioxide is toxic. A high level of carbon dioxide has three main effects.

- The majority of carbon dioxide is carried in the blood as hydrogencarbonate ions. However, from your AS work you will recall that forming hydrogencarbonate ions also forms hydrogen ions (see below). This occurs inside the red blood cells, under the influence of the enzyme carbonic anhydrase. The hydrogen ions combine with haemoglobin. They compete with oxygen for space on the haemoglobin. If there is too much carbon dioxide in the blood it can reduce oxygen transport.

- The carbon dioxide also combines directly with haemoglobin to form carbaminohaemoglobin. This molecule has a lower affinity for oxygen than normal haemoglobin.

- Excess carbon dioxide can also cause respiratory acidosis. The carbon dioxide dissolves in the blood plasma. Once dissolved it can combine with water to produce carbonic acid:

$$CO_2 + H_2O \rightarrow H_2CO_3$$

The carbonic acid dissociates to release hydrogen ions:

$$H_2CO_3 \rightarrow H^+ + HCO_3^-$$

The hydrogen ions lower the pH and make the blood more acidic. Proteins in the blood act as buffers to resist the change in pH. If the change in pH is small then the extra hydrogen ions are detected by the respiratory centre in the medulla oblongata of the brain. This causes an increase in breathing rate to help remove the excess carbon dioxide. However, if the blood pH drops below 7.35 it results in slowed or difficult breathing, headache, drowsiness, restlessness, tremor and confusion. There may also be a rapid heart rate and changes in blood pressure. This is respiratory acidosis. It can be caused by diseases or conditions that affect the lungs themselves, such as emphysema, chronic bronchitis, asthma or severe pneumonia. Blockage of the airway due to swelling, a foreign object, or vomit can also induce respiratory acidosis.

Nitrogenous compounds

The body cannot store proteins or amino acids. However, amino acids contain almost as much energy as carbohydrates. Therefore it would be wasteful to simply excrete excess amino acids. Instead they are transported to the liver and the potentially toxic amino group is removed (deamination). The amino group initially forms the very soluble and highly toxic compound, ammonia. This is converted to a less soluble and less toxic compound called urea, which can be transported to the kidneys for excretion. The remaining keto acid can be used directly in respiration to release its energy or it may be converted to a carbohydrate or fat for storage.

Deamination: amino acid + oxygen → keto acid + ammonia

Formation of urea: ammonia + carbon dioxide → urea + water

$$2NH_3 + CO_2 \rightarrow CO(NH_2)_2 + H_2O$$

Questions

1 Explain the difference between excretion and egestion.
2 Explain what causes respiratory acidosis.
3 Suggest why fish can excrete ammonia but mammals must convert it to urea for excretion.

Examiner tip

Review the production of and transport of carbon dioxide in the AS student book, spread 1.2.12.

By the end of this spread, you should be able to ...

* **Describe, with the aid of diagrams and photographs, the histology and gross structure of the liver.**

The structure of the liver

The liver cells (**hepatocytes**) carry out many hundreds of metabolic processes and the liver has an important role in **homeostasis**. It is therefore essential that it has a very good supply of blood. The internal structure of the liver is arranged to ensure that as much blood as possible flows past as many liver cells as possible.

Blood flow to and from the liver

Unusually, the liver is supplied with blood from two sources.

* Oxygenated blood from the heart. Blood travels from the aorta via the hepatic artery into the liver. This supplies the oxygen that is essential for aerobic respiration (see Module 4). The liver cells are very active as they carry out many metabolic processes. Many of these processes require energy, in the form of ATP, so it is important that there is a good supply of oxygen.
* Deoxygenated blood from the digestive system. This enters the liver via the **hepatic portal vein**. This blood is rich in the products of digestion. The concentrations of various compounds will be uncontrolled, and the blood may contain toxic compounds that have been absorbed in the intestine.

Key definition

The **hepatic portal vein** is an unusual blood vessel that has capillaries at both ends – it carries blood from the digestive system to the liver.

Blood leaves the liver via the hepatic vein. This rejoins the vena cava and the blood returns to normal circulation.

A fourth vessel connected to the liver is not a blood vessel. It is the bile duct. Bile is a secretion from the liver. It has both a digestive function and an excretory function. The bile duct carries bile from the liver to the gall bladder where it is stored until required to aid the digestion of fats in the small intestines.

The arrangement of cells inside the liver

The cells, blood vessels and chambers inside the liver are arranged to ensure the best possible contact between the blood and the liver cells. The liver is divided into lobes, which are further divided into cylindrical lobules.

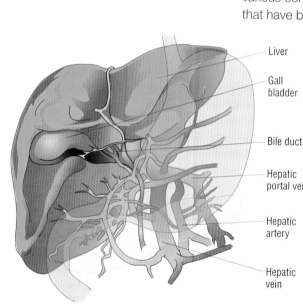

Liver

Gall bladder

Bile duct

Hepatic portal vein

Hepatic artery

Hepatic vein

Figure 1 The liver and its connections to the blood system

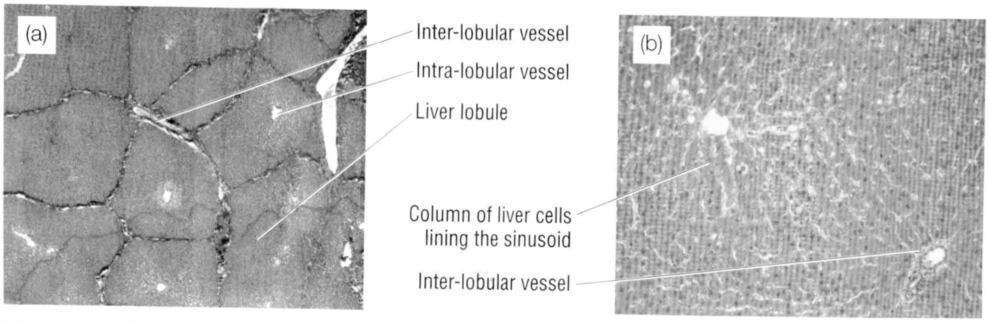

(a)

Inter-lobular vessel

Intra-lobular vessel

Liver lobule

Column of liver cells lining the sinusoid

Inter-lobular vessel

(b)

Liver cells   Sinusoid   Branch of hepatic vein

Branch of hepatic portal vein   Branch of bile duct   Branch of hepatic artery

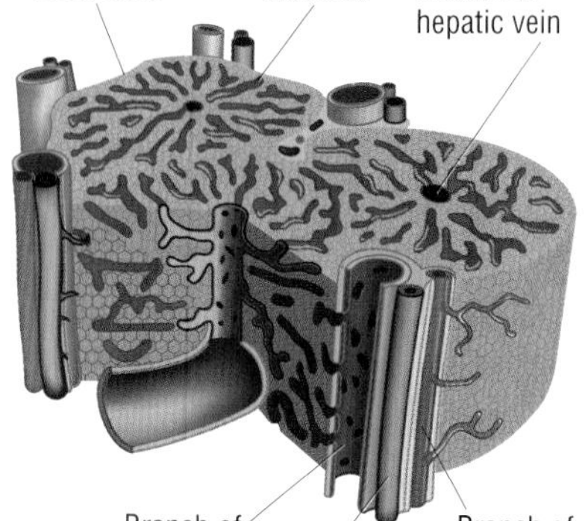

Figure 3 The arrangement of liver cells into cylindrical lobules

Figure 2 (a) Liver lobules ×40; **(b)** A section through a lobule ×35

As the hepatic artery and hepatic portal vein enter the liver they split into smaller and smaller vessels. These vessels run between, and parallel to, the lobules, and are known as inter-lobular vessels. At intervals, branches from the hepatic artery and the hepatic portal vein enter the lobules. The blood from the two blood vessels is mixed and passes along a special chamber called a sinusoid. The sinusoid is lined by liver cells. The sinusoids empty into the intra-lobular vessel, a branch of the hepatic vein. The branches of the hepatic vein from different lobules join together to form the hepatic vein, which drains blood from the liver.

As the blood flows along the sinusoid it is in very close contact with the liver cells. They are able to remove molecules from the blood and pass molecules into the blood.

One of the many functions of the liver cells is to manufacture bile. This is released into the bile canaliculi (meaning small canals). These join together to form the bile duct, which transports the bile to the gall bladder.

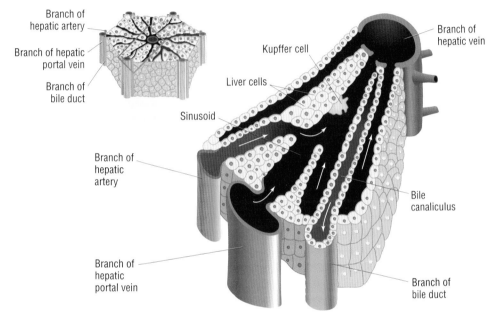

Figure 4 The arrangement of liver cells in a lobule

Liver cells

Liver cells, or hepatocytes, appear to be relatively unspecialised. They have a simple cuboidal shape with many **microvilli** on their surface. However, their many metabolic functions include protein synthesis, transformation and storage of carbohydrates, synthesis of cholesterol and bile salts, detoxification and many other processes. This means that their cytoplasm must be very dense and is specialised in the amounts of certain **organelles** that it contains.

Kupffer cells

Kupffer cells are specialised macrophages. They move about within the sinusoids and are involved in the breakdown and recycling of old red blood cells. One of the products of haemoglobin breakdown is **bilirubin**, which is excreted as part of the bile and in faeces. Bilirubin is the brown pigment in faeces.

Questions

1 Explain why the liver has two supplies of blood.
2 Describe how liver structure ensures that the blood flows past as many liver cells as possible.
3 Suggest which organelles may be particularly common in the cytoplasm of liver cells.

Key definitions

The primary function of **Kupffer cells** appears to be the breakdown and recycling of old red blood cells.

Bilirubin is one of the waste products from the breakdown of haemoglobin.

By the end of this spread, you should be able to . . .

* Describe the formation of urea in the liver, including an outline of the ornithine cycle.
* Describe the roles of the liver in detoxification.

A wide range of functions

The liver is very metabolically active and carries out a wide range of functions, including:

- *control of*: blood glucose levels, amino acid levels, lipid levels
- *synthesis of*: red blood cells in the fetus, bile, plasma proteins, cholesterol
- *storage of*: vitamins A, D and B_{12}, iron, glycogen
- *detoxification of*: alcohol, drugs
- breakdown of hormones
- destruction of red blood cells.

Formation of urea

Every day we each need 40–60 g of protein. However, most people in developed countries eat far more than this. Excess amino acids cannot be stored, as the amine groups make them toxic. However, the amino acid molecules contain a lot of energy, so it would be wasteful to excrete the whole molecule. Therefore it undergoes treatment in the liver before the amino component is excreted. This treatment consists of two processes, **deamination** (a) and the **ornithine cycle** (b).

$$\text{amino acid} \longrightarrow \text{ammonia} \longrightarrow \textbf{urea}$$
$$\uparrow \quad + \text{keto acid} \quad \uparrow$$
$$\text{(a) deamination} \quad \text{(b) ornithine cycle}$$

Deamination

The process of deamination produces ammonia, which is very soluble and highly toxic. This ammonia must not be allowed to accumulate. It also produces an organic compound, a keto acid, which can enter respiration directly (see spread 1.4.9) to release its energy.

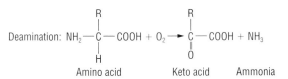

Deamination: $NH_2 - \overset{R}{\underset{H}{C}} - COOH + O_2 \rightarrow \overset{R}{\underset{O}{C}} - COOH + NH_3$

Amino acid   Keto acid   Ammonia

The ornithine cycle

The ammonia is very soluble and highly toxic. It must be converted to a less toxic form very quickly. The ammonia is combined with carbon dioxide to produce urea. This occurs in the ornithine cycle. Urea is both less soluble and less toxic than ammonia. It can be passed back into the blood and transported around the body to the kidneys. In the kidneys the urea is filtered out of the blood and concentrated in the urine. Urine can be stored in the bladder until it is released from the body. The ornithine cycle can be summarised as:

$$2NH_3 \quad + \quad CO_2 \quad \rightarrow CO(NH_2)_2 + H_2O$$
$$\text{ammonia} + \text{carbon dioxide} \rightarrow \quad \text{urea} \quad + \text{water}$$

Key definitions

Urea is an excretory product formed from the breakdown of excess amino acids.

The **ornithine cycle** is the process in which ammonia is converted to urea. It occurs partly in the cytosol and partly in mitochondria, as ATP is used.

Ammonia NH_3 CO_2

H_2O

Citrulline NH_3

Ornithine   **Ornithine cycle**

Arginine H_2O

H_2O

Urea $CO(NH_2)_2$

Figure 1 The ornithine cycle

Detoxification

The liver is able to detoxify many compounds. Some of these compounds, such as hydrogen peroxide, are produced in the body. Some compounds, such as alcohol, may be consumed as a part of our diet. Others, such as drugs, may be taken for health reasons or for recreational purposes. Toxins can be rendered harmless by oxidation, reduction, methylation or combination with another molecule.

Liver cells contain many enzymes that render toxic molecules less toxic. These include catalase, which converts hydrogen peroxide to oxygen and water. Catalase has a particularly high turnover number – the number of molecules of hydrogen peroxide that one molecule of catalase can render harmless in one minute – of five million.

Key definition

Detoxification is the conversion of toxic molecules to less toxic or non-toxic molecules.

Detoxification of alcohol

Alcohol, or ethanol, is a drug that depresses nerve activity. In addition, it contains chemical potential energy, which can be used for respiration. It is broken down in the hepatocytes (liver cells) by the action of the enzyme ethanol dehydrogenase. The resulting compound is ethanal. This is dehydrogenated further by the enzyme ethanal dehydrogenase. The final compound produced is ethanoate (acetate), which is combined with coenzyme A to form acetyl coenzyme A, which enters the process of respiration. The hydrogen atoms released in this process are combined with another coenzyme called NAD to form reduced NAD (see spread 1.4.2).

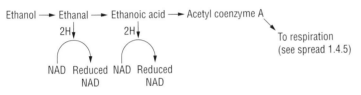

NAD is also required to oxidise and break down fatty acids for use in respiration (see spread 1.4.9). If the liver has to detoxify too much alcohol it has insufficient NAD to deal with the fatty acids. These fatty acids are then converted back to lipids and are stored in hepatocytes, causing the liver to become enlarged. This is a condition known as 'fatty liver' which can lead to alcohol-related hepatitis or to cirrhosis.

STRETCH and CHALLENGE

Liver cells contain a large group of enzymes called the cytochrome P450 enzymes. These are responsible for the breakdown of some toxic molecules, such as cocaine and other drugs (recreational and medicinal). The P450s are most concentrated in the endoplasmic reticulum of liver cells. As a result of variation these enzymes can be more effective in some people than in others.
Many drugs can be more effective in some people than in others and may also cause variable side effects.

Questions

A Suggest why the P450s are most concentrated in the endoplasmic reticulum.
B Suggest why many medicinal drugs have different side effects in different people.
C Explain why the P450s are not identical in every person.

Questions

1 Why must ammonia be converted to urea?
2 Explain why excess amino acids and alcohol should not be excreted.
3 Suggest why the liver cells have large numbers of mitochondria and ribosomes.

By the end of this spread, you should be able to . . .

✳ Describe, with the aid of diagrams and photographs, the histology and gross structure of the kidney.

✳ Describe, with the aid of diagrams and photographs, the detailed structure of a nephron and its associated blood vessels.

The structure of the kidney

Most people have two kidneys. These are positioned on each side of the spine just below the lowest rib. Each kidney is supplied with blood from a renal artery and is drained by a renal vein. The role of the kidney is to remove waste products from the blood and to produce urine. The urine passes out of the kidney down the ureter to the bladder where it can be stored before release.

In longitudinal section we can see that the kidney consists of easily identified regions surrounded by a tough capsule.

- The outer region is called the cortex.
- The inner region is called the medulla.
- In the centre is the pelvis which leads into the ureter.

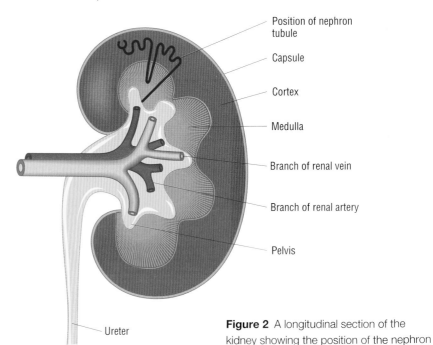

Figure 2 A longitudinal section of the kidney showing the position of the nephron

The nephron

The bulk of each kidney actually consists of tiny tubules called **nephrons**. There are about one million nephrons in each kidney. The nephrons are closely associated with many tiny blood capillaries.

Each nephron starts in the cortex. In the cortex the capillaries form a knot called the **glomerulus**. This is surrounded by a cup-shaped structure called the **Bowman's capsule**. Fluid from the blood is pushed into the Bowman's capsule by the process of **ultrafiltration**.

The capsule leads into the nephron which is divided into four parts:

- proximal convoluted tubule
- loop of Henle
- distal convoluted tubule
- collecting duct.

Figure 1 A kidney in longitudinal section

Cortex

Medulla

Pelvis

Key definitions

The **nephron** is the functional unit of the kidney. It is a microscopic tubule that receives fluid from the blood capillaries in the cortex and converts this to urine, which drains into the ureter.

The **glomerulus** is a fine network of capillaries that increases the local blood pressure to squeeze fluid out of the blood. It is surrounded by a cup- or funnel-shaped capsule which collects the fluid and leads into the nephron.

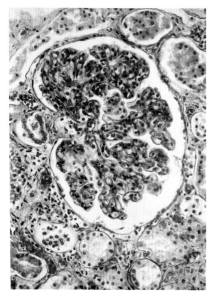

Figure 3 A section of the cortex showing a glomerulus (×500)

As the fluid moves along the nephron its composition is altered. This is achieved by **selective reabsorption**. Substances are reabsorbed back into the tissue fluid and blood capillaries surrounding the nephron tubule. The final product in the collecting duct is urine. This passes into the pelvis and down the ureter to the bladder.

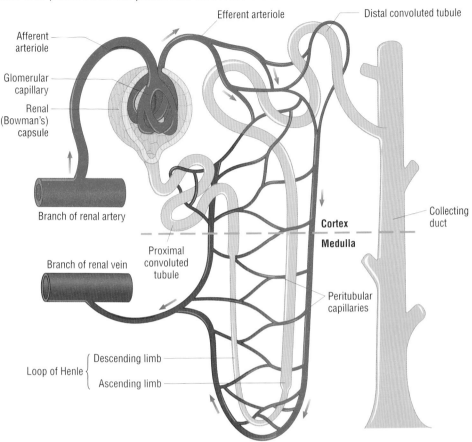

Figure 4 Diagram of the nephron and associated blood capillaries

Parts of the nephron

- The proximal tubule is the part closest to the glomerulus.
- The distal tubule is the part furthest from the glomerulus.
- A convoluted tubule is one that is bent and coiled.

Key definition

In **selective reabsorption** useful substances are reabsorbed from the nephron into the bloodstream while other excretory substances remain in the nephron.

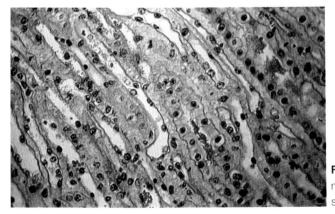

Figure 5 A micrograph of the medulla in longitudinal section, showing tubules (×400)

How does the composition of the fluid change?

- In the proximal convoluted tubule the fluid is altered by the reabsorption of all the sugars, most salts and some water.
- In total about 85% of the fluid is reabsorbed here.
- In the descending limb of the loop of Henle the water potential of the fluid is decreased by the addition of salts and the removal of water.
- In the ascending limb the water potential is increased as salts are removed by active transport.
- In the collecting duct the water potential is decreased again by the removal of water. This ensures that the final product (urine) has a low water potential. This means that the urine has a higher concentration of solutes than is found in the blood and tissue fluid.

Questions

1 Suggest why the nephrons are convoluted.
2 Why are there many capillaries around each nephron?
3 Explain why reabsorption from the nephron must be selective.

By the end of this spread, you should be able to ...

✳ Describe and explain the production of urine, with reference to the processes of ultrafiltration and selective reabsorption.

Key definitions

All organs have **afferent** vessels – they bring blood into the organ. Similarly, **efferent** vessels carry blood away from the organ. In a glomerulus the efferent vessel is an arteriole – which is muscular and can constrict to raise the blood pressure in the glomerulus. In most organs a venule carries blood away.

Ultrafiltration is filtration at a molecular level – as in the glomerulus where large molecules and cells are left in the blood and smaller molecules pass into the Bowman's capsule.

Podocytes are specialised cells that make up the lining of the Bowman's capsule.

Examiner tip

The endothelium of the capillary and the epithelium of Bowman's capsule contain gaps or pores. These two layers of cells do little to filter out larger molecules. It is the basement membrane that is actually involved in ultrafiltration.

Ultrafiltration

Blood flows into the **glomerulus** from the **afferent** arteriole. This is wider than the **efferent** arteriole, which carries blood away from the glomerulus. The difference in diameters ensures that the blood in the capillaries of the glomerulus is under increased pressure. The pressure in the glomerulus is higher than the pressure in the **Bowman's capsule**. This pressure difference tends to push fluid from the blood into the Bowman's capsule that surrounds the glomerulus.

The barrier between the blood in the capillary and the lumen of the Bowman's capsule consists of three layers: the *endothelium* of the capillary, a *basement membrane* and the *epithelial* cells of the Bowman's capsule. These are each adapted to allow ultrafiltration:

• The endothelium of the capillaries has narrow gaps between its cells that blood plasma, and the substances dissolved in it, can pass through.

• The basement membrane consists of a fine mesh of collagen fibres and glycoproteins. These act as a filter to prevent the passage of molecules with a relative molecular mass of greater than 69 000. This means that most proteins (and all blood cells) are held in the capillaries of the glomerulus.

• The epithelial cells of Bowman's capsule, called **podocytes**, have a very specialised shape. Podocytes have many finger-like projections called major processes. These ensure that there are gaps between the cells. Fluid from the blood in the glomerulus can pass between these cells into the lumen of the Bowman's capsule.

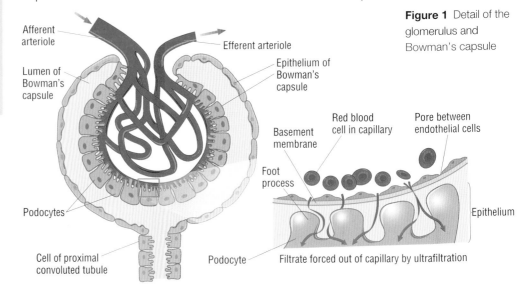

Figure 1 Detail of the glomerulus and Bowman's capsule

What is filtered out of the blood?

Blood plasma containing dissolved substances is pushed under pressure from the capillary into the lumen of the Bowman's capsule. This includes the following substances:

• water
• amino acids
• glucose
• urea
• inorganic ions (sodium, chloride, potassium).

Substance	Concentration in blood plasma/g dm⁻³	Concentration in glomerular filtrate/g dm⁻³
Water	900	900
Proteins	80.0	0.005
Amino acids	0.5	0.5
Glucose	1.0	1.0
Urea	0.3	0.3
Inorganic ions	7.2	7.2

Table 1 A comparison of blood plasma and glomerular filtrate

What is left in the capillary?

The blood cells and proteins are left in the capillary. The presence of the proteins means that the blood has a very low (very negative) water potential. This low water potential ensures that some of the fluid is retained in the blood, and this contains some of the water and dissolved substances listed above. The very low water potential of the blood in the capillaries is important to help reabsorb water at a later stage. The total volume of fluid filtered out of the blood by both kidneys is about 125 cm^3 min^{-1}. This is about 180 dm^3 day^{-1}. You can see why it is essential to reabsorb much of that water!

Selective reabsorption

As fluid moves along the nephron, substances are removed from the fluid and reabsorbed into the blood. Most reabsorption occurs from the proximal convoluted tubule, where about 85% of the filtrate is reabsorbed. All the glucose and amino acids and some salts are reabsorbed along with some of the water. Reabsorption is achieved by a combination of processes as described below. The cells lining the proximal convoluted tubule are specialised to achieve this reabsorption:

- The cell surface membrane in contact with the tubule fluid is highly folded to form **microvilli**. These increase the surface area for reabsorption.
- This membrane also contains special **co-transporter proteins** that transport glucose or amino acids, in association with sodium ions, from the tubule into the cell. This is **facilitated diffusion**.
- The opposite membrane of the cell, close to the tissue fluid and blood capillaries, is also folded to increase its surface area. This membrane contains **sodium–potassium pumps** that pump sodium ions out of the cell and potassium ions into the cell.
- The cell cytoplasm has many mitochondria. This indicates that an active, or energy-requiring, process is involved, as many mitochondria will produce a lot of ATP.

How does reabsorption occur?

- The sodium–potassium pumps remove sodium ions from the cells lining the proximal convoluted tubule. This reduces the concentration of sodium ions in the cell cytoplasm.
- Sodium ions are transported into the cell along with glucose or amino acid molecules by facilitated diffusion (see above).
- As the glucose and amino acid concentrations rise inside the cell, these substances are able to diffuse out of the opposite side of the cell into the tissue fluid. This process may be enhanced by active removal of glucose and amino acids from the cells.
- From the tissue fluid these substances diffuse into the blood and are carried away.
- Reabsorption of salts, glucose and amino acids reduces the water potential in the cells and increases the water potential in the tubule fluid. This means that water will enter the cells and then be reabsorbed into the blood by osmosis.
- Larger molecules, such as small proteins that may have entered the tubule, will be reabsorbed by **endocytosis**.

Questions

1 Explain what is meant by ultrafiltration.
2 Suggest what might happen if water is not reabsorbed from the nephron.
3 Explain why the concentrations of glucose and amino acids are the same in the glomerular filtrate as in the blood plasma.

STRETCH and CHALLENGE

High blood pressure can damage the capillaries of the glomerulus and the epithelium of Bowman's capsule.

Question

A Explain why the presence of protein in urine can be a sign of hypertension.

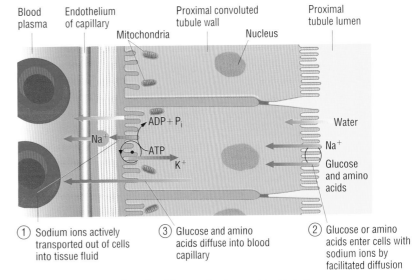

Figure 2 Reabsorption in cells of the proximal convoluted tubule

Labels: Blood plasma; Endothelium of capillary; Mitochondria; Proximal convoluted tubule wall; Nucleus; Proximal tubule lumen; ADP + P$_i$; Na$^+$; ATP; K$^+$; Water; Na$^+$; Glucose and amino acids

① Sodium ions actively transported out of cells into tissue fluid
③ Glucose and amino acids diffuse into blood capillary
② Glucose or amino acids enter cells with sodium ions by facilitated diffusion

Key definitions

Microvilli are microscopic folds of the cell surface membrane that increase the surface area of the cell.

Co-transporter proteins are proteins in the cell surface membrane that allow the facilitated diffusion of simple ions to be accompanied by transport of a larger molecule such as glucose.

Facilitated diffusion is diffusion that is enhanced by the action of proteins in the cell membrane.

Sodium–potassium pumps are special proteins in the cell surface membrane that actively transport sodium and potassium ions against their concentration gradients.

By the end of this spread, you should be able to . . .

✳ Explain, using water potential terminology, the control of the water content of the blood, with reference to the roles of the kidney, osmoreceptors in the hypothalamus and the posterior pituitary gland.

Reabsorption of water

Each minute about 125 cm³ of fluid is filtered from the blood and enters the nephron tubules. After selective reabsorption in the proximal convoluted tubule about 45 cm³ is left. The role of the loop of Henle is to create a low (very negative) water potential in the tissue of the medulla. This ensures that even more water can be reabsorbed from the fluid in the collecting duct.

The loop of Henle

The loop of Henle consists of a *descending limb* that descends into the medulla and an *ascending limb* that ascends back out to the cortex. The arrangement of the loop of Henle allows salts (sodium and chloride ions) to be transferred from the ascending limb to the descending limb. The overall effect is to increase the concentration of salts in the tubule fluid and consequently they diffuse out from the thin-walled ascending limb into the surrounding medulla tissue, giving the tissue fluid in the medulla a very low (very negative) water potential.

How is this achieved?

As the fluid in the tubule descends deeper into the medulla its water potential becomes lower (more negative). This is due to:

* loss of water by osmosis to the surrounding tissue fluid
* diffusion of sodium and chloride ions into the tubule from the surrounding tissue fluid.

As the fluid ascends back up towards the cortex its water potential becomes higher (less negative). This is because:

* at the base of the tubule, sodium and chloride ions diffuse out of the tubule into the tissue fluid
* higher up the tubule, sodium and chloride ions are actively transported out into the tissue fluid
* the wall of the ascending limb (particularly near the top) is impermeable to water, so water cannot leave the tubule
* the fluid loses salts but not water as it moves up the ascending limb.

The arrangement of the loop of Henle is known as a **hairpin countercurrent multiplier** system. The overall effect of this arrangement is to increase the efficiency of salt transfer from the ascending limb to the descending limb. This causes a build-up of salt concentration in the surrounding tissue fluid.

The movement of salts from the ascending limb into the medulla creates a high salt concentration in the tissue fluid of the medulla so that there is a low (very negative) water potential. The water potential becomes increasingly negative deeper into the medulla.

The removal of ions from the ascending limb means that at the top of the ascending limb the urine is dilute. Water may then be reabsorbed from the urine in the distal tubules and collecting ducts. The amount of water reabsorbed depends on the needs of the body, and so the kidney is also an organ of **osmoregulation**.

Key definitions

A **hairpin countercurrent multiplier** is the arrangement of a tubule in a sharp hairpin so that one part of the tubule passes close to another part of the tubule with the fluid flowing in opposite directions. This allows exchange between the contents and can be used to create a very high concentration of solutes.

Osmoregulation is the control and regulation of the water potential of the blood and body fluids. In humans the kidney controls the water potential of the blood.

STRETCH and CHALLENGE

Question: Why do camels have humps? Answer: to allow space for an extra long loop of Henle! This old teachers' joke is obviously not accurate. The hump stores fat which can be metabolised to release energy and water. But why do camels need an extra long loop of Henle? All mammals adapted to living in arid regions share this feature. It provides them with a longer countercurrent mechanism that can increase the salt concentration in the medulla more than in other mammals.

Question

A Explain why it is beneficial to mammals living in arid regions to have higher salt concentrations in their medullas.

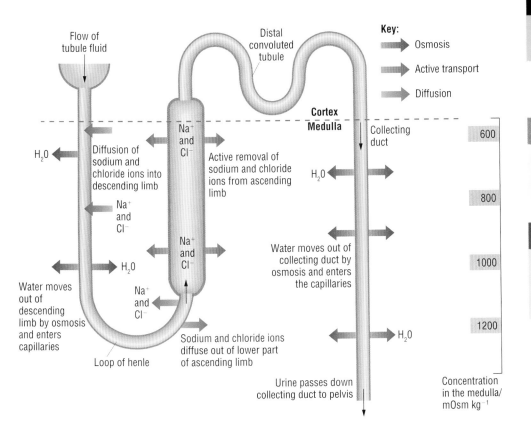

Figure 1 The loop of Henle and the collecting duct

Key definition

The **distal convoluted tubule** is the coiled portion of the nephron between the loop of Henle and the collecting duct.

Examiner tip

Milliosmoles per kilogram (mOsm kg^{-1}) are units of concentration, thousandths of a mole of molecules, ions or both in 1 kg (about 1 dm^3) of solution.

The collecting duct

From the top of the ascending limb the tubule fluid passes along a short **distal convoluted tubule** where active transport is used to adjust the concentrations of various salts. From here the fluid flows into the collecting duct. At this stage the tubule fluid still contains a lot of water – it has a high water potential. The collecting duct carries the fluid back down through the medulla to the pelvis. Remember that the tissue fluid in the medulla has a low water potential that becomes even lower deeper into the medulla. As the tubule fluid passes down the collecting duct water moves, by osmosis, from the tubule fluid into the surrounding tissue. It then enters the blood capillaries, by osmosis, and is carried away.

The amount of water that is reabsorbed depends on the permeability of the walls of the collecting duct. Only about 1.5–2.0 dm^3 of fluid (urine) reaches the pelvis each day. By the time the urine reaches the pelvis it has a low (very negative) water potential and the concentration of urea and salts in urine is higher than that of the blood plasma.

STRETCH and CHALLENGE

Question

B The tissue fluid in the medulla has a low water potential, so how can water pass from the tissue fluid into the blood plasma by osmosis?
Suggest an arrangement of the blood vessels that could create blood plasma with an even lower water potential than the tissue fluid.

Questions

1 Why must the collecting duct pass back through a region of low water potential?
2 Why is it important for terrestrial mammals to reabsorb as much water as possible?
3 Suggest why beavers have short loops of Henle.

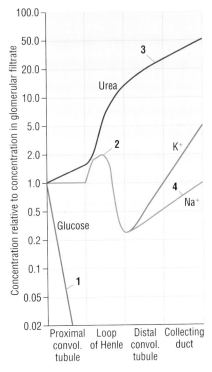

1. Glucose actively transported out of tubule
2. Sodium ions enter descending limb and leave ascending limb
3. Urea concentration rises as water is removed from tubule
4. Water is absorbed from the tubule, increasing the relative concentration of sodium ions, and potassium ions are actively transported into the tubule to be removed in urine

Figure 2 Changes in the relative concentrations of certain substances as fluid passes along the nephron

By the end of this spread, you should be able to ...

* Explain, using water potential terminology, the control of the water content of the blood, with reference to the roles of the kidney, osmoreceptors in the hypothalamus and the posterior pituitary gland.

Osmoregulation

Osmoregulation is the control of water levels and salt levels in the body. The correct water balance between cells and the surrounding fluids must be maintained to prevent problems with osmosis.

Water is gained from three sources:

- food
- drink
- metabolism (e.g. respiration).

Water is lost in:

- urine
- sweat
- water vapour in exhaled air
- faeces.

If it is a cool day and you have drunk a lot of fluid you will produce a large volume of dilute urine. Alternatively, on a hot day when you have drunk very little you will produce smaller volumes of more concentrated urine. Controlling the loss of water in urine is just one part of the osmoregulation process.

The walls of the collecting duct can be made more or less permeable according to the needs of the body. On a cool day when you need to conserve less water the walls of the collecting duct are less permeable. This means that less water is reabsorbed and more urine will be produced. On a hot day you need to conserve more water. The collecting duct walls are made more permeable so that more water can be reabsorbed into the blood. You will produce a smaller volume of urine.

Altering the permeability of the collecting duct

The walls of the collecting duct respond to the level of **antidiuretic hormone** (**ADH**) in the blood. Cells in the wall have membrane-bound receptors for ADH. The ADH binds to these receptors and causes a chain of enzyme-controlled reactions inside the cell. The end result of these reactions is to insert vesicles containing water-permeable channels (aquaporins) into the cell surface membrane. This makes the walls more permeable to water. If there is more ADH in the blood, more water-permeable channels are inserted. This allows more water to be reabsorbed, by osmosis, into the blood. Less urine, of a lowered water potential, passes out of the body.

If there is less ADH in the blood then the cell surface membrane folds inwards to create new vesicles that remove water-permeable channels from the membrane. This makes the walls less permeable and less water is reabsorbed, by osmosis, into the blood. More water passes out in the (dilute) urine.

Adjusting the concentration of ADH in the blood

- The water potential of the blood is monitored by **osmoreceptors** in the **hypothalamus** of the brain.
- These cells probably respond to the effects of osmosis. When the water potential of the blood is low (very negative) the osmoreceptor cells lose water by osmosis. This causes them to shrink and stimulate **neurosecretory cells** in the hypothalamus.
- The neurosecretory cells are specialised neurones (nerve cells) that produce and release ADH. The ADH is manufactured in the cell body of these cells, which lies in the hypothalamus. ADH flows down the axon to the terminal bulb in the **posterior pituitary gland**. It is stored there until needed.
- When the neurosecretory cells are stimulated they send action potentials down their axons and cause the release of ADH.

Key definitions

Antidiuretic hormone (ADH) is released from the pituitary gland and acts on the collecting ducts in the kidneys to increase their reabsorption of water.

Osmoreceptors are receptor cells that monitor the water potential of the blood. If the blood has a low water potential then water is moved out of the osmoreceptor cells by osmosis, causing them to shrink. This causes stimulation of the neurosecretory cells.

The **hypothalamus** is a part of the brain that contains neurosecretory cells and various receptors that monitor the blood.

Neurosecretory cells are specialised cells that act like nerve cells but release a hormone into the blood. ADH is manufactured in the cell body and passes down the axon to be stored in the terminal bulb. If an action potential passes down the axon then ADH is released from the terminal bulb.

The **posterior pituitary gland** is the hind part of the pituitary gland, which releases ADH.

Examiner tip

Don't say that the ADH receptors are in cell walls. They are in the membranes of the cells that make up the walls of the collecting ducts.

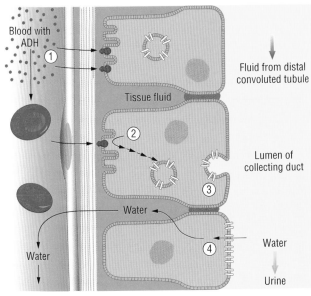

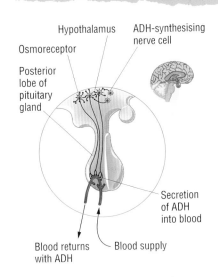

Figure 2 The hypothalamus and posterior pituitary

1. ADH detected by cell surface receptors
2. Enzyme-controlled reactions
3. Vesicles containing water-permeable channels (aquaporins) fuse to membrane
4. More water can be reabsorbed

Figure 1 The effect of ADH on the collecting duct wall

- The ADH enters the blood capillaries running through the posterior pituitary gland. It is transported around the body and acts on the cells of the collecting ducts (its target cells).
- Once the water potential of the blood rises again, less ADH is released.
- ADH is slowly broken down – it has a **half-life** of about 20 minutes. Therefore, the ADH present in the blood is broken down and the collecting ducts will receive less stimulation.

Key definition

The **half-life** of a substance is the time taken for its concentration to drop to half its original value.

STRETCH and CHALLENGE

A number of drugs have an effect on urine production. Some have effects that are unwanted or may be a nuisance. Alcohol inhibits the production of ADH, and certain antibiotics, such as tetracycline, can cause renal failure through a variety of mechanisms, including direct toxicity to the nephron tubules.

Other drugs have effects that may be useful. Diuretic drugs increase urine production and antidiuretic drugs do the opposite by increasing the reabsorption of water at the distal tubule and collecting ducts without significantly modifying the rate of glomerular filtration.

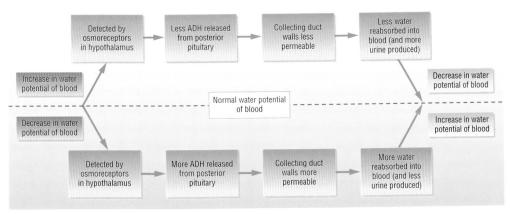

Figure 3 Control of water potential in the blood by negative feedback

Questions

1. How do neurosecretory cells differ from normal nerve cells?
2. Explain what is meant by negative feedback.
3. Why is it important that ADH is broken down?

Questions

A. Explain why drinking too much alcohol can cause a hangover.
B. Suggest what symptoms may be relieved by the use of **(i)** diuretics, and **(ii)** antidiuretics.

By the end of this spread, you should be able to ...

* Outline the problems that arise from kidney failure and discuss the use of renal dialysis and transplants for the treatment of kidney failure.
* Describe how urine samples can be used to test for pregnancy and detect misuse of anabolic steroids.

Key definitions

Dialysis is the use of a partially permeable membrane to filter the blood.

The **dialysis membrane** is a partially permeable membrane that separates the dialysis fluid from the patient's blood in a dialysis machine. **Dialysis fluid** is a complex solution that matches the composition of body fluids.

In **haemodialysis** blood is taken from a vein and passed through a dialysis machine so that exchange can occur across an artificial partially permeable membrane. In **peritoneal dialysis**, dialysis fluid is pumped into the body cavity so that exchange can occur across the peritoneal membrane.

Kidney failure

Kidney failure can occur for a number of reasons. The most common causes are:

* **diabetes mellitus** (both type I and type II sugar diabetes)
* hypertension
* infection.

Once the kidneys fail completely the body is unable to remove excess water and certain waste products from the blood. This includes urea and excess salts. It is also unable to regulate the levels of water and salts in the body. This will rapidly lead to death.

Treatment of kidney failure

There are two main treatments for kidney failure: dialysis and transplant.

Dialysis

Dialysis is the most common treatment for kidney failure. It removes wastes, excess fluid and salt from blood by passing the blood over a **dialysis membrane**. The dialysis membrane is a **partially permeable membrane** that allows the exchange of substances between the blood and **dialysis fluid**. This fluid contains the correct concentrations of salts, urea, water and other substances in blood plasma. Any substances in excess in the blood diffuse across the membrane into the dialysis fluid. Any substances that are too low in concentration diffuse into the blood from the dialysis fluid. Dialysis must be combined with a carefully monitored diet.

In **haemodialysis** blood from a vein is passed into a machine that contains an artificial dialysis membrane. Heparin is added to avoid clotting, and any bubbles are removed before the blood returns to the body. Haemodialysis is usually performed at a clinic three times a week for several hours at each session, but some patients learn to carry it out at home.

In **peritoneal dialysis** (PD) the filter is the body's own abdominal membrane (peritoneum). First, a surgeon implants a permanent tube in the abdomen. Dialysis solution is poured through the tube and fills the space between the abdominal wall and organs. After several hours, the used solution is drained from the abdomen. PD is usually performed in several consecutive sessions daily at home or work. As the patient can walk around having dialysis, the method is sometimes called ambulatory PD.

Kidney transplant

In a kidney transplant the old kidneys are left in place unless they are likely to cause infection or are cancerous. The donor kidney can be from a living relative who is willing to donate one of their healthy kidneys or from someone who has died.

(a)

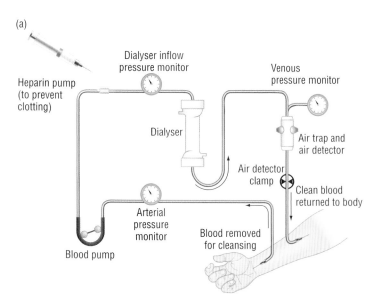

(b)

Blood pumped in at a higher pressure than the dialysis fluid

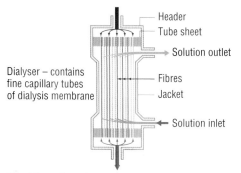

Blood flows through artificial capillaries, while the dialysis fluid flows along the outside in the opposite direction

Figure 1 Haemodialysis

A kidney transplant is major surgery. While the patient is under anaesthesia, the surgeon implants the new organ into the lower abdomen and attaches it to the blood supply and the bladder. Many patients feel much better immediately after the transplant, which is the best life-extending treatment for kidney failure. However, the patient's immune system will recognise the new organ as a foreign object and produce a reaction. Patients are given immunosuppressant drugs to help prevent rejection.

Advantages	Disadvantages
Freedom from time-consuming dialysis	Need immunosuppressants for the life of the kidney
Diet is less limited	Need major surgery under a general anaesthetic
Feeling better physically	Risks of surgery include infection, bleeding and damage to surrounding organs
A better quality of life, e.g. able to travel	Frequent checks for signs of organ rejection
No longer seeing oneself as chronically ill	Side effects: anti-rejection medicines cause fluid retention and high blood pressure; immunosuppressants increase susceptibility to infections

Table 1 Advantages and disadvantages of kidney transplants

Testing urine samples

Substances or molecules with a relative molecular mass of less than 69 000 can enter the nephron. This means that any metabolic product or other substance that is in the blood can be passed into the urine – as long as it is small enough. If these substances are not reabsorbed further down the nephron they can be detected in urine.

Pregnancy testing

Once implanted in the uterine lining, a human embryo starts secreting a pregnancy hormone called **human chorionic gonadotrophin (hCG)**. hCG is a relatively small glycoprotein with a molecular mass of 36 700. It can be found in urine as early as 6 days after conception. The pregnancy tests on the market today are manufactured with **monoclonal antibodies**. The antibody is specific – it will bind only to hCG, not to other hormones. When someone takes a home pregnancy test, she soaks a portion of the test strip in her urine. Any hCG in the urine attaches to an antibody that is tagged with a blue bead. This hCG–antibody complex moves up the strip until it sticks to a band of immobilised antibodies. As a result all the antibodies carrying a blue bead and attached to hCG are held in one place forming a blue line. There is always one control blue line to use for comparison; a second blue line indicates pregnancy.

Testing for anabolic steroids

Anabolic steroids increase protein synthesis within cells. This results in the build-up of cell tissue, especially in the muscles. Non-medical uses for anabolic steroids are controversial because they can give advantage in competitive sports and they have dangerous side effects. The use of anabolic steroids is banned by all major sporting bodies.

Anabolic steroids have a half-life of about 16 hours and remain in the blood for many days. They are relatively small molecules and can enter the nephron easily. Testing for anabolic steroids involves analysing a urine sample in a laboratory using **gas chromatography** or mass spectrometry.

In gas chromatography the sample is vaporised in the presence of a gaseous solvent and passed down a long tube lined by an absorption agent. Each substance dissolves differently in the gas and stays there for a unique, specific time, the retention time. Eventually the substance comes out of the gas and is absorbed onto the lining. This is then analysed to create a **chromatogram**. Standard samples of drugs, as well as the urine samples, are run so that drugs can be identified and quantified in the chromatograms.

Key definitions

Human chorionic gonadotrophin (hCG) is a hormone released by human embryos; its presence in the mother's urine confirms pregnancy.

Monoclonal antibodies are identical because they have been produced by cells that are clones of one original cell.

Anabolic steroids are drugs that mimic the action of steroid hormones that increase muscle growth.

Gas chromatography is a technique used to separate substances in a gaseous state. A **chromatogram** is a chart produced when substances are separated by movement of a solvent along a permeable material such as paper or gel.

Questions

1. What components of the diet must be carefully monitored in someone who undergoes dialysis?
2. Explain why haemodialysis fluid has to be sterile and at 37 °C.
3. Create a table of advantages and disadvantages of dialysis as a treatment for kidney failure.
4. Explain why standard samples of drugs must be run alongside a urine sample in gas chromatography.

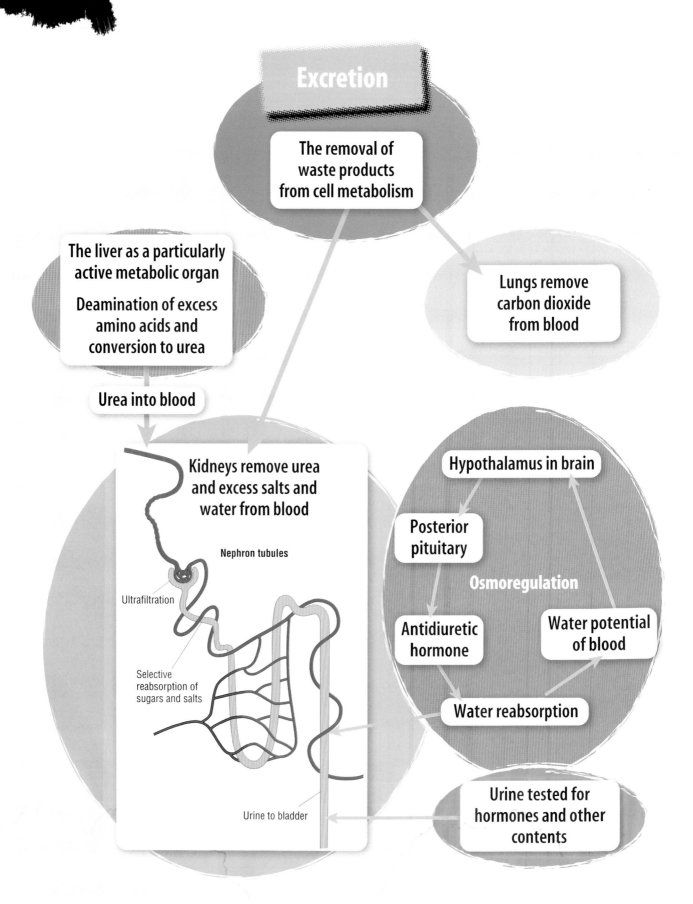

Excretion

The removal of waste products from cell metabolism

Lungs remove carbon dioxide from blood

The liver as a particularly active metabolic organ

Deamination of excess amino acids and conversion to urea

Urea into blood

Kidneys remove urea and excess salts and water from blood

Nephron tubules

Ultrafiltration

Selective reabsorption of sugars and salts

Urine to bladder

Hypothalamus in brain

Posterior pituitary

Osmoregulation

Antidiuretic hormone

Water potential of blood

Water reabsorption

Urine tested for hormones and other contents

Practice questions

1. What is meant by the term *excretion*? [2]

2. (a) Name the two main excretory products produced in mammals. [2]
   (b) Name the two organs that remove these products from the body. [2]

3. (a) Name the blood vessels that carry blood to the liver. [2]
   (b) What is a sinusoid and where is it found? [2]

4. What is a hepatocyte? [1]

5. (a) What is meant by the term deamination? [2]
   (b) What occurs during the ornithine cycle? [2]

6. (a) Name the three sections of the kidney as seen in longitudinal section. [3]
   (b) Name the tubules found in the kidney. [1]
   (c) Name the tube that leads from the kidney down to the bladder. [1]

7. Suggest two ways in which liver cells are specialised. [2]

8. (a) Explain why it is important to remove carbon dioxide from the blood. [3]
   (b) Explain why most mammals cannot excrete nitrogenous waste in the form of ammonia. [2]

9. (a) Where is antidiuretic hormone (ADH) made? [1]
   (b) Where is ADH released into the blood? [1]
   (c) What are the target cells for ADH? [1]

10. Name the part of the kidney tubule in which:
    (a) ultrafiltration occurs [1]
    (b) selective reabsorption occurs [1]
    (c) urine is concentrated. [1]

11. Suggest why the cells lining the proximal convoluted tubule have the following features:
    (a) microvilli [2]
    (b) many mitochondria. [2]

12. Complete the table.

| Substance | Concentration relative to blood plasma | |
	In urine	In Bowman's capsule
Glucose	Less	
Amino acids	Less	Same
Urea		Same
Protein		Much less/none

[3]

13. (a) Which part of the brain acts as a receptor to monitor the water potential of the blood? [1]
    (b) Describe how the body uses negative feedback to maintain the water potential of the blood. [4]

14. (a) Describe how alcohol is detoxified. [3]
    (b) Explain why excess alcohol consumption can lead to a fatty liver. [2]

15. (a) What is meant by a countercurrent multiplier? [2]
    (b) Describe how such an arrangement is used to increase salt concentrations in the kidney medulla. Explain why the salt concentration of the tissue fluid in the medulla needs to be increased. [3]

16. Describe how glucose is selectively reabsorbed from the tubule fluid and returned to the blood. [3]

17. State two advantages of a kidney transplant compared to dialysis. [2]

1 An investigation was conducted into the filtration and reabsorption of glucose in the kidney of a mammal as the glucose concentration in the plasma of the renal artery was increased.

The glucose concentrations were measured in the following fluids:

- glomerular filtrate
- urine.

From the measurements obtained, the concentration of glucose in the fluid reabsorbed from the glomerular filtrate was calculated.

The results of this investigation are shown in Figure 1.1.

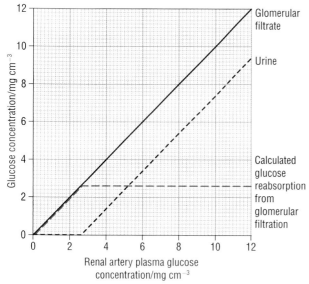

Figure 1.1

(a) Use the data in Figure 1.1 to answer the following questions.
   (i) Describe the relationship between plasma glucose concentration in the renal artery and the concentration of glucose in the glomerular filtrate. [1]
   (ii) State the plasma glucose concentration in the renal artery above which the kidney is unable to reabsorb all the glucose from the glomerular filtrate.
      Answer = ... mg cm^{-3} [1]
   (iii) Explain why plasma glucose concentrations in the renal artery greater than the figure you have given in (ii) would result in the presence of glucose in the urine. [3]
(b) Explain why proteins with a relative molecular mass (RMM) greater than 69 000 do not pass from the blood plasma into the glomerular filtrate. [2]
(c) The kangaroo rat, *Dipodomys spectabilis*, is common in the deserts of North America. It does not need to drink water and feeds mostly on seeds and other dry plant material. It produces very little urine.

(i) Suggest how the kidney of this mammal is adapted to reduce the volume of urine produced. [3]
[Total: 10]
(OCR 2804 Jun06)

2 Figure 2.1 shows diagrams of the nephrons from the kidneys of three different animals, **X**, **Y** and **Z**.

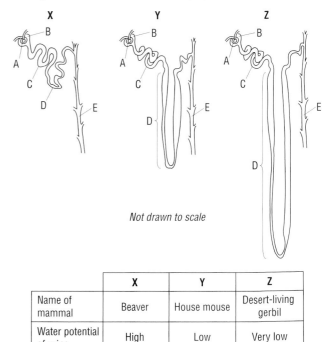

Not drawn to scale

	X	Y	Z
Name of mammal	Beaver	House mouse	Desert-living gerbil
Water potential of urine	High	Low	Very low

Figure 2.1

(a) Name parts **A** to **E**. [5]
(b) Explain the relationship between the length of part **D** and the water potential of the urine in the three mammals. [4]

Figure 2.2 is a drawing of a cell from part **C** of the diagram of the nephrons shown in Figure 2.1.

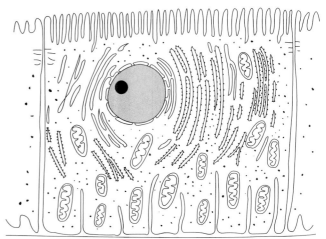

Figure 2.2

Answers to examination questions will be found on the Exam Café CD.

(c) State three structural features visible in this cell which help in selective reabsorption from the glomerular filtrate. [3]

[Total: 12]

(OCR 2804 Jun03)

3 (a) Define the term *excretion*. [2]

(b) Name two groups of macromolecules that are broken down to form nitrogenous excretory products in mammals. [2]

(c) Table 3.1 shows the amount of different substances excreted by a volunteer during two 24-hour periods. During the first 24-hour period the volunteer was fed a protein-deficient diet; during the second 24-hour period the volunteer was fed a protein-rich diet. All other variables were kept constant.

Substance excreted/g	Protein-deficient diet	Protein-rich diet
Urea	2.20	14.70
Uric acid	0.09	0.18
Ammonium ions	0.04	0.49
Creatine	0.60	0.58

Table 3.1

(i) Calculate the percentage increase in urea excreted when the volunteer switched from a protein-deficient diet to a protein-rich diet. Show your working. [2]

(ii) Explain why more urea is produced when eating a protein-rich diet. [2]

(d) Explain why the main nitrogenous excretory product in humans is urea rather than ammonia. [2]

(e) Table 3.2 shows the concentrations of glucose and urea in the renal artery and renal vein.

	Concentration/mg 100 cm^{-3} plasma	
	Renal artery	**Renal vein**
Glucose	90	80
Urea	30	16

Table 3.2

Both substances are present in lower concentrations in the renal vein than in the renal artery. However, urea appears in the urine of a healthy person but glucose does not. Explain why this is so. [5]

[Total: 15]

(OCR 2804 Jan06)

4 The mammalian liver is made up of lobules that consist of liver cells (hepatocytes) arranged in plates.
Figure 4.1 shows a section of a liver lobule and its associated blood vessels.

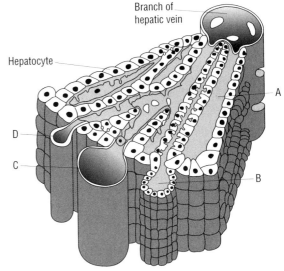

Figure 4.1

(a) Name structures **A** to **D**. [4]

(b) Figure 4.2 is an outline of an important biochemical process performed by the liver.

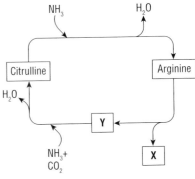

Figure 4.2

With reference to Figure 4.2:

(i) Name **X** and **Y**. [2]

(ii) Outline the role of this process in the mammal. [2]

(iii) State the name given to this process. [1]

[Total: 9]

(OCR 2805 Jan06 and Jun02)

Module 3
Photosynthesis

Introduction

We all depend on photosynthesis for our existence. All the food we eat has originated from plants. They use sunlight energy and turn it into chemical energy that other life forms (like us) can use. We eat the plants, products made from them, products from microorganisms often fed on plant waste, and animals that have eaten plants or herbivores. Without photosynthesis we wouldn't be here; indeed, some scientists think that oxygen produced by photosynthesis changed the course of evolution.

You will already have learnt a bit about photosynthesis. It is interesting to understand how our detailed knowledge of photosynthesis has come about. In the early seventeenth century Jan Baptist van Helmont, a Flemish chemist and medic, was wondering how plants grew, so he grew a willow tree in a tub for 5 years. He found that the tree gained about 80 kg but the mass of soil (minerals) lost was small, and he thought the increase came from the water added. In 1772, Joseph Priestley discovered that plants produce oxygen, and in 1779 Jan Ingenhousz found that they need light to do this. In 1781 carbon dioxide gas in the air was identified. In 1804 it was shown that the gain in mass of a plant grown in a carefully weighed pot was the sum of the carbon, entirely from absorbed carbon dioxide, and water taken up by the roots. In the late nineteenth century, Theodor Engelmann shone different colours of light onto *Spirogyra*, an aquatic alga (picture), and used motile oxygen-loving bacteria to detect high concentrations of oxygen. They accumulated near chloroplasts lit by red and blue light. He concluded that chloroplasts were also needed and that red and blue light were most effective. In 1931 Cornelius van Niel put forward a hypothesis that light energy caused a proton donor to release hydrogen ions that were used to reduce carbon dioxide and produce organic molecules. In 1939 Robert Hill showed that isolated chloroplasts could produce oxygen. We now know that some bacteria also photosynthesise (Engelmann showed that purple bacteria use ultraviolet light to manufacture food).

In this module you will learn about the stages of photosynthesis and the factors that influence it. Some of the biochemistry you will need is covered in more detail in the next module, Respiration.

Test yourself

1 Where in plant cells does photosynthesis take place?
2 What are the raw materials needed for photosynthesis?
3 What is the energy source for photosynthesis?
4 Name two phyla, other than plants, that contain photosynthetic organisms.
5 Besides being a source of food, how else are plants necessary to humans and other animals?
6 Draw a flow diagram showing how energy from sunlight is used to produce muscle contractions in your arm as you draw the diagram.

Module contents

By the end of this spread, you should be able to ...

* Define the terms *autotroph* and *heterotroph*.
* State that light energy is used during photosynthesis to produce complex organic molecules.
* Explain how respiration in plants and animals depends upon the products of photosynthesis.
* State that in plants photosynthesis is a two-stage process that takes place in chloroplasts.

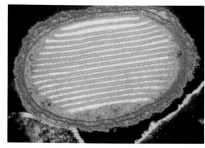

Figure 1 Photosynthetic bacterium *Ectothiorhodospira* lives in hot salty pools; it has inwardly folded cell membranes where a type of chlorophyll is situated; ×5000

Key definitions

Autotrophs are organisms that use light energy or chemical energy and inorganic molecules (carbon dioxide and water) to synthesise complex organic molecules.

Heterotrophs are organisms that ingest and digest complex organic molecules, releasing the chemical potential energy stored in them.

What is photosynthesis?

* Photosynthesis is the process whereby light energy from the Sun is transformed into chemical energy and used to synthesise large organic molecules from inorganic substances.
* Scientists think that this process first evolved in some *prokaryotes* (ancestors of modern-day cyanobacteria) at least 2500 million years ago and possibly earlier than that.
* It is perhaps the most important biochemical process, since nearly all life on Earth depends on it.
* It transforms light energy into chemical potential energy that is then available to **consumers** and **decomposers**.
* It also releases oxygen, from water, into the atmosphere, so all **aerobes** depend on it for their respiration.

Autotrophs and heterotrophs

Autotrophs are organisms that can synthesise complex organic molecules, such as carbohydrates, lipids, proteins, nucleic acids and vitamins, from inorganic molecules and an energy source. The first life forms on Earth were **chemoautotrophs** – prokaryotes that synthesised complex organic molecules, using energy derived from **exergonic** chemical reactions. Many bacteria are chemoautotrophs, for example the nitrifying bacteria that are so important in the recycling of nitrogen. They obtain their energy from **oxidising** ammonia to nitrite or oxidising nitrite to nitrate. There are also chemoautotrophs living in darkness, near thermal oceanic vents, supporting very specific food chains in those areas.

Organisms that can photosynthesise are described as **photoautotrophs**. Their source of energy is sunlight and the raw materials are inorganic molecules – carbon dioxide and water. Plants, some bacteria and some **protoctists** (such as algae) are photoautotrophs. The majority of food chains on Earth have producers that are photoautotrophs.

Animals, fungi and some bacteria are **heterotrophs**. They cannot make their own food but **digest** complex organic molecules into simpler soluble ones from which they synthesise complex molecules such as lipids, proteins and nucleic acids.

Why does respiration in autotrophs and heterotrophs depend on photosynthesis?

Both photoautotrophs and heterotrophs can release the chemical potential energy in complex organic molecules (made during photosynthesis). This is respiration. They can also use the oxygen (which first appeared in the atmosphere after being released from photosynthesis) for aerobic respiration.

The equation below summarises the process of photosynthesis:

$$6CO_2 + 6H_2O \text{ (+ light energy)} \rightarrow C_6H_{12}O_6 + 6O_2$$

Once the Earth's atmosphere contained free oxygen, organisms evolved that could use the oxygen for aerobic respiration. This releases carbon dioxide back into the atmosphere and produces water.

$$C_6H_{12}O_6 + 6O_2 \rightarrow 6CO_2 + 6H_2O \text{ (+ energy, some as ATP)}$$

Figure 2 The giant panda, *Ailuropoda melanoleuca*, a heterotroph, feeding on the shoots and leaves of bamboo, an autotroph

Using radioactive isotopes

You can see from the equation that photosynthesis releases free oxygen. When scientists used water containing radioactive isotopes of oxygen they found that the oxygen produced during photosynthesis was radioactive. When they gave the plant carbon dioxide containing radioactive oxygen, the oxygen produced was not radioactive. This showed that the oxygen is released from water.

Where does photosynthesis take place?

Photoautotrophs have special **organelles** within their cells called **chloroplasts**. The equation above only summarises the process. It happens in two main stages. Each stage consists of many smaller steps, which will be considered in the following spreads.

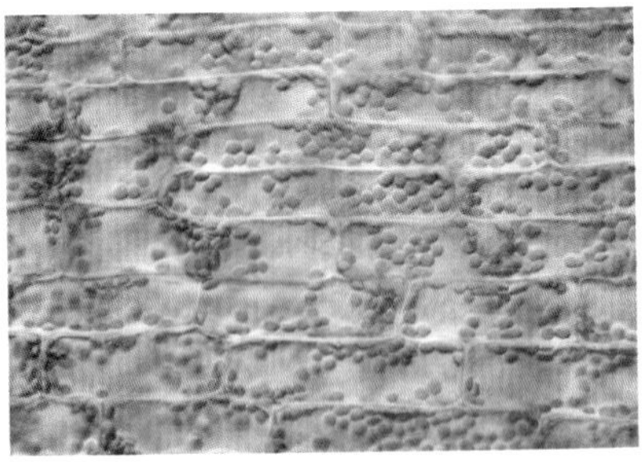

Figure 3 Chloroplasts inside *Elodea* leaf ×250

STRETCH and CHALLENGE

Chloroplasts have many similarities with photosynthetic bacteria. They are about the same size, have circular DNA, prokaryote-type ribosomes and similar pigments. The endosymbiont theory suggests that photosynthetic bacteria were acquired, possibly by **endocytosis**, by early **eukaryotic** cells, to form the first algal or plant cells. Chloroplasts may be primitive photosynthetic bacteria adapted to living inside plant cells.

Two species of marine molluscs, *Elysia viridis* and *Elysia chlorotica*, incorporate chloroplasts from algae into their cells. However, these chloroplasts do not get transferred to the next generation.

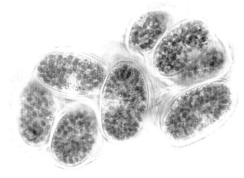

Figure 4 Endosymbiotic cyanobacteria inside protoctist cells ×450

Question

A Suggest an advantage to *E. viridis* and *E. chlorotica* of incorporating chloroplasts into their cells.

Examiner tip

Remember that autotroph means self-feeding. Photoautotrophs make food using energy from sunlight.

Questions

1 Which process evolved first on Earth – aerobic respiration or photosynthesis? Give reasons for your answer.
2 Complete the following table.

	Autotrophs	Heterotrophs
Do they respire?		
Can they synthesise complex organic molecules from simple inorganic molecules?		
Do they use light energy?		
Can they hydrolyse complex organic molecules?		
Examples		

By the end of this spread, you should be able to . . .

* Explain, with the aid of diagrams and electron micrographs, how the structure of chloroplasts enables them to carry out their functions.
* Define the term *photosynthetic pigment*.
* Explain the importance of photosynthetic pigments in photosynthesis.

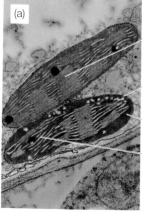

(a) Granum, consisting of thylakoids — Envelope — Stroma — Intergranal lamella

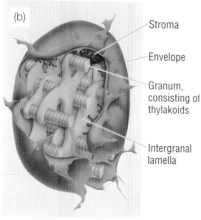

(b) Stroma — Envelope — Granum, consisting of thylakoids — Intergranal lamella

Figure 1 Chloroplast structure: **(a)** false colour TEM of chloroplast: **(b)** cutaway illustration of chloroplast structure ×12000

Key definitions

A **nanometre (nm)** is one thousandth of a micrometre, or one millionth of a millimetre.

Photosynthetic pigments are molecules that absorb light energy. Each pigment absorbs a range of wavelengths in the visible region and has its own distinct peak of absorption. Other wavelengths are reflected.

The structure of chloroplasts

Photosynthesis takes place within **organelles** called **chloroplasts**.

* They vary in shape and size but most are disc-shaped and between 2–10 μm long.
* Each chloroplast is surrounded by a double membrane – an **envelope**.
* There is an intermembrane space, about 10–20 nm wide, between the inner and outer membrane.
* The outer membrane is permeable to many small ions.
* The inner one is less permeable and has transport proteins embedded in it. It is folded into **lamellae** (thin plates), which are stacked up like piles of pennies. Each stack of lamellae is called a **granum** (plural: grana).
* Between the grana are intergranal lamellae.

There are two distinct regions inside each chloroplast – the **stroma** and the **grana**. Both of these regions can be seen when chloroplasts are viewed under a light microscope.

* The stroma is a fluid-filled matrix. The reactions of the **light-independent stage** of photosynthesis occur in the stroma where the necessary enzymes are located. Within the stroma are starch grains and oil droplets, as well as DNA and prokaryote-type ribosomes.
* The grana are stacks of flattened membrane compartments, called **thylakoids**. These are the sites of light absorption and ATPsynthesis during the **light-dependent stage** of photosynthesis. The thylakoids can only be seen using an electron microscope.

How chloroplasts are adapted for their role

* The inner membrane, with its transport proteins, can control entry and exit of substances between the cytoplasm and the stroma inside the chloroplasts.
* The many grana, consisting of stacks of up to 100 thylakoid membranes, provide a large surface area for the photosynthetic pigments, electron carriers and ATP synthase enzymes, all of which are involved in the light-dependent reaction.
* The **photosynthetic pigments** are arranged into special structures called **photosystems**, which allow maximum absorption of light energy.
* Proteins embedded in the grana hold the photosystems in place.
* The fluid-filled stroma contains the enzymes needed to catalyse the reactions of the light-independent stage of photosynthesis.
* The grana are surrounded by the stroma so the products of the light-dependent reaction, which are needed for the light-independent reaction, can readily pass into the stroma.
* Chloroplasts can make some of the proteins they need for photosynthesis, using genetic instructions in the chloroplast DNA, and the chloroplast ribosomes to assemble the proteins.

Photosynthetic pigments

Photosynthetic pigments are substances that absorb certain wavelengths of light and reflect others. They appear to us as the colour of the light wavelengths that they are reflecting. There are many different pigments that act together, to capture as much light energy as possible. They are in thylakoid membranes, arranged in funnel-shaped structures called photosystems, held in place by proteins.

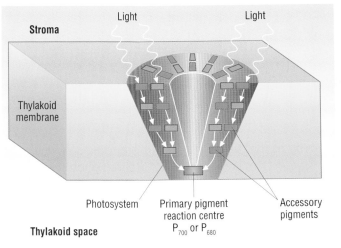

Figure 2 A photosystem. A funnel-shaped light-harvesting cluster of photosynthetic pigments, held in place in the thylakoid membrane of a chloroplast. Only a few pigment molecules are shown. The primary pigment reaction centre is a molecule of chlorophyll a. The accessory pigments consist of molecules of chlorophyll b and carotenoids

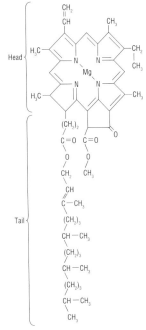

Figure 3 Structure of chlorophyll a

Chlorophylls

Chlorophyll is a mixture of pigments. All have a similar molecular structure, consisting of a long phytol (hydrocarbon) chain and a porphyrin group. The porphyrin group is similar to haem, found in haemoglobin, but contains a magnesium atom instead of an iron atom.

- Light hitting chlorophyll causes a pair of electrons associated with the magnesium to become excited.
- There are two forms of chlorophyll a – P_{680} and P_{700}. Both appear yellow-green.
- Each absorbs red light at a slightly different wavelength (absorption peak).
- Both are found at the centre of photosystems and are known as the **primary pigment reaction centre**.
- P_{680} is found in photosystem II and its peak of absorption is light at a wavelength of 680 nm.
- P_{700} is found in photosystem I and its peak of absorption is light at a wavelength of 700 nm.
- Chlorophyll a also absorbs blue light, of wavelength around 450 nm.
- Chlorophyll b absorbs light of wavelengths around 500 nm and 640 nm. It appears blue-green.

Accessory pigments

- Carotenoids reflect yellow and orange light and absorb blue light.
- They do not contain a porphyrin group and are not directly involved in the light-dependent reaction.
- They absorb light wavelengths that are not well absorbed by chlorophylls and pass the energy associated with that light to the chlorophyll a at the base of the photosystem.
- Carotene (orange) and xanthophyll (yellow) are the main carotenoid pigments.

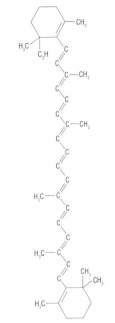

Figure 4 β-carotene – a carotenoid

Questions

1 What range of wavelengths do you think chlorophylls reflect?
2 Do carotenoids absorb light of wavelength 500 nm?
3 Explain how the structure of the grana enables them to carry out their function.
4 Explain how the structure of the stroma enables chloroplasts to carry out their functions.

Examiner tip

When answering questions about light absorption, remember that pigments absorb a range of wavelengths. Refer to the wavelengths or the wavelength of peak absorption and not just to the colour of the light.

61

By the end of this spread, you should be able to ...

* Outline how light energy is converted to chemical energy (ATP and reduced NADP) in the light-dependent stage.
* Explain the role of water in the light-dependent stage.

Naming terms

You may notice that during non-cyclic photophosphorylation, photosystem II is used before photosystem I. This may seem strange – why not name them the other way around? They are named in this way because photosystem I was the first to be isolated and identified.

Key definitions

Photophosphorylation is the making of ATP from ADP and P_i in the presence of light.

Electron carriers are molecules that transfer electrons.

Electron acceptors are chemicals that accept electrons from another compound. They are reduced while acting as oxidising agents.

The **light-dependent** stage of photosynthesis takes place on the thylakoid membranes of the chloroplasts. The photosystems, with the photosynthetic pigments, are embedded in these membranes. Photosystem I (PSI) occurs *mainly* on the intergranal lamellae and PSII occurs *almost exclusively* on the granal lamellae. These pigments trap light energy (usually from sunlight) so that it can be converted to chemical energy in the form of ATP.

The role of water

Photosystem II also contains an enzyme that, in the presence of light, can split water into H^+ ions (protons), electrons and oxygen. This splitting of water is called **photolysis**.

$$2H_2O \rightarrow 4H^+ + 4e^- + O_2$$

Some of the oxygen produced in this way is used by the plant for its aerobic respiration but much of it diffuses out of the leaves, through **stomata**, into the air.

Water is a source of:

- hydrogen ions, which are used in **chemiosmosis** (see spreads 1.4.4, 1.4.6 and 1.4.7) to produce ATP. These protons are then accepted by a **coenzyme** (see spread 1.4.2), **NADP** (nicotinamide adenine dinucleotide phosphate), which becomes **reduced NADP**, to be used during the light-independent stage to reduce carbon dioxide and produce organic molecules
- electrons to replace those lost by the oxidised chlorophyll.

Water is one of the raw materials used in photosynthesis. It also keeps plant cells turgid, enabling them to function. The by-product of photosynthesis, oxygen, comes from water.

Photophosphorylation

Light has a dual nature. We usually think of it as travelling in waves. However, we can also think of it as travelling in particles that we call photons. When a photon hits a chlorophyll molecule the energy of the photon is transferred to two electrons and they become excited. These electrons are captured by **electron acceptors** and passed along a series of **electron carriers** embedded in the thylakoid membranes. The electron carriers are proteins that contain iron atoms.

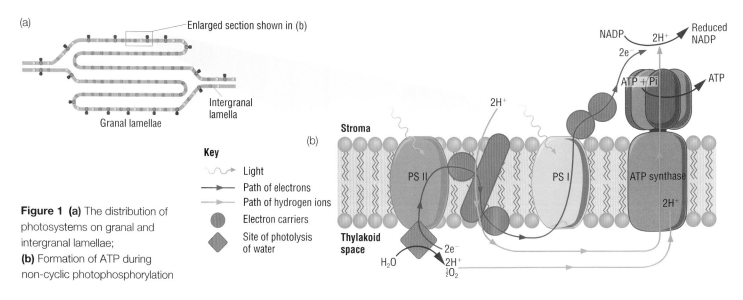

Figure 1 (a) The distribution of photosystems on granal and intergranal lamellae;
(b) Formation of ATP during non-cyclic photophosphorylation

Energy is released as electrons pass along the chain of electron carriers. This pumps protons across the thylakoid membranes into the thylakoid space where they accumulate. A proton gradient is formed across the thylakoid membrane and the protons flow down their gradient, through channels associated with ATP synthase enzymes. This flow of protons is called chemiosmosis. It produces a force that joins ADP and P_i to make ATP. The kinetic energy from the proton flow is converted to chemical energy in the ATP molecules, which is used in the **light-independent stage** of photosynthesis. The making of ATP using light energy is called **photophosphorylation**. There are two types – **cyclic** and **non-cyclic**. Both are illustrated in Figure 2.

Cyclic photophosphorylation

This uses only photosystem I (P_{700}). The excited electrons pass to an electron acceptor and back to the chlorophyll molecule from which they were lost. There is no photolysis of water and no generation of reduced NADP, but small amounts of ATP are made. This may be used in the light-independent reaction of photosynthesis or it may be used in guard cells (their chloroplasts contain only photosystem I) to bring in potassium ions, lowering the water potential and causing water to follow by osmosis. This causes the guard cells to swell and opens the stomata.

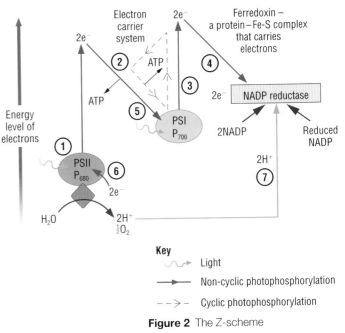

Figure 2 The Z-scheme

Non-cyclic photophosphorylation

This involves both photosystems – PSI and PSII.

1 Light strikes photosystem II, exciting a pair of electrons that leave the chlorophyll molecule from the **primary pigment reaction centre** (see spread 1.3.2).
2 The electrons pass along a chain of electron carriers and the energy released is used to synthesise ATP.
3 Light has also struck photosystem I and a pair of electrons has been lost.
4 These electrons, along with protons (produced at photosystem II by photolysis of water), join NADP, which becomes reduced NADP.
5 The electrons from the oxidised photosystem II replace the electrons lost from PSI.
6 Electrons from photolysed water replace those lost by the oxidised chlorophyll in PSII.
7 Protons from photolysed water take part in chemiosmosis to make ATP and are then captured by NADP, in the stroma. They will be used in the light-independent stage.

Questions

1 Complete the table below to compare cyclic and non-cyclic photophosphorylation.

	Cyclic	Non-cyclic
Photosystems involved		I and II
Is photolysis involved?		
Fate of electrons released from chlorophyll		
Products	ATP	

2 Outline the role of water in photosynthesis.
3 Explain how light causes stomata to open. What is the significance of this?
4 Suggest why a lack of iron in soil can reduce growth in plants.

STRETCH and CHALLENGE

If leaves are ground up in ice-cold, buffered 2% sucrose solution and then centrifuged, chloroplasts can be extracted. These can be suspended in the buffered sucrose solution. A dye, DCPIP, is then added and the suspension is illuminated. DCPIP is a redox agent. It is blue when oxidised and colourless when reduced. In the presence of illuminated chloroplasts, DCPIP changes from blue to colourless, so the suspension turns green.

Questions

A Suggest how the DCPIP becomes reduced when in the presence of illuminated chloroplasts.
B Suggest why the isolated chloroplasts are suspended in ice-cold, buffered 2% sucrose solution.

By the end of this spread, you should be able to ...

✳ **Outline how the products of the light-dependent stage are used in the light-independent stage (Calvin cycle) to produce triose phosphate (TP), referring also to ribulose bisphosphate (RuBP), ribulose bisphosphate carboxylase (rubisco) and glycerate 3-phosphate (GP).**

✳ **Explain the role of carbon dioxide in the light-independent stage.**

✳ **State that TP (and GP) can be used to make carbohydrates, lipids and amino acids.**

✳ **State that most TP is recycled to RuBP.**

Key definition

The **light-independent stage** of photosynthesis is where carbon dioxide is fixed and used to build complex organic molecules.

Working out the sequence

The sequence of events was worked out by Melvin Calvin and his associates (Andrew Benson and James Bassham) between 1946 and 1953. They grew algae in light with carbon dioxide containing radioactive (labelled) carbon. At differing time intervals they dropped some of the algae into boiling alcohol (to arrest any reactions) and identified the compounds present.

The **light-independent stage** of photosynthesis takes place in the **stroma** of chloroplasts.

It is also called the Calvin cycle. Although light is not directly used, the products of the light-dependent stage are used, and the light-independent stage soon ceases if light is not available.

The role of carbon dioxide

Carbon dioxide is the source of carbon (and oxygen) for the production of all large organic molecules. These molecules are used as structures, or act as energy stores or sources, for all the (carbon-based) life forms on this planet.

The Calvin cycle

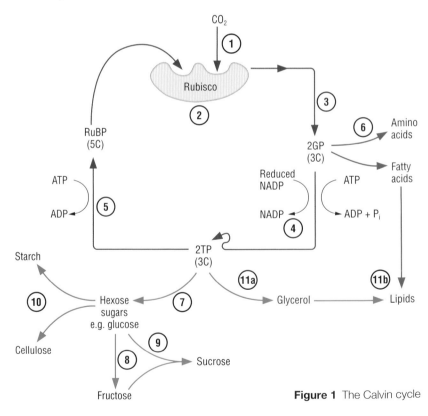

Figure 1 The Calvin cycle

1 Carbon dioxide from the air **diffuses** into the leaf through open **stomata**, most of which are on the underside of the leaf. It then diffuses throughout the air spaces in the spongy mesophyll and reaches the palisade mesophyll layer. Here it diffuses through the thin cellulose walls, the cell surface membrane, the cytoplasm and the **chloroplast envelope**, into the stroma.

2 In the stroma, carbon dioxide combines with a 5-carbon compound, **ribulose bisphosphate** (**RuBP**) (a carbon dioxide acceptor). The reaction is catalysed by the enzyme, ribulose bisphosphate carboxylase-oxygenase, usually called **rubisco**. RuBP becomes carboxylated (combined with carbon dioxide so it now has a carboxyl group).

3 The product of this reaction is two molecules of a 3-carbon compound, **glycerate 3-phosphate** (**GP**). The carbon dioxide has now been fixed.

4 GP is reduced and phosphorylated to another 3-carbon compound, **triose phosphate** (**TP**). ATP and reduced NADP from the light-dependent reaction are used in this process.

5 Five out of every six molecules of TP (3C) are recycled by phosphorylation, using ATP from the light-dependent reaction, to three molecules of RuBP (5C).

How the products of the Calvin cycle are used

6 Some GP can be used to make amino acids and fatty acids.

7 Pairs of TP molecules combine to form hexose (6C) sugars, such as glucose.

8 Some glucose molecules may be isomerised to form another hexose sugar, fructose.

9 Glucose and fructose molecules may be combined to form the disaccharide sucrose – the sugar translocated in phloem sieve tubes.

10 Hexose sugars can be polymerised into other carbohydrates (polysaccharides) such as cellulose and starch.

11 (a) TP can also be converted to glycerol and (b) this may be combined with fatty acids formed from GP, to make lipids.

Examiner tip

When writing about chemicals in the Calvin cycle, use the full name first, show the abbreviation, and then use the abbreviation for all further references to that chemical.

STRETCH and CHALLENGE

Rubisco has been described as the most important enzyme on Earth. It is also the most abundant as it makes up 50% of leaf protein.

When chloroplasts are illuminated, protons are pumped from the stroma into the thylakoid space. The pH of the stroma changes from pH 7 to pH 8. The optimum pH for the enzyme rubisco is pH 8.

Rubisco is short for ribulose bisphosphate carboxylase-oxygenase. Oxygen can also fit onto the active site of this enzyme resulting in a reaction called photorespiration. It undoes a lot of the work of photosynthesis and wastes some of the ATP made during the light-dependent stage. It also leads to the formation of hydrogen peroxide, which may be toxic.

As temperature increases, the oxygenase activity of rubisco increases more than its carboxylase activity.

When photosynthesis evolved, the Earth's atmosphere contained very little or no free oxygen and, therefore, a relatively higher proportion of carbon dioxide.

Questions

A Explain why illumination of chloroplasts will lead to optimum conditions for the enzyme rubisco, involved in the light-independent stage.

B Suggest why there has been no selection of plants with rubisco enzymes having reduced oxygenase activity.

C Suggest how genetically modified plants, with an altered tertiary structure of rubisco, could improve the efficiency of photosynthesis.

D It has been suggested that with the increased carbon dioxide in the atmosphere, plants will carry out photosynthesis more efficiently. However, some studies of the greenhouse effect show that the opposite occurs and the rate of photosynthesis is reduced. Suggest a reason for this.

E Why do plant tissues contain the enzyme catalase?

F Why is rubisco described as the most important enzyme on Earth?

Questions

1 Explain why there are only small amounts of RuBP in the stroma of chloroplasts.

2 Discuss the role of carbon dioxide in photosynthesis.

3 Explain how a lack of nitrates in soil can lead to reduced growth in plants.

4 Which of the following contain nitrogen: NADP, DNA, GP, RuBP, ATP, RNA, sucrose, electron carriers, chlorophyll?

By the end of this spread, you should be able to . . .

✳ **Discuss the limiting factors in photosynthesis, with reference to carbon dioxide concentration, light intensity and temperature.**

We can see from the equation summarising photosynthesis that certain factors are needed.

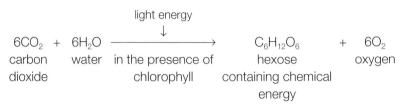

$$6CO_2 + 6H_2O \xrightarrow[\text{in the presence of chlorophyll}]{\text{light energy}} C_6H_{12}O_6 + 6O_2$$

carbon dioxide   water   in the presence of chlorophyll   hexose containing chemical energy   oxygen

The **chlorophyll** is present in the plants' **chloroplasts**. The other factors, light and supplies of carbon dioxide and water, are present in the environment. They may influence the rate at which photosynthesis proceeds and therefore the rate at which food for nearly all organisms on the planet becomes available.

Limiting factors

Photosynthesis is a complex process and the factors that affect its rate are all operating simultaneously. However, the factor that is present at the least favourable level is the one that *limits* the process, just as if you wished to make a cake and you had lots of flour, sugar and butter but only three eggs, the size of cake you could make would be limited by the small number of eggs you had available.

The factor present in the least favourable amount is called the **limiting factor**.

In the early 1900s Blackman investigated the effect of varying light intensity and temperature on the rate of photosynthesis.
- At constant temperature, the rate of photosynthesis varies with light intensity. At zero light intensity there will be no photosynthesis.
- At low light intensities, as light intensity increases so does the rate of photosynthesis, so light intensity is *limiting* the process.
- However at higher light intensities, the rate of photosynthesis plateaus. Light intensity is no longer limiting the process because changing the light intensity does *not* alter the rate of photosynthesis.
- Some other factor, such as carbon dioxide availability, must now be limiting the process.
- Increasing the carbon dioxide concentration will increase the rate of photosynthesis but not indefinitely. At some point the rate will plateau again as temperature is now limiting the process.
- Increasing the temperature can increase the rate of photosynthesis, although it will reach a plateau where some other factor is limiting the process. At too high a temperature, proteins such as enzymes involved in the Calvin cycle may be denatured.

The law of limiting factors states that at any given moment, the rate of a metabolic process is limited by the factor that is present at its least favourable (lowest) value.

The effect of carbon dioxide concentration on the rate of photosynthesis

Carbon dioxide constitutes around 0.03–0.06% of the Earth's atmosphere, depending on the location on the planet. The average figures are 0.039% by volume or 0.058% by mass. 500 million years ago there was 20 times this amount of carbon dioxide in the atmosphere.

Key definition

The **limiting factor** for a metabolic process is the factor that is present at the lowest or least favourable value.

Figure 1 The rate of photosynthesis at constant temperature and different light intensities

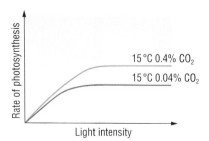

Figure 2 The rate of photosynthesis at constant temperature, with raised concentrations of carbon dioxide and different light intensities

Figure 3 Artist's impression of a Carboniferous forest

This fell during the Carboniferous Period, when fossil fuels were made. It rose again in the Triassic and Jurassic then continued to gradually decrease until the beginning of industrialisation, when it began to slowly rise again, because of the burning of fossil fuels by humans. This releases carbon dioxide that was fixed during the Carboniferous Period. The oceans act as carbon sinks and have absorbed about one third of the carbon dioxide produced by human activities. Growing forests can absorb carbon dioxide from the atmosphere but mature forests produce, by respiration and decomposition of dead leaves and wood, as much as they take in for photosynthesis.

In greenhouses, where many plants are grown in close proximity, even with ventilation the levels of carbon dioxide drop to around 0.02%. Growers can introduce more carbon dioxide by burning methane- or oil-fired heaters. Enhanced levels of carbon dioxide will increase the rate of photosynthesis, provided that no other factor is limiting the process.

The effect of light intensity on the rate of photosynthesis
When light is the limiting factor then the rate of photosynthesis is directly proportional to the light intensity. As light intensity increases, the rate of photosynthesis increases.

Light has three main effects:
- It causes stomata to open so that carbon dioxide can enter leaves.
- It is trapped by chlorophyll where it excites electrons.
- It splits water molecules to produce protons.

The electrons and protons are involved in **photophosphorylation**, producing ATP for the fixation of carbon dioxide.

The rate of photosynthesis will vary throughout the day as light intensity increases and then decreases.

The effect of temperature on the rate of photosynthesis
The photochemical reactions of the light-dependent stage of photosynthesis are not very much influenced by temperature. However, the enzyme-catalysed reactions of the Calvin cycle are. Between the temperatures of 0 °C and 25 °C, the rate of photosynthesis approximately doubles for each 10 °C rise in temperature. Above 25 °C the rate of photosynthesis levels off and then falls as enzymes work less efficiently and as oxygen more successfully competes for the active site of rubisco and prevents it from accepting carbon dioxide. High temperatures will also cause more water loss from stomata, leading to a stress response in which the stomata close, limiting the availability of carbon dioxide.

Questions
1 Explain why an increase in temperature from 20 °C to 25 °C increases the rate of photosynthesis.
2 Explain how burning an oil-fired stove in a greenhouse will increase the growth of the plants in the greenhouse.
3 Explain why cutting softwood trees for making paper, and replacing them by planting more trees, is considered to be a sustainable (not damaging to the environment) agricultural practice.
4 Explain how reduced light intensity can affect the rate of the light-independent stage of photosynthesis.

By the end of this spread, you should be able to ...

* Describe the effect on the rate of photosynthesis of changing the light intensity.
* Describe how to investigate experimentally the factors that affect the rate of photosynthesis.

Examiner tip

When answering data-based questions about measuring the rate of photosynthesis, don't refer to 'amount' of anything, be specific. Refer to *volume* of gas, *length* of time, *concentration* of a solution/chemical and *intensity* of light.

Measuring photosynthesis

There are many ways to measure photosynthesis. We could measure the uptake of substrates or the appearance of products. If we measure the photosynthesis happening per unit time (per second or per minute), then we have measured the rate of photosynthesis.

We could measure the:

* volume of oxygen produced per unit time (i.e. rate of oxygen production)
* rate of uptake of carbon dioxide
* rate of increase in dry mass of plants.

The rate of photosynthesis is usually found by measuring the volume of oxygen (the by-product) produced per minute, by an aquatic plant. This method has some limitations as:

* some of the oxygen produced by photosynthesis will be used by the plant for its respiration
* there may be some dissolved nitrogen in the gas collected.

Investigating the effect of light intensity on the rate of photosynthesis

The apparatus, known as a **photosynthometer** or **Audus microburette**, shown in Figure 1 is set up so that it is air-tight and there are no air bubbles in the capillary tubing. Gas, given off by the plant over a known period of time, collects in the flared end of the capillary tube. As the experimenter manipulates the syringe, this gas bubble can be moved into the part of the capillary tube against the scale and its length can be measured. This length can be converted to volume if we know the radius of the capillary bore.

volume of gas collected = length of bubble × πr^2

Key definition

A **photosynthometer** is used to measure the rate of photosynthesis by collecting and measuring the volume of oxygen produced in a certain time.

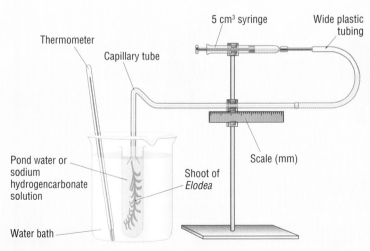

Figure 1 A photosynthometer – apparatus used to measure the rate of photosynthesis under various conditions

However, if the same apparatus is used throughout the investigation, the diameter (and therefore the radius) is constant and we can compare rates of photosynthesis by using just the length of gas bubble evolved per unit time. The water bath keeps the temperature constant. Sodium hydrogencarbonate solution, added to the water in the tube, provides carbon dioxide. The investigation has to be carried out in a darkened room, so that the only light available to the plant is from the light source. Ideally a low-energy bulb should be used as these do not release as much heat.

1 To fill the apparatus with tap water, remove the plunger from the syringe and allow a *gentle* stream of tap water into the barrel of the syringe, until the whole barrel and plastic tube are full of water. Replace the syringe plunger and *gently* push water out of the flared end of the capillary tubing, until the plunger is nearly at the end of the syringe and there are no air bubbles in the water in the capillary tube.

2 Cut a piece of well-illuminated *Elodea* (Canadian pondweed) about 7 cm long and make sure that bubbles of gas are emerging from the cut stem. Place this, cut end upwards, into a test tube containing the same water that the pondweed has been kept in. Add two drops of hydrogencarbonate solution to the water of the test tube. Stand the test tube in a beaker of water at about 20 °C. Use the thermometer to measure the temperature of the beaker at intervals during the investigation, and add cold water to adjust if necessary.

3 Place a light source as close to the beaker as possible. Measure the distance (*d*) from the piece of pondweed to the light source and note this. Light intensity (*I*) follows the inverse square law and is given by the formula

$$I = 1/d^2$$

Leave the apparatus, with the capillary tube positioned so that it is *not* collecting gas given off by the plant, for 5–10 minutes so that the plant acclimatises to these conditions.

4 Position the capillary tube over the cut end of the plant and after a known period of time (5–10 minutes) gently pull the syringe plunger so that the bubble of gas collected is in the capillary tube near the scale. Read and note the length of the bubble. Now gently push in the plunger so that the bubble is expelled.

5 Reposition the capillary tube to collect more gas from the plant and repeat step 4 twice more.

6 Move the light source further from the plant. Measure the distance and calculate the light intensity. Or you could use a light meter to measure light intensity. Allow a 5–10 minute acclimatisation period and then repeat steps 4 and 5.

7 Continue the investigation with different light intensities. Tabulate your data and plot a graph of rate of photosynthesis, calculated by volume of oxygen evolved per minute, against light intensity ($1/d^2$).

Light intensity $1/d^2$ (where *d* is in metres)	Volume of gas collected in one minute/cm³			Mean rate of photosynthesis/ cm³ min⁻¹
	1	2	3	

Questions

1 From your results, at what range of light intensities is light the limiting factor?

2 From your results, when is light no longer the main limiting factor?

3 List the main limitations of this experimental method.

4 Suggest how the method could be improved to overcome these limitations.

By the end of this spread, you should be able to . . .

* Describe the effect on the rate of photosynthesis of changing the carbon dioxide concentration, light intensity and temperature.
* Describe how to investigate experimentally the factors that affect the rate of photosynthesis.

Investigating the effects of temperature and carbon dioxide concentration on the rate of photosynthesis

The investigation described in spread 1.3.6 can be adapted to measure the effect of other factors on the rate of photosynthesis.

Temperature

Keep all other factors constant. Use a light intensity that produces a high rate of photosynthesis. Alter the temperature of the water bath and measure the volume of gas produced, in a known period of time, at each temperature. Note that this is not wholly accurate as warmer water will reduce the solubility of oxygen gas.

Carbon dioxide concentration

Keep all other factors constant and vary the number of drops of sodium hydrogencarbonate solution. Measure the volume of gas produced, in a known period of time, at each concentration of carbon dioxide.

STRETCH and CHALLENGE

Investigate different wavelengths of light. Keep all other factors constant and place different filters in front of the light source. Measure the volume of gas produced, in a known period of time, with each wavelength of light.

Investigating the rate of photosynthesis using changes in density of leaf discs

1 Use a drinking straw to cut several leaf discs from cress **cotyledons**.
2 Place five or six leaf discs in a 10 cm³ syringe and half-fill the syringe with dilute sodium hydrogencarbonate solution.
3 Hold the syringe upright, place your finger over the end of the syringe and gently pull on the plunger. This pulls the air out of the air spaces of spongy mesophyll in the leaf discs. The air is replaced by sodium hydrogencarbonate solution. As the density of the discs increases they sink to the bottom of the syringe.
4 Once all the discs have sunk, transfer the contents of the syringe to a small beaker. Illuminate from above using a bright light and time how long it takes for one leaf disc to float to the top of the solution. The reciprocal of the time taken ($1/t$) is a measure of the rate of photosynthesis.
5 Repeat the procedure twice more at this light intensity and find the mean rate of photosynthesis.
6 Repeat at different light intensities.
7 Tabulate your results.

This method can also be adapted to investigate the effect of temperature, light wavelength or carbon dioxide concentration on the rate of photosynthesis.

Key definition

The seeds of dicotyledonous plants store food in the **cotyledons**. These sometimes appear above the soil after germination and act as the first leaves, as in cress.

Examiner tip

Always refer to changes in *density* for the leaf discs. *Don't* describe them as becoming lighter.

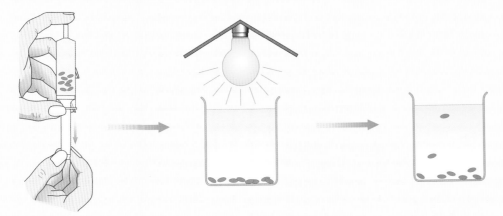

Figure 1 Using leaf discs to measure the rate of photosynthesis

STRETCH and CHALLENGE

Investigating the rate of photosynthesis by measuring uptake of carbon dioxide

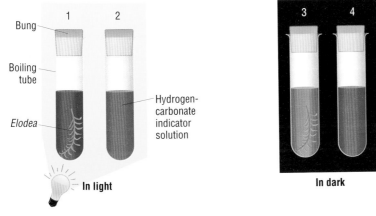

Figure 2 Using an indicator solution

Some indicator solutions are sensitive to small changes in pH. Hydrogencarbonate indicator solution is red when neutral, yellow at pH 6 and purple/red at pH slightly above 7.

Changes in colour of the indicator solution could be measured more objectively using a colorimeter with either a blue or green filter.

The change in absorption divided by time taken can give an indication of the rate of uptake of carbon dioxide by a piece of aquatic plant, such as *Elodea*, or by photosynthetic algae such as *Spirogyra*.

Questions

A Explain why hydrogencarbonate indicator solution will change from red to purple/red as the aquatic plant or alga in it photosynthesises.

B Explain why the solution in tube 3 becomes yellow.

C Design an investigation, using hydrogencarbonate indicator solution, to study the effect of light intensity or light wavelength on the rate of photosynthesis in *Spirogyra*.

Questions

1 Explain why the leaf discs float after they have been illuminated for a few minutes.

2 Explain why replicates are carried out for each light intensity.

3 Describe the pathway followed by the air that leaves the leaf discs at the beginning of the experiment.

4 State at which stage of photosynthesis each of the three factors – **(a)** light wavelength, **(b)** temperature and **(c)** carbon dioxide concentration – will have most effect. Explain your answers.

5 Write an outline plan for how you could use increase in dry mass to measure the effects of factors, such as light intensity, on the rate of photosynthesis.

By the end of this spread, you should be able to ...

✳ Describe the effect on the levels of GP, RuBP and TP of changing the carbon dioxide concentration, light intensity and temperature.

You have seen that changes in carbon dioxide concentration, light intensity and temperature all alter the rate of photosynthesis.

Light intensity

Light intensity gives a measure of how much energy is associated with the light. Light from a source spreads out and so if the distance from the source is doubled, the light intensity is quartered. It follows the inverse square law and $I = 1/d^2$.

An increase in light intensity will alter the rate of the **light-dependent reaction**.
- More light energy is available to excite more electrons.
- The electrons take part in **photophosphorylation**, so increased light intensity means that more ATP and more reduced NADP will be produced.
- These are both used in the **light-independent stage** (Calvin cycle), as sources of hydrogen and energy, to reduce **glycerate phosphate** (**GP**) to **triose phosphate** (**TP**). ATP is also used to phosphorylate five out of every six molecules of TP to regenerate **RuBP**.

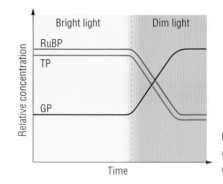

Figure 1 Effect on relative concentratons of GP, TP and RuBP, of reducing light intensity

If there is no or very little light available, the light-dependent stage will cease. This will stop the light-independent stage as it needs the products of the light-dependent stage.
- GP cannot be changed to TP, so GP will accumulate and levels of TP will fall.
- This will lower the amount of RuBP, reducing the fixation of carbon dioxide and the formation of more GP.

Carbon dioxide concentration

An increase in carbon dioxide concentration may lead to an increase in carbon dioxide fixation in the Calvin cycle, provided that light intensity is not limiting the process.
- More carbon dioxide fixation leads to more molecules of GP (some of which may be converted to amino acids or fatty acids) and hence more molecules of TP (to be converted to hexose sugar, disaccharides and polysaccharides, and glycerol) and more regeneration of RuBP.
- However, the number of **stomata** that open to allow gaseous exchange leads to increased transpiration and may lead to the plant wilting, if its water uptake from the soil cannot exceed water loss by transpiration. This, in turn, leads to a stress response and, following release of a plant growth regulator (abscisic acid), stomata close (see spread 1.3.5).
- This will reduce carbon dioxide uptake and reduce the rate of photosynthesis.

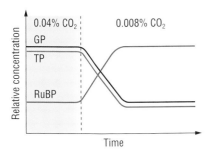

Figure 2 Effect on relative concentrations of GP, TP, and RuBP of reducing carbon dioxide concentration

If carbon dioxide concentration is reduced below 0.01% then RuBP, the carbon dioxide acceptor, will accumulate. As a result, levels of GP, and subsequently of TP, will fall.

Temperature

- Increasing the temperature will have little effect upon the rate of the light-dependent reaction as, apart from photolysis of water, it is not dependent upon enzymes. However, it will alter the rate of the light-independent reaction as that is a series of biochemical steps, each catalysed by a specific enzyme. Increasing temperature will, at first, increase the rate of photosynthesis.
- However, as temperatures rise above 25 °C, the oxygenase activity of rubisco increases more than its carboxylase activity increases.
- This means that photorespiration exceeds photosynthesis.
- As a result ATP and reduced NADP, from the light-dependent reaction, are dissipated and wasted.
- This reduces the overall rate of photosynthesis.
- Very high temperatures may also damage proteins involved in photosynthesis.
- Increased temperatures cause an increase in water loss from leaves by transpiration. This may lead to closure of stomata and subsequent reduction in the rate of photosynthesis (see spread 1.3.5).

Examiner tip

You may well get questions that expect you to interpret, from graphs, what happens to levels (relative concentrations) of GP, TP or RuBP when factors are changed. Make a quick sketch of the Calvin cycle and then you can more easily explain why levels of chemicals rise or fall.

STRETCH and CHALLENGE

What other factors may alter the rate of photosynthesis?

Questions

Suggest and explain how each of the following factors could alter the rate of photosynthesis:

A lack of magnesium in soil
B lack of water in soil
C the presence of DCMU (dichlorophenyl dimethylurea), a herbicide that blocks electron flow between PSII and PSI

Questions

1 Which proteins, besides enzymes, involved in photosynthesis, may be damaged by high temperatures? Explain how high temperatures may damage these proteins.
2 Explain how stomatal closure leads to a reduction in the rate of photosynthesis.
3 Explain why, at low levels of atmospheric carbon dioxide, levels of RuBP will rise and RuBP accumulates in chloroplasts, but levels of GP and TP fall.
4 Explain why levels of RuBP in chloroplasts drop when light intensity falls.

Photosynthesis summary

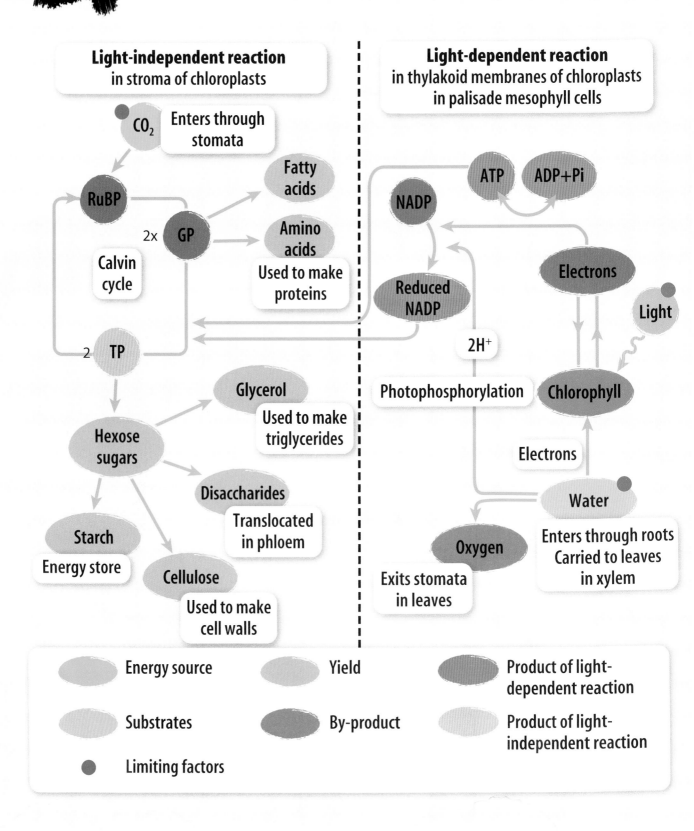

Light-independent reaction
in stroma of chloroplasts

Light-dependent reaction
in thylakoid membranes of chloroplasts
in palisade mesophyll cells

CO_2 Enters through stomata

RuBP

GP 2x

Fatty acids

Amino acids

Used to make proteins

Calvin cycle

2 TP

ATP   ADP+Pi

NADP

Reduced NADP

Electrons

Light

$2H^+$

Photophosphorylation

Chlorophyll

Electrons

Glycerol

Used to make triglycerides

Hexose sugars

Disaccharides

Translocated in phloem

Starch

Energy store

Cellulose

Used to make cell walls

Oxygen

Exits stomata in leaves

Water

Enters through roots
Carried to leaves in xylem

Energy source

Yield

Product of light-dependent reaction

Substrates

By-product

Product of light-independent reaction

Limiting factors

Practice questions

1. What do the following terms mean?
   (a) autotroph
   (b) heterotroph   [2]

2. Why does respiration in plants and animals depend on the products of photosynthesis?   [2]

3. Where in chloroplasts do the following take place?
   (a) light-dependent stage of photosynthesis   [1]
   (b) light-independent stage of photosynthesis   [1]
   (c) protein synthesis   [1]

4. What is meant by the term 'photosynthetic pigment'?   [4]

5. What are the products of non-cyclic photophosphorylation?
                                                              [3]

6. Explain the roles, in photosynthesis, of each of the following:
   (a) water   [3]
   (b) ribulose bisphosphate   [3]
   (c) NADP   [3]
   (d) ATP   [3]
   (e) carbon dioxide   [3]
   (f) rubisco   [3]

7. Some plants that are adapted to living in the shade produce many red pigments, as well as some green chlorophyll pigments. Their leaves appear red.
   (a) Explain why the leaves appear red.   [2]
   (b) Suggest the advantage to these plants of having a lot of this pigment in their leaves.   [2]

8. Many plants that live in the Arctic Circle have very dark, almost black, leaves. Suggest the advantage to these plants of having such pigmentation.   [4]

9. Ribulose bisphosphate carboxylase (rubisco) is said to be the most important enzyme on Earth. It comprises up to 50% of leaf proteins and is therefore the most abundant protein in the biosphere. Rubisco from plants consists of eight large subunits encoded by chloroplast DNA, and eight small subunits encoded by nuclear DNA. It is sometimes given the symbol L_8S_8. The enzyme will work without the small subunits.

   The active site of rubisco allows carbon dioxide and a pentose sugar, ribulose bisphosphate, to sit side by side and form an unstable 6-carbon compound that immediately breaks down to two molecules of triose phosphate. The unstable intermediate has not been isolated or identified. There is a competitive inhibitor of rubisco, called CABP (2-carboxyarabinitol-1,5-bisphosphate), the structure of which is shown below.

   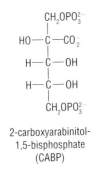

   2-carboxyarabinitol-
   1,5-bisphosphate
   (CABP)

   (a) Suggest why rubisco is described as the most important enzyme on Earth.   [3]
   (b) Explain why rubisco is described as having a quaternary structure.   [2]
   (c) Scientists think that the active site for this enzyme involves only the large subunits. What evidence is described in the passage above to support this hypothesis?   [1]
   (d) Explain how the competitive inhibition of rubisco by CABP supports the hypothesis that an intermediate 6-carbon compound is formed when ribulose bisphosphate reacts with carbon dioxide.   [3]

10. (a) Explain how the structure of chloroplasts enables them to carry out their functions.   [8]
    (b) Plant cells contain protein microtubules. These can move organelles within the cells. Chloroplasts can be moved around within palisade mesophyll cells. On a dull day many will be positioned at the top of the palisade cells, just below the upper epidermis of the leaf. What is the advantage to the plants of chloroplasts being moved to this position on a dull day?   [3]

1 The rate of photosynthesis at different wavelengths of light can be measured and plotted on a graph. This is called an action spectrum and is shown on Figure 1.1.

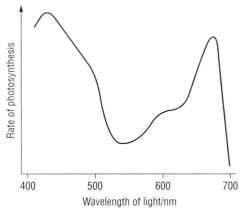

Figure 1.1

(a) Describe the effects of different wavelengths of light on the rate of photosynthesis. [3]

(b) (i) Name **two** pigment molecules found in chloroplasts. [2]

   (ii) State **one** role of pigment molecules in a photosystem. [1]

   (iii) State the location of photosystems within the chloroplast. [1]

(c) The production of ATP in chloroplasts is known as photophosphorylation. There are two types of photophosphorylation, cyclic and non-cyclic.

   (i) Complete the following table comparing the two types of photophosphorylation.

	Cyclic	Non-cyclic
Photosystem(s) involved		
End product(s)		

[4]

   (ii) Explain the role of ATP in the light-**independent** stage of photosynthesis. [3]

(d) Some chloroplasts have been shown to possess photosystem 1 only. Explain why these chloroplasts are unable to form sugars. [3]

[Total: 17]
(OCR 2804 Jan03)

2 An investigation was carried out into photosynthesis and respiration in a leaf. The net uptake of carbon dioxide by the leaf in bright light, and the mass of carbon dioxide released in the dark, were determined at different temperatures. The results are shown in Table 2.1.

Temperature/°C	5	10	15	20	25	30
Net uptake of CO_2 in bright light/mg g^{-1} dry mass h^{-1}	1.3	2.4	3.0	3.3	3.0	2.2
Release of CO_2 in dark/mg g^{-1} dry mass h^{-1}	0.4	0.7	1.0	1.4	1.9	2.8
True rate of photosynthesis/ mg CO_2 g^{-1} dry mass h^{-1}						

Table 2.1

(a) (i) State **two** types of tissue in a leaf where there is a net uptake of carbon dioxide in bright light. [2]

   (ii) Assuming the rate of respiration in the light is equal to the rate of respiration in the dark, calculate the true rate of carbon dioxide uptake in photosynthesis at each temperature and add the figures to Table 2.1. [1]

   (iii) The term temperature coefficient (Q_{10}) is used to express the effect of a 10 °C rise in temperature on the rate of a chemical reaction. It is calculated in the following way:
   $$Q_{10} = \frac{\text{rate of reaction at } t + 10\,°C}{\text{rate of reaction at } t\,°C}$$
   where t = any given temperature
   Between 5 °C and the optimum temperature for enzyme-catalysed reactions, the Q_{10} is approximately 2.
   Discuss whether the data in Table 2.1 supports this statement for both respiration and photosynthesis. [7]

(b) The carbon dioxide taken up by a leaf enters the chloroplasts.
   Name and describe the **biochemical pathway** which fixes the carbon dioxide into hexose sugars in the chloroplasts.
   In this question, one mark is available for the quality of written communication. [5]

(c) Figure 2.1 shows the population curve of unicellular photosynthetic organisms (algae) in a freshwater lake in southern England.

Answers to examination questions will be found on the Exam Café CD.

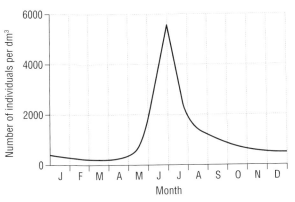

Figure 2.1

Explain how a change in environmental factors could account for the sudden rise and fall of this population between May and September. [5]

[Total: 20]
(OCR 2804 Jan03 and Jun06)

3 The light-dependent stage of photosynthesis takes place on thylakoid membranes in chloroplasts. These membranes surround the thylakoid space (lumen) and are arranged into stacks known as grana. Figure 3.1 is a diagram summarising the processes that take place at the thylakoid membrane.

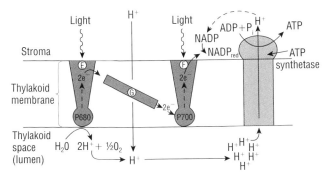

Figure 3.1

(a) State the general name of the pigment complexes shown as **E** and **F** on the diagram. [1]

(b) Name the pigment represented by P680 and P700. [1]

(c) Name the type of molecule represented by **G**. [1]

(d) State, using the information in Figure 3.1, why the pH of the thylakoid space (lumen) is lower than that of the stroma. [1]

(e) Explain the function of this pH gradient. [3]

(f) Herbicides (weedkillers), such as diquat and paraquat, act on the chloroplast thylakoids. They interfere with electron transport by accepting electrons and prevent the light-dependent stage of photosynthesis from taking place. Explain how this causes plants to die. [5]

(g) Some weed species are **not** killed when herbicides are applied. Suggest why. [2]

[Total: 14]
(OCR 2804 Jan04)

4 In the majority of photosynthetic organisms, fixation of carbon dioxide occurs using the Calvin cycle. Figure 4.1 is an outline of this cycle.

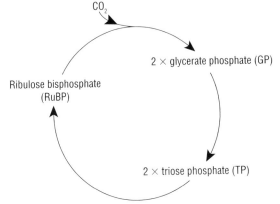

Figure 4.1

(a) (i) Name the five-carbon sugar in the cycle. [1]

(ii) Name the enzyme that fixes the carbon dioxide. [1]

(iii) Mark an **A** on Figure 4.1 to show where reduced NADP from the light-dependent reaction is used.[1]

(iv) State where in the chloroplast the Calvin cycle occurs. [1]

(v) Name another compound produced in the light-dependent stage that is used in this cycle. [1]

(b) Figure 4.2 shows the changes in relative amounts of RuBP and GP production in the Calvin cycle before and after a light source is switched off. All other conditions are constant.

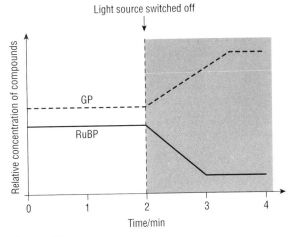

Figure 4.2

Explain the changes in the relative amounts of GP and RuBP after the light source is switched off. [3]

[Total: 8]
(OCR 2804 Jun03)

Module 4
Respiration

Introduction

All living organisms need energy to do work. This work includes movement, active transport, bulk transport, nerve conduction, synthesis of large molecules such as cellulose and proteins, and replication of DNA. Energy is also needed for synthesis of new organelles before a cell divides.

When life developed on Earth around 3500 million years ago, the atmosphere did not contain any free oxygen. The earliest forms of life on Earth used various metabolic pathways to obtain energy from chemicals in their environment. After about 1500 million years, cyanobacteria evolved that could trap sunlight energy for the process of photosynthesis. They used water as a source of electrons and protons, releasing free oxygen into the air. This changed the previously reducing atmosphere to an oxidising one and led eventually to vast biodiversity, as organisms that could use this oxygen for respiration evolved.

Cyanobacteria still exist today. The photograph shows a group of stromatolites in Shark Bay, Western Australia. These are living, rocky lumps formed over the last 4000 years from cyanobacteria, the calcium carbonate they secrete, and trapped sediment.

In this module you will learn some of the details of the process of respiration.

Test yourself

1 What is aerobic respiration?
2 What is anaerobic respiration?
3 Which organelle carries out most of the stages of respiration in eukaryotic cells?
4 What are the products of aerobic respiration?
5 What is the universal energy currency molecule?
6 What is energy?
7 Why do living organisms need energy?
8 Do plants respire?

By the end of this spread, you should be able to ...

✳ Outline why living organisms need to respire.
✳ Describe the structure of ATP.
✳ State that ATP provides the immediate source of energy for biological processes.

What is respiration?

Respiration is the process whereby **energy** stored in complex organic molecules (carbohydrates, fats and proteins) is used to make **ATP**. It occurs in living cells.

What is energy?

Energy exists as **potential** (stored) energy and **kinetic** energy (the energy of movement). Moving molecules have kinetic energy that allows them to diffuse down a concentration gradient. Large organic molecules contain chemical potential energy.

Energy:
- cannot be created or destroyed but can be converted from one form to another
- is measured in joules or kilojoules
- has many forms, e.g. sound (mechanical), light, heat, electrical, chemical and atomic.

Why do we need it?

All living organisms need energy to drive their biological processes. All the reactions that take place within organisms are known collectively as **metabolism**. Metabolic reactions that build large molecules are described as **anabolic** and those that break large molecules into smaller ones are **catabolic**.

Metabolic processes that need energy include:
- Active transport – moving ions and molecules across a membrane against a concentration gradient. Much of an organism's energy is used for this. All cell membranes have sodium–potassium pumps and these maintain the resting potential. When this pump momentarily stops in neurone membranes, sodium ions enter the neurone and an action potential occurs.
- Secretion – large molecules made in some cells are exported by exocytosis.
- Endocytosis – bulk movement of large molecules into cells.
- Synthesis of large molecules from smaller ones, such as proteins from amino acids, steroids from cholesterol and cellulose from β-glucose. These are all examples of anabolism.
- Replication of DNA and synthesis of organelles before a cell divides.
- Movement – such as movement of bacterial flagella, eukaryotic cilia and undulipodia, muscle contraction and microtubule motors that move organelles around inside cells.
- Activation of chemicals – glucose is phosphorylated at the beginning of respiration so that it is more unstable and can be broken down to release energy.

Some of the energy from catabolic reactions is released in the form of heat. This is useful as metabolic reactions are controlled by enzymes, so organisms need to maintain a suitable temperature that allows enzyme action to proceed at a speed that will sustain life.

Key definitions

Energy is the ability to do work.

ATP is a phosphorylated nucleotide and is the universal energy currency.

Key definitions

Anabolic reactions are biochemical reactions where large molecules are synthesised from smaller ones.

In **catabolic** reactions larger molecules are hydrolysed to produce smaller molecules.

Where does the energy come from?

Plants, some protoctists and some bacteria are **photoautotrophs**. They use sunlight energy in photosynthesis to make large, organic molecules that contain chemical potential energy, which they and consumers and decomposers can then use. Respiration releases the energy, which is used to phosphorylate (add inorganic phosphate to) ADP, making ATP. This phosphorylation also transfers energy to the ATP molecule.

The role of ATP

ATP is a phosphorylated nucleotide. It is a high-energy intermediate compound, found in both prokaryotic and eukaryotic cells. Each molecule consists of adenosine (adenine and ribose sugar) plus three phosphate (more correctly, phosphoryl) groups. It can be hydrolysed to ADP and P_i (inorganic phosphate), releasing 30.6 kJ energy per mol. So, energy is immediately available to cells in small, manageable amounts that will not damage the cell and will not be wasted. ATP is described as the universal energy currency.

Respiration occurs in many small steps. The energy released at each stage joins ADP and P_i to make ATP. You probably use 25–50 kg ATP each day, depending on your level of activity, but you will only have about 5 g of ATP in your body at any one point in time. It is continually being hydrolysed and resynthesised.

The hydrolysis of ATP is coupled with a synthesis reaction, such as DNA replication or protein synthesis, in cells. Such synthesis reactions require energy. The energy released from ATP hydrolysis is an immediate source of energy for these biological processes.

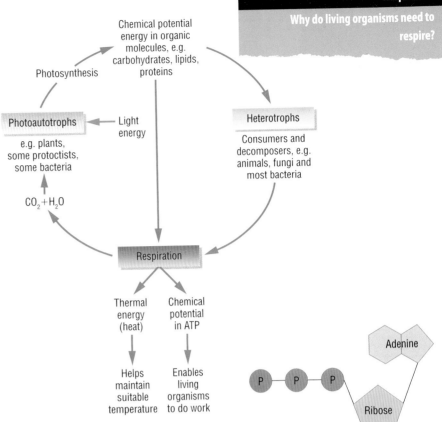
Figure 1 Energy transfer

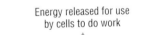
Figure 2 The structure of ATP

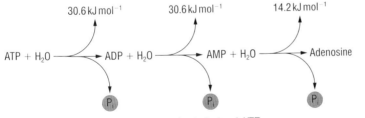
Figure 3 The energy released from hydrolysis of ATP

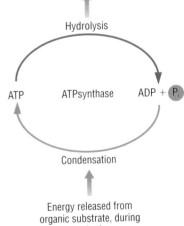
Figure 4 The ATP cycle

Questions

1 Fireflies can produce light in a process called bioluminescence. Outline the energy transformations that occur in fireflies as they use energy from their food to produce bioluminescence.
2 ATP is a nucleic acid/nucleotide derivative. Do you think it is derived from DNA or RNA nucleotides? Give reasons for your answer.
3 Decide whether each of the following is an anabolic or catabolic reaction:
   (a) synthesis of spindle microtubules during mitosis
   (b) digestion of starch to maltose
   (c) formation of insulin in cells of the pancreas
   (d) conversion of glycogen to glucose in liver cells
   (e) digestion of a pathogen inside a phagolysosome of a macrophage.
4 Explain why ATP is known as the universal energy currency.

Examiner tip

Remember that energy cannot be created or destroyed. So, never refer to energy being *produced*. Respiration *releases* energy to *produce* ATP.

② Coenzymes

By the end of this spread, you should be able to ...

✳ **Explain the importance of coenzymes in respiration, with reference to NAD and coenzyme A.**

The stages of respiration

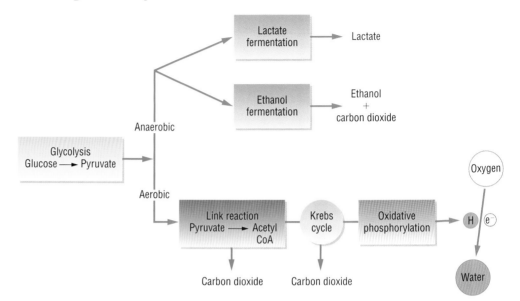

Figure 1 The stages of respiration

Respiration of glucose can be described in four stages:

- **Glycolysis** – this happens in the cytoplasm of all cells. It is an ancient biochemical pathway. It doesn't need oxygen and can take place in aerobic or anaerobic conditions. During glycolysis, glucose (a 6-carbon sugar) is broken down to two molecules of pyruvate (a 3-carbon compound).
- **The link reaction** – this happens in the matrix of mitochondria. Pyruvate is dehydrogenated (hydrogen removed) and decarboxylated (carboxyl removed) and converted to acetate.
- **Krebs cycle** – also takes place in the matrix of mitochondria. Acetate is decarboxylated and dehydrogenated.
- **Oxidative phosphorylation** – takes place on the folded inner membranes (cristae) of mitochondria. This is where ADP is phosphorylated to ATP.

The last three stages will only take place under aerobic conditions. Under anaerobic conditions, pyruvate is converted to either ethanol or lactate.

Why are coenzymes needed?

During glycolysis, the link reaction and Krebs cycle, hydrogen atoms are removed from substrate molecules in **oxidation** reactions. These reactions are catalysed by dehydrogenase enzymes. Although enzymes catalyse a wide variety of metabolic reactions, they are not very good at catalysing oxidation or **reduction** reactions. Coenzymes are needed to help them carry out the oxidation reactions of respiration. The hydrogen atoms are combined with coenzymes such as **NAD**. These carry the hydrogen atoms, which can later be split into hydrogen ions and electrons, to the inner mitochondrial membranes. Here, they will be involved in the process of oxidative phosphorylation (see spread 1.4.6), which produces a lot of ATP. Delivery of the hydrogens to the cristae reoxidises the coenzymes so they can combine with (or 'pick up') more hydrogen atoms from the first three stages of respiration.

Key definitions

Oxidation reactions involve loss of electrons.

Reduction reactions involve addition of electrons.

These reactions are coupled – one substrate becomes oxidised and another becomes reduced. In the reactions of respiration where coenzymes are involved, the coenzymes become reduced as substrate becomes oxidised. Later the reduced coenzyme becomes reoxidised so that it can be used again.

NAD

This is an organic, non-protein molecule that helps dehydrogenase enzymes to carry out oxidation reactions. Nicotinamide adenine dinucleotide (NAD) is made of two linked nucleotides. It is made in the body from nicotinamide (Vitamin B_3), the 5-carbon sugar ribose, adenine and two phosphate (or, more accurately, phosphoryl) groups. One nucleotide contains the nitrogenous base adenine. The other contains a nicotinamide ring that can accept hydrogen atoms – each of which can later be split into a hydrogen ion and an electron.

When a molecule of NAD has accepted two hydrogen atoms with their electrons, it is reduced. When it loses the electrons it is oxidised. NAD operates during glycolysis (see spread 1.4.3), the link reaction (see spread 1.4.5), Krebs cycle (see spread 1.4.5) and during the anaerobic ethanol and lactate pathways (see spread 1.4.8).

Coenzyme A (CoA)

This coenzyme is made from pantothenic acid (a B-group vitamin), adenosine (ribose and adenine), three phosphate (phosphoryl) groups and cysteine (an amino acid). Its function is to carry ethanoate (acetate) groups, made from pyruvate during the link reaction, onto Krebs cycle. It can also carry acetate groups that have been made from fatty acids or from some amino acids (see spread 1.4.9) onto Krebs cycle.

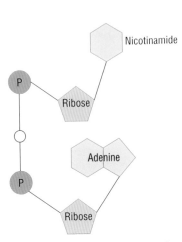

Figure 2 Molecular structure of NAD

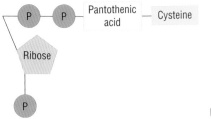

Figure 3 The structure of coenzyme A

Examiner tips

Don't confuse NAD and NADP. We have met NADP in photosynthesis (think of the P in NADP as a reminder of photosynthesis!).

We have met NAD involved in respiration.

When these coenzymes become reduced they carry hydrogen atoms (which later become protons and electrons). *Don't* say that they carry hydrogen ions or molecules.

STRETCH and CHALLENGE

In the early part of the twentieth century, a dietary deficiency disease called *pellagra* (diarrhoea, dermatitis and dementia) was endemic in rural parts of the Southern United States of America, where the diet consisted mainly of corn products. It could be treated with nicotinamide (vitamin B_3). Humans, like many animals, can synthesise nicotinamide from the amino acid tryptophan. Corn contains very little tryptophan. It contains a lot of nicotinamide but in a form that needs to be treated with a base (alkaline substance) before it can be absorbed from the intestine.

Mexican Indians are thought to have domesticated the corn plant and their diet has always contained a lot of corn. They soak the corn in limewater (calcium hydroxide solution) before using it to make tortillas. They do not suffer from pellagra.

Question

A Explain why Mexican Indians do not suffer from pellagra, whilst people living in rural Southern states of the US did, although both ate a diet rich in corn.

Questions

1 Explain why living organisms do not have very much NAD or CoA in their cells.
2 Alcohol is metabolised in the liver. It is oxidised to ethanal by dehydrogenation, and then to ethanoate (acetate). Suggest why people who drink large amounts of alcohol may be deficient in NAD.
3 Explain why NAD is called a nucleic acid/nucleotide derivative.

1.4 ③ Glycolysis

By the end of this spread, you should be able to ...

* **State that glycolysis occurs in the cytoplasm of cells.**
* **Outline the process of glycolysis.**
* **State that in aerobic respiration, pyruvate is actively transported into mitochondria.**

Key definitions

Glycolysis is a metabolic pathway where each glucose molecule is broken down to two molecules of pyruvate. It occurs in the cytoplasm of all living cells and is common to anaerobic (without oxygen) and aerobic (with oxygen) respiration.

Hexose sugars have six carbon atoms in each molecule.

Hydrolysis is the breaking down of large molecules to smaller molecules by the addition of water.

Triose sugars have three carbon atoms in each molecule.

Glycolysis is a very ancient biochemical pathway, occurring in the cytoplasm of all living cells that respire. This means it happens in prokaryotic and eukaryotic cells. It has been studied extensively by biochemists and is probably the best-understood metabolic pathway.

This pathway involves a sequence of ten reactions, each catalysed by a different enzyme. The coenzyme NAD (see spread 1.4.2) is also involved. You only need to know this pathway in outline so we will consider it as just four stages.

Stage 1: Phosphorylation

Glucose is a **hexose** sugar – it contains six carbon atoms. Its molecules are stable and need to be activated before they can be split into two.

* One ATP molecule is **hydrolysed** and the phosphate group released is attached to the glucose molecule at carbon 6.
* Glucose 6-phosphate is changed to fructose 6-phosphate.
* Another ATP is hydrolysed and the phosphate group released is attached to fructose 6-phosphate at carbon 1. This activated hexose sugar is now called fructose 1,6-bisphosphate.
* The energy from the hydrolysed ATP molecules activates the hexose sugar and prevents it from being transported out of the cell. We can refer to the activated, phosphorylated sugar as **hexose 1,6-bisphosphate**. (This name tells us that it is a hexose sugar with two phosphates attached, one at carbon 1 and the other at carbon 6.)
* Note that this stage has *used* two molecules of ATP for each molecule of glucose.

Stage 2: Splitting of hexose 1,6-bisphosphate

* Each molecule of hexose bisphosphate is split into two molecules of **triose** phosphate (3-carbon sugar molecules each with one phosphate group attached).

Stage 3: Oxidation of triose phosphate

* Although this process is anaerobic, it involves oxidation.
* Two hydrogen atoms (with their electrons) are removed from each triose phosphate molecule (the substrate).
* This involves dehydrogenase enzymes.
* These are aided by the coenzyme NAD (nicotinamide adenine dinucleotide), which is a hydrogen acceptor (see spread 1.4.2). NAD combines with the hydrogen atoms, becoming reduced NAD.
* So, at this stage of glycolysis, two molecules of NAD are reduced per molecule of glucose.
* Also, at this stage, two molecules of ATP are formed. This is called **substrate-level phosphorylation**.

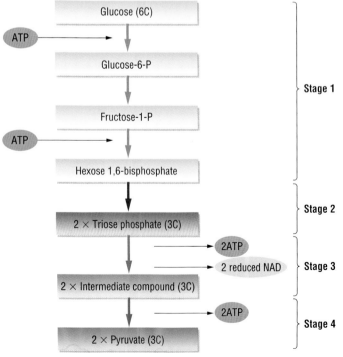

Figure 1 Summary of glycolysis

Stage 4: Conversion of triose phosphate to pyruvate

- Four enzyme-catalysed reactions convert each triose phosphate molecule to a molecule of pyruvate. Pyruvate is also a 3-carbon compound.
- In the process another two molecules of ADP are phosphorylated (an inorganic phosphate group, P_i, is added) to two molecules of ATP (by substrate-level phosphorylation).

What are the products of glycolysis?

From each molecule of glucose at the beginning of this pathway, at the end of glycolysis there are:

- two molecules of ATP. Four have been made but two were used to 'kick-start' the process, so the *net gain* is two molecules of ATP
- two molecules of reduced NAD. These will carry hydrogen atoms, indirectly via a shunt mechanism, to the inner mitochondrial membranes and be used to generate more ATP during oxidative phosphorylation (see spread 1.4.6)
- two molecules of pyruvate. This will normally be actively transported into the mitochondrial matrix for the next stage of aerobic respiration (see spread 1.4.4). In the absence of oxygen it will be changed, in the cytoplasm, to lactate or ethanol (see spread 1.4.8).

Examiner tip

Learn the stages of glycolysis where:
ATP is used
ATP is produced
NAD is reduced

Fermentation and glycolysis

For thousands of years humans used the process of fermentation of glucose to ethanol, by yeast, without understanding that glycolysis was involved. In the second half of the nineteenth century, scientists investigated the mechanism. Pasteur established that alcoholic fermentation is caused by microorganisms. Buchner showed that extracts from yeast cells could also cause fermentation. He used the word enzyme – the word means 'in yeast'. By 1940 many biochemists had helped to analyse and work out the pathway. They had used cells and tissues from many living organisms in their studies and found that, with very few exceptions (such as some Archaea), all living things have this metabolic pathway.

STRETCH and CHALLENGE

Enzymes that cause the shape of a molecule to change (without changing the proportions of atoms in that molecule) are called *isomerases*.

Questions

A At which stage of glycolysis are isomerase enzymes involved?
B How does the fact that nearly all living things use the glycolysis pathway support the theory of evolution?

Questions

1 What was in Buchner's cell-free extract (made from yeast) that enabled the fermentation of glucose to alcohol?
2 Outline the role of coenzymes (spread 1.4.2) in the glycolysis pathway.
3 Explain why the net gain of ATP during glycolysis is two, not four, molecules.
4 Explain how oxidation occurs during glycolysis, although no oxygen is involved.

By the end of this spread, you should be able to ...

✳ **Explain, with the aid of diagrams and electron micrographs, how the structure of mitochondria enables them to carry out their functions.**

Key definition

Mitochondria are organelles found in eukaryote cells. They are the sites of the link reaction, Krebs cycle and oxidative phosphorylation – the aerobic stages of respiration.

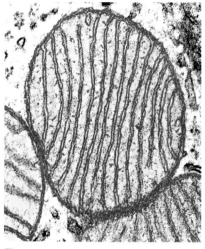

Figure 1 Electron micrograph of a mitochondrion from an intestinal cell (×32 000)

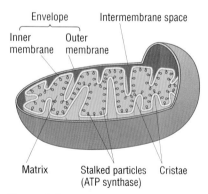

Envelope   Intermembrane space

Inner membrane   Outer membrane

Matrix   Stalked particles (ATP synthase)   Cristae

Figure 2 Structure of a mitochondrion

Examiner tip

Always say that protons flow *down* the (electrochemical/pH/proton) gradient through ATP synthase enzyme. *Don't* say they flow along the gradient.

Mitochondrial ultrastructure

Mitochondria were first identified in animal cells, using light microscopy, in 1840. Plant mitochondria were observed about 60 years later. In 1953 the first extensive electron microscope studies of mitochondria were made.

* All mitochondria have an inner and outer phospholipid membrane. These two membranes make up the *envelope*.
* The outer membrane is smooth and the inner membrane is folded into **cristae** (singular **crista**) that give the inner membrane a large surface area.
* The two membranes enclose and separate the two compartments within the mitochondrion. Between the inner and outer membranes is the **intermembrane space**.
* The **matrix** is enclosed by the inner membrane. It is semi-rigid and gel-like, consisting of a mixture of proteins and lipids. It also contains looped mitochondrial DNA, mitochondrial ribosomes and enzymes.

Shape, size and distribution

Mitochondria may be rod-shaped or thread-like. Their shape can change but most are between 0.5–1.0 μm in diameter and 2–5 μm long, although some can be 10 μm long. A trained athlete may have larger mitochondria in his/her muscle tissue. Metabolically active cells (large demand for ATP) have more mitochondria. These mitochondria usually have longer and more densely packed cristae to house more electron transport chains and more ATP synthase enzymes. Mammalian liver cells may each contain up to 2500 mitochondria, occupying up to 20% of the cell's volume.

Mitochondria can be moved around within cells by the cytoskeleton (microtubules). In some types of cells the mitochondria are permanently positioned near a site of high ATP demand, for example at the synaptic knobs of nerve cells. However, they have been moved to that position by microtubules.

How does their structure enable them to carry out their functions?

The matrix

The matrix is where the **link reaction** and **Krebs cycle** take place. It contains:

* the enzymes that catalyse the stages of these reactions
* molecules of coenzyme NAD
* oxaloacetate – the 4-carbon compound that accepts acetate from the link reaction
* mitochondrial DNA, some of which codes for mitochondrial enzymes and other proteins
* mitochondrial ribosomes (structurally the same as prokaryote ribosomes) where these proteins are assembled.

The outer membrane

The phospholipid composition of the outer membrane is similar to membranes around other organelles. It contains proteins, some of which form channels or carriers that allow the passage of molecules such as pyruvate. Other proteins in this membrane are enzymes.

The inner membrane

The inner membrane:

- has a different lipid composition from the outer membrane and is impermeable to most small ions, including hydrogen ions (protons)
- is folded into many cristae to give a large surface area
- has embedded in it many **electron carriers** and **ATP synthase** enzymes.

The electron carriers are protein complexes, arranged in **electron transport chains**.

- Each electron carrier is an enzyme. Each is associated with a cofactor. The cofactors are non-protein groups. They are haem groups and contain an iron atom.
- The cofactors can accept and donate electrons because the iron atoms can become reduced (to Fe^{2+}) by accepting an electron and oxidised (to Fe^{3+}) by donating an electron to the next electron carrier.
- They are oxidoreductase enzymes as they are involved in oxidation and reduction reactions.
- Some of the electron carriers also have a coenzyme that *pumps* (using energy released from the passage of electrons) protons from the matrix to the intermembrane space.
- Because the inner membrane is impermeable to small ions, protons accumulate in the intermembrane space, building up a proton gradient – a source of potential energy.

The ATP synthase enzymes:

- are large and protrude from the inner membrane into the matrix
- are also known as stalked particles
- allow protons to pass through them.

Protons flow down a proton gradient, through the ATP synthase enzymes, from the intermembrane space into the matrix. This flow is called **chemiosmosis**. The force of this flow (the proton motive force) drives the rotation of part of the enzyme and allows ADP and P_i (inorganic phosphate) to be joined to make ATP.

The coenzyme FAD, which becomes reduced during one stage of Krebs cycle, is tightly bound to a dehydrogenase enzyme that is embedded in the inner membrane. The hydrogen atoms accepted by FAD do *not* get pumped into the inert membrane space. Instead they pass back into the mitochondrial matrix.

FAD is flavine adenine dinucleotide, derived from vitamin B_2 (riboflavin), adenine, ribose and two phosphate groups.

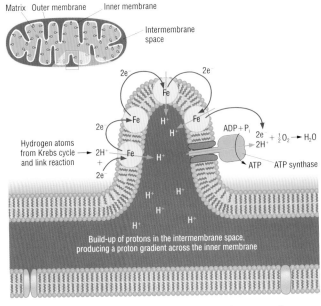

Figure 3 Diagram showing the structure of the inner membrane and the flow of electrons between electron carriers and the flow of protons into the intermembrane space

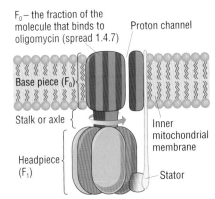

Figure 4 The structure of ATP synthase

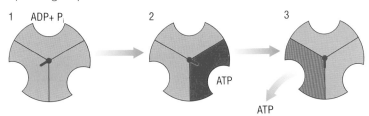

Figure 5 ATPsynthesis occurs in three steps. The axle (stalk) rotates the head, shown here. ADP and P_i join to form ATP, which is then released as that section of the headpiece undergoes a conformational (shape) change

STRETCH and CHALLENGE

It has been suggested that mitochondria are derived from prokaryotes.

Question

A What features of their structure support this suggestion?

Questions

1 Suggest how the structure of a mitochondrion from a skin cell would differ from that of a mitochondrion from heart muscle tissue.
2 Suggest why synaptic knobs of nerve cells have many mitochondria.
3 Explain the following terms: chemiosmosis; proton motive force; oxidoreductase enzyme.

By the end of this spread, you should be able to ...

* ✳ Outline the link reaction, with reference to decarboxylation of pyruvate to acetate and the reduction of NAD, and state that it takes place in the mitochondrial matrix.
* ✳ Explain that coenzyme A carries acetate from the link reaction to Krebs cycle.
* ✳ Outline the Krebs cycle, including the roles of NAD and FAD, and substrate-level phosphorylation, and state that it takes place in the mitochondrial matrix.

Key definitions

The **link reaction** converts pyruvate to acetate. NAD is reduced.

Krebs cycle oxidises acetate to carbon dioxide. NAD and FAD are reduced. ATP is made by **substrate-level phosphorylation**.

Both of these reactions occur in the mitochondrial matrix.

Pyruvate produced during glycolysis is transported across the inner and outer mitochondrial membranes to the matrix. It is changed into a 2-carbon compound, acetate, during the **link reaction**. Acetate is then oxidised during Krebs cycle.

The link reaction

Decarboxylation and dehydrogenation of pyruvate to acetate are enzyme-catalysed reactions.

* **Pyruvate dehydrogenase** removes hydrogen atoms from pyruvate.
* **Pyruvate decarboxylase** removes a carboxyl group, which eventually becomes carbon dioxide, from pyruvate.
* The coenzyme NAD (spread 1.4.2) accepts the hydrogen atoms.
* Coenzyme A (CoA) (spread 1.4.2) accepts acetate, to become **acetyl coenzyme A**. The function of CoA is to carry acetate to Krebs cycle.

The following equation summarises the link reaction:

$$2\text{pyruvate} + 2\text{NAD}^+ + 2\text{CoA} \rightarrow 2\text{CO}_2 + 2\text{reduced NAD} + 2\text{acetyl CoA}$$

NAD^+ indicates NAD in the oxidised state. Two molecules of pyruvate are considered in the equation as two molecules of pyruvate are derived from each molecule of glucose.

Note that *no* ATP is produced. However each reduced NAD will take a pair of hydrogen atoms to the inner mitochondrial membrane and they will be used to make ATP during oxidative phosphorylation (spread 1.4.6).

The Krebs cycle

The Krebs cycle also takes place in the mitochondrial matrix. It is a series of enzyme-catalysed reactions that oxidise the acetyl group of acetyl CoA to two molecules of carbon dioxide. It also produces one molecule of ATP by **substrate-level phosphorylation**, and reduces three molecules of NAD and one molecule of FAD. These reduced coenzymes have the potential to produce more ATP during oxidative phosphorylation.

1 The acetate is offloaded from coenzyme A (which is then free to collect more acetate) and joins with a 4-carbon compound, called oxaloacetate, to form a 6-carbon compound, called citrate.
2 Citrate is decarboxylated (one molecule of carbon dioxide removed) and dehydrogenated (a pair of hydrogen atoms removed) to form a 5-carbon compound. The pair of hydrogen atoms is accepted by a molecule of NAD, which becomes reduced.

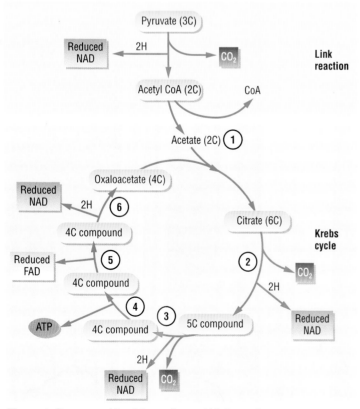

Figure 1 Summary of the link reaction and Krebs cycle

3   The 5-carbon compound is decarboxylated and dehydrogenated to form a 4-carbon compound and another molecule of reduced NAD.

4   The 4-carbon compound is changed into another 4-carbon compound. During this reaction a molecule of ADP is phosphorylated to produce a molecule of ATP. This is substrate-level phosphorylation.

5   The second 4-carbon compound is changed into another 4-carbon compound. A pair of hydrogen atoms is removed and accepted by the coenzyme FAD, which is reduced.

6   The third 4-carbon compound is further dehydrogenated and regenerates oxaloacetate. Another molecule of NAD is reduced.

How many turns of the cycle?

There is one turn of the cycle for each molecule of acetate, which was made from one molecule of pyruvate. Therefore there are *two* turns of the cycle for each molecule of glucose.

What are the products of the link reaction and Krebs cycle?

For each molecule of glucose (i.e. two turns of the cycle):

Product per molecule of glucose	Link reaction	Krebs cycle
Reduced NAD	2	6
Reduced FAD	0	2
Carbon dioxide	2	4
ATP	0	2

Table 1 The products of the link reaction and Krebs cycle

Although oxygen is not used in these stages of respiration, they won't occur in the absence of oxygen so they are aerobic.

*   Other food substrates besides glucose can be respired.
*   Fatty acids are broken down to acetates and can enter Krebs cycle via coenzyme A.
*   Amino acids can be deaminated (NH_2 group removed) and the rest of the molecule may enter Krebs cycle directly or be changed to pyruvate or acetate, depending on the type of amino acid (spread 1.4.9).

Examiner tip

You may be asked why an enzyme has a particular name. The answer is that the name describes its role. For example pyruvate decarboxylase is so called because it removes carboxyl groups from its substrate, pyruvate.

STRETCH and CHALLENGE

Questions

A Explain why mature erythrocytes (red blood cells) cannot carry out the link reaction or Krebs cycle.

B The inner mitochondrial membranes are impermeable to reduced NAD. For this reason a shunt mechanism moves hydrogen atoms from reduced NAD made during glycolysis, to the matrix side of the inner mitochondrial membrane. The hydrogens are carried in by another chemical that then becomes reoxidised, reducing NAD that is already in the mitochondrial matrix. Explain why such a shunt mechanism is not needed for NAD reduced during the link reaction and Krebs cycle.

C Aerobic prokaryotes can carry out the link reaction, Krebs cycle and oxidative phosphorylation. Suggest where in the prokaryotic cell these reactions take place.

Questions

1   Suggest why living organisms have only small amounts of oxaloacetate in their cells.

2   Explain why each stage of Krebs cycle needs to be catalysed by its own specific enzyme.

3   State the role of pyruvate dehydrogenase.

4   Describe how amino acids that are converted to pyruvate enter Krebs cycle.

⑥ Oxidative phosphorylation and chemiosmosis

By the end of this spread, you should be able to ...

✳ Outline the process of oxidative phosphorylation, with reference to the roles of electron carriers, oxygen and mitochondrial cristae.
✳ Outline the process of chemiosmosis, with reference to the electron transport chain, proton gradients and ATP synthase.
✳ State that oxygen is the final electron acceptor in aerobic respiration.
✳ Explain that the theoretical yield of ATP per glucose molecule is rarely, if ever, achieved.

Key definition

Oxidative phosphorylation is the formation of ATP by adding a phosphate group to ADP, in the presence of oxygen, which is the final electron acceptor.

The final stage of aerobic respiration

- The final stage of aerobic respiration involves electron carriers embedded in the inner mitochondrial membranes (spread 1.4.4).
- These membranes are folded into **cristae**, increasing the surface area for electron carriers and ATP synthase enzymes.
- Reduced NAD and reduced FAD are reoxidised when they donate hydrogen atoms, which are split into protons and electrons, to the electron carriers.
- The first electron carrier to accept electrons from reduced NAD is a protein complex, complex I, called NADH – coenzyme Q reductase (also known as NADH dehydrogenase).
- The protons go into solution in the matrix.

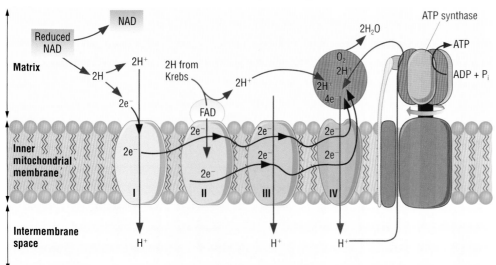

Figure 1 The electron transport chain and chemiosmosis

The electron transport chain

The electrons are passed along a chain of electron carriers and then donated to molecular oxygen, the final electron acceptor.

Chemiosmosis

- As electrons flow along the electron transport chain, energy is released and used, by coenzymes associated with some of the electron carriers (complexes I, III and IV), to *pump* the protons across to the intermembrane space.
- This builds up a proton gradient, which is also a pH gradient and an electrochemical gradient.

- Thus, potential energy builds up in the intermembrane space.
- The hydrogen ions cannot diffuse through the lipid part of the inner membrane but can diffuse through ion channels in it. These channels are associated with the enzyme ATP synthase. This flow of hydrogen ions (protons) is chemiosmosis.

Oxidative phosphorylation

Oxidative phosphorylation is the formation of ATP by the addition of inorganic phosphate to ADP in the presence of oxygen. This is how it happens:

- As protons flow through an ATP synthase enzyme, they drive the rotation of part of the enzyme and join ADP and P_i (inorganic phosphate) to form ATP.
- The electrons are passed from the last electron carrier in the chain to molecular oxygen, which is the final electron acceptor.

- Hydrogen ions also join so that oxygen is reduced to water.

$$4H^+ + 4\,e^- + O_2 \rightarrow 2H_2O$$

How much ATP is made before oxidative phosphorylation?

So far, for each glucose molecule:
- two molecules of ATP have been gained, during glycolysis, by substrate-level phosphorylation
- two molecules of ATP have been made, during Krebs cycle, by substrate-level phosphorylation.

How much ATP is made during oxidative phosphorylation?

- More ATP will be made during oxidative phosphorylation, where the reduced NAD and FAD molecules are reoxidised.

The number of molecules made from one molecule of glucose			
Name of molecule produced	Stage of respiration		
	Glycolysis	Link	Krebs cycle
Reduced NAD	2	2	6
Reduced FAD	0	0	2

Table 1 Number of molecules of reduced NAD and FAD per molecule of glucose

- The reduced NAD and reduced FAD will both provide electrons to the electron transport chain, to be used in oxidative phosphorylation.
- Reduced NAD also provides hydrogen ions that contribute to the build-up of the proton gradient for chemiosmosis. The hydrogens from reduced FAD stay in the matrix but can combine with oxygen to form water.
- The 10 molecules of reduced NAD can theoretically produce 26 molecules of ATP during oxidative phosphorylation.
- Therefore for each molecule of reduced NAD that is reoxidised, up to 2.6 molecules of ATP should be made.
- Together with the ATP made during glycolysis and Krebs cycle, the total yield of ATP molecules, per molecule of glucose respired, should be 30.

However this is rarely achieved for the following reasons:
- Some protons leak across the mitochondrial membrane, reducing the number of protons to generate the proton motive force.
- Some ATP produced is used to actively transport pyruvate into the mitochondria.
- Some ATP is used for the shuttle to bring hydrogen from reduced NAD made during glycolysis, in the cytoplasm, into the mitochondria.

STRETCH and CHALLENGE

In the cytoplasm, reduced NAD from glycolysis reduces oxaloacetate to malate. In the process the coenzyme NAD is reoxidised. Malate passes into the mitochondria, through the outer and inner membranes, to the matrix.

Malate dehydrogenase catalyses the oxidation of malate back to oxaloacetate, with the formation of reduced NAD, which goes to the inner membrane. The oxaloacetate is changed to aspartate which can pass from the mitochondria back into the cytoplasm, where it is converted to oxaloacetate.

Question

A Suggest why malate and aspartate can pass through the inner mitochondrial membrane but oxaloacetate and reduced NAD cannot.

Examiner tip

Always refer to protons being *pumped* into the intermembrane space. Don't say that they are actively transported as this implies that ATP is used and, in this case, the energy is from the electron flow, *not* from ATP.

Questions

1 Explain why oxygen is known as the final electron acceptor.
2 Explain why the proton gradient across the inner membrane is a source of potential energy.
3 Describe the pathway taken by an oxygen molecule from a red blood cell in a capillary to the matrix of a mitochondrion in a respiring cell.
4 Suggest how the formation of water from hydrogen ions, from reduced FAD, and oxygen in the matrix can *indirectly* contribute to the proton gradient across the inner mitochondrial membrane.

By the end of this spread, you should be able to ...

* Evaluate the experimental evidence for the theory of chemiosmosis.

Key definition

Chemiosmosis is the diffusion of ions through a partially permeable membrane. It relates specifically to the flow of hydrogen ions (protons) across a membrane, which is coupled to the generation of ATP during respiration. In eukaryotic cells the membrane is the inner mitochondrial membrane and in prokaryotes it is the cell surface membrane, which may be invaginated to increase surface area.

Early studies

By the early 1940s the link between oxidation of sugars and the formation of ATP, the universal energy currency of cells, was made. By the end of that decade, scientists knew that reduced NAD linked metabolic pathways, such as Krebs cycle, with the production of ATP.

However, they did not know the biochemical mechanism by which the ATP was made and thought that the energy associated with reduced NAD was first stored in a high-energy intermediate chemical before being used to make ATP. Investigations did not find such a high-energy intermediate.

By the early 1960s, research teams were extracting mitochondria from cells and examining them, using electron microscopes and special staining techniques. They could identify an outer and inner membrane with a space between them, and could see that the inner membrane was folded into cristae covered on the inner surface with many small (9 nm diameter), mushroom-shaped particles.

Peter Mitchell's theory

In 1961, Peter Mitchell realised that the build-up of hydrogen ions on one side of a membrane would be a source of potential energy and that the movement of ions across the membrane, down an electrochemical gradient, could provide the energy needed to power the formation of ATP from ADP and P_i. He called this **chemiosmosis theory**.

The inner mitochondrial membrane is therefore an energy-transducing membrane. He postulated that the energy released from the transfer of electrons along the electron transport chain was used to *pump* hydrogen ions from the matrix to the intermembrane space and that these protons then flowed through protein channels, attached to enzymes. The kinetic energy or the force of this flow, the **proton motive force**, drove the formation of ATP.

At first this theory was greeted with great scepticism as it was radically different from the idea of a high-energy intermediate compound. However, by 1978 there was much evidence

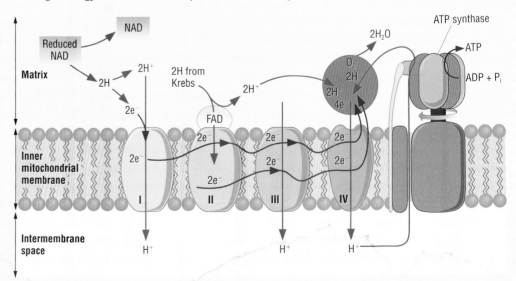

Figure 1 The electron transport chain and chemiosmosis

supporting the theory and Mitchell was awarded the Nobel Prize for chemistry. Since then scientists have established that the stalked particles are ATP synthase enzymes and have discovered how they function. It is also now known that some of the complexes in the electron transport chain have coenzymes that can use the energy released from electron transport to *pump* hydrogen ions across the membrane, into the intermembrane space, where a proton or electrochemical gradient builds up.

Evidence from other studies

Some researchers treated isolated mitochondria by placing them in solutions of very low water potential so that the outer membrane ruptured, releasing the contents of the intermembrane space. By further treating the resulting mitoblasts (mitochondria stripped of their outer membranes) with strong detergent, they could rupture the inner membrane and release the contents of the matrix.

All this allowed them to identify where various enzymes are in the mitochondria, and to work out that the link reaction and Krebs cycle take place in the matrix, whilst the electron transfer chain enzymes are embedded in the inner mitochondrial membrane.

Electron transfer in mitoblasts did not produce any ATP, so they concluded that the intermembrane space was also involved. ATP was not made if the mushroom-shaped parts of the stalked particles were removed from the inner membrane of intact mitochondria. ATP was not made in the presence of oligomycin, an antibiotic, now known to block the flow of protons through the ion channel part of the stalked particles.

In intact mitochondria:
- the potential difference across the inner membrane was −200 mV, being more negative on the matrix side of the membrane than on the intermembrane space side of the membrane
- the pH of the intermembrane space was also lower than that of the matrix.

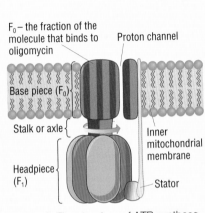

Figure 2 The structure of ATP synthase

F_0 – the fraction of the molecule that binds to oligomycin

Proton channel

Base piece (F_0)

Stalk or axle

Inner mitochondrial membrane

Headpiece (F_1)

Stator

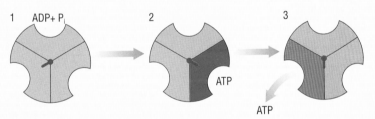

Figure 3 ATP synthesis occurs in three steps. The axle (stalk) rotates the head. ADP and P_i join to form ATP, which is then released as that section of the headpiece undergoes a conformational (shape) change

Notice that there is quite a large time lag between making a discovery and being awarded the Nobel Prize. During this time other scientists repeat the work or carry out further research, gathering more evidence to support the theory. The more the studies are replicated, with other scientists coming to the same conclusion, the more reliable the evidence is. By this time, the scientific community is able to accept a new theory and to judge just how significant the discovery is.

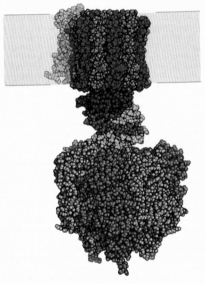

Figure 4 Molecular structure of ATP synthase

Question

1 Explain how each of the following pieces of evidence supports the chemiosmosis theory:
   (a) lower pH in intermembrane space than in mitochondrial matrix
   (b) the more negative potential on the matrix side of the inner mitochondrial membrane
   (c) no ATP made in mitoblasts
   (d) no ATP made if headpieces are removed from the stalked particles
   (e) no ATP made in the presence of oligomycin
   (f) coenzymes within complexes I, III and IV can use energy released from the transfer of electrons to pump hydrogen ions across the inner mitochondrial membrane to the intermembrane space.

By the end of this spread, you should be able to . . .

* Explain why anaerobic respiration produces a much lower yield of ATP than aerobic respiration.
* Compare and contrast anaerobic respiration in mammals and in yeast.

Key definition

Anaerobic respiration is the release of energy from substrates, such as glucose, in the absence of oxygen.

Figure 1 Zebra running from a predator

What happens if there is no oxygen?

We have seen that oxygen acts as the final electron acceptor in oxidative phosphorylation. If oxygen is absent, the electron transport chain cannot function, so Krebs cycle and the link reaction also stop. This leaves only the anaerobic process of glycolysis as a source of ATP. The reduced NAD, generated during the oxidation of glucose, has to be reoxidised so that glycolysis can keep operating. This increases the chances of the organism surviving under temporary adverse conditions.

For eukaryote cells there are two pathways to reoxidise NAD:
* Fungi, such as yeast, use ethanol (alcohol) fermentation (plant cells, such as root cells under waterlogged conditions, can also use this pathway).
* Animals use lactate fermentation.

Neither of these pathways produce any ATP but two molecules of ATP, per molecule of glucose, are made by substrate-level phosphorylation during glycolysis.

Glycolysis (see spread 1.4.3) produces two molecules of ATP, two molecules of reduced NAD and two molecules of pyruvate per molecule of glucose.

Lactate fermentation

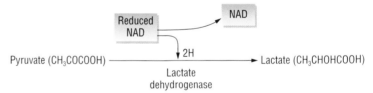

Figure 2 The fate of pyruvate under anaerobic conditions in mammals – the lactate pathway. Pyruvate accepts hydrogen atoms from reduced NAD, which is reoxidised. Pyruvate is reduced to lactate

Lactate fermentation occurs in mammalian muscle tissue during vigorous activity, such as when running to escape a predator, when the demand for ATP (for muscle contraction) is high and there is an oxygen deficit.
* Reduced NAD must be reoxidised to NAD^+.
* Pyruvate is the hydrogen acceptor.
* It accepts hydrogen atoms from reduced NAD.
* NAD is now reoxidised and is available to accept more hydrogen atoms from glucose.
* Glycolysis can continue, generating enough ATP to sustain muscle contraction.
* The enzyme lactate dehydrogenase catalyses the oxidation of reduced NAD, together with the reduction of pyruvate to lactate.

The lactate is carried in the blood away from muscles, to the liver. When more oxygen is available the lactate can be converted back to pyruvate, which may then enter Krebs cycle via the link reaction, or it may be recycled to glucose and glycogen. It is not a build-up of lactate that causes muscle fatigue (muscles can still function in the presence of lactate if their pH is kept constant by buffers), but it is specifically the reduction in pH that will reduce enzyme activity in the muscles.

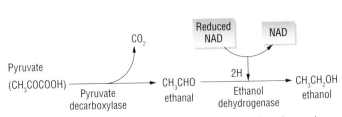

Alcoholic fermentation

Under anaerobic conditions in yeast cells:

- each pyruvate molecule loses a carbon dioxide molecule; it is decarboxylated and becomes ethanal
- this reaction is catalysed by the enzyme pyruvate decarboxylase (not present in animals), which has a coenzyme (thiamine diphosphate) bound to it
- ethanal accepts hydrogen atoms from reduced NAD, which becomes reoxidised as ethanal is reduced to ethanol (catalysed by ethanol dehydrogenase)
- the reoxidised NAD can now accept more hydrogen atoms from glucose, during glycolysis.

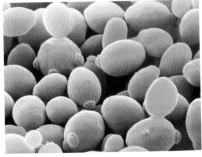

Figure 3 The fate of pyruvate under anaerobic conditions in yeast – ethanol fermentation. Pyruvate is decarboxylated to ethanal. Ethanal accepts hydrogen atoms from reduced NAD, which is reoxidised. Ethanal is reduced to ethanol

Yeast is a facultative anaerobe – it can live without oxygen, although it is killed when the concentration of ethanol builds up to around 15%. However, the rate of growth is faster under aerobic conditions (with equal concentrations of glucose). At the beginning of the brewing process, yeast is grown under aerobic conditions and then placed in anaerobic conditions to undergo alcoholic fermentation.

STRETCH and CHALLENGE

Pasteur observed that yeast consumes far more glucose when growing under anaerobic conditions than when growing under aerobic conditions. Scientists now know that the rate of ATP production by anaerobic glycolysis can be up to 100 times faster than that of oxidative phosphorylation, but a lot of glucose is consumed and the end product, ethanol, still has a lot of potential chemical energy.

Question

A When mammalian muscle tissues are rapidly using ATP, they can regenerate it almost entirely by anaerobic glycolysis and lactate fermentation. A great deal of glucose is used but this process is not as wasteful as ethanol fermentation. Suggest why this is.

Figure 4 Scanning electron micrograph of yeast cells, *Saccharomyces cerevisiae*, ×3000

Questions

1 Complete the table, comparing anaerobic respiration in yeast and mammals.

	Yeast	**Mammals**
Hydrogen acceptor		
Is carbon dioxide produced?		
Is ATP produced?		
Is NAD reoxidised?		
End products		
Enzymes involved		

2 Aerobic respiration can theoretically produce a maximum of 30 molecules of ATP per molecule of glucose. How many molecules of ATP are produced per molecule of glucose during anaerobic respiration?

3 Why can't mammalian tissues carry out alcoholic fermentation of pyruvate under anaerobic conditions?

4 Explain how a build-up of acid during glycolysis leads to muscle fatigue in mammals.

5 Suggest how diving mammals, such as seals, whales and dolphins can swim below water without suffering muscle fatigue.

Examiner tip

Remember that the main significance or purpose of the anaerobic pathways in mammals and yeast is to reoxidise NAD and allow glycolysis to continue, thereby generating some ATP.

By the end of this spread, you should be able to ...

✳ Define the term *respiratory substrate*.
✳ Explain the difference in relative energy values of carbohydrate, lipid and protein respiratory substrates.

Energy values of different respiratory substrates

We have seen that the majority of ATP made during respiration is produced during oxidative phosphorylation when hydrogen ions (protons) flow through channels associated with ATP synthase enzymes, on the inner mitochondrial membranes. The hydrogen ions and electrons then combine with oxygen to produce water.

The more protons, the more ATP is produced. It follows, then, that the more hydrogen atoms there are in a molecule of **respiratory substrate**, the more ATP can be generated when that substrate is respired. It also follows that if there are more hydrogen atoms per **mole** of respiratory substrate, then more oxygen is needed to respire that substrate.

Carbohydrate

You may remember from AS that the general formula for carbohydrate is $C_n(H_2O)_n$. Glucose is the chief respiratory substrate and some mammalian cells, e.g. brain cells and red blood cells, can use only glucose for respiration. Animals store glucose as glycogen and plants store it as starch. Both can be hydrolysed to glucose for respiration.

Other monosaccharides, such as fructose and galactose, are changed to glucose for respiration.
- The theoretical maximum energy yield for glucose is 2870 kJ mol^{-1}.
- It takes 30.6 kJ to produce 1 mol ATP.
- So, theoretically the respiration of 1 mol of glucose should produce nearly 94 mol ATP.
- The actual yield is more like 30 mol ATP, an efficiency of about 32%.
- The remaining energy is released as heat, which helps maintain a suitable body temperature, thus allowing enzyme-controlled reactions to proceed.

Protein

Excess amino acids, released after protein digestion, may be deaminated. This involves removal of the amine group and its conversion to urea – see spread 1.2.3. The rest of the molecule is changed into glycogen or fat. These can be stored and later respired to release energy.
- When an organism is undergoing fasting, starvation or prolonged exercise, protein from muscle can be hydrolysed to amino acids, which can be respired.
- Some can be converted to pyruvate, or to acetate, and be carried to Krebs cycle.
- Some enter Krebs cycle directly.
- The number of hydrogen atoms per mole accepted by NAD and then used in oxidative phosphorylation is slightly more than the number of hydrogen atoms per mole of glucose, so proteins release slightly more energy than equivalent masses of carbohydrate.

Key definitions

A **respiratory substrate** is an organic substance that can be used for respiration.

One **mole** is the gram molecular mass of a substance. 180 g glucose is one mole of glucose (mol for short).

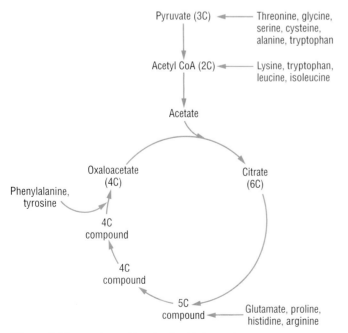

Figure 1 How amino acids enter the Krebs cycle

Lipids

Lipids are an important respiratory substrate for many tissues, particularly muscle. Triglycerides are hydrolysed by lipase to fatty acids and glycerol. Glycerol can be converted to glucose, and then respired, but fatty acids cannot.

Figure 2 Hydrolysis of triglyceride to fatty acids and glycerol

Figure 3 Palmitic acid, a fatty acid

Fatty acids are long-chain hydrocarbons with a carboxylic acid group. Hence, in each molecule there are many carbon atoms and even more hydrogen atoms. These molecules are a source of many protons for oxidative phosphorylation so they produce a lot of ATP.

- Each fatty acid is combined with CoA. This requires energy from the hydrolysis of a molecule of ATP to AMP (adenosine monophosphate) and two inorganic phosphate groups.
- The fatty acid–CoA complex is transported into the mitochondrial matrix where it is broken down into 2-carbon acetyl groups that are attached to CoA.
- During this breakdown, by the β-oxidation pathway, reduced NAD and reduced FAD are formed.
- The acetyl groups are released from CoA and enter Krebs cycle, where three molecules of reduced NAD, one molecule of reduced FAD and one molecule of ATP (by substrate-level phosphorylation) are formed for each acetate.
- The large amount of reduced NAD is reoxidised at the electron transport chain, during oxidative phosphorylation, producing large amounts of ATP by chemiosmosis.

Respiratory substrate	Mean energy value/kJ g^{-1}
Carbohydrate	15.8
Lipid	39.4
Protein	17.0

Table 1 Energy values per gram of different respiratory substrates. These are mean values as lipids and proteins vary in their compositions of fatty acids or amino acids respectively

Examiner tip

Remember that fats and proteins can only be respired aerobically. They cannot undergo glycolysis.

STRETCH and CHALLENGE

Palmitic acid produces eight 2-carbon fragments. This requires seven turns of the β-oxidation cycle.

For each turn of the β-oxidation cycle one reduced NAD and one reduced FAD are produced. The seven FAD and seven NAD are reoxidised via oxidative phosphorylation; the hydrogen atoms from reduced NAD are involved in chemiosmosis and ATP synthesis.

Each acetyl group enters the Krebs cycle and produces one reduced FAD and three reduced NAD, as well as one ATP, by substrate-level phosphorylation. Eight turns of the Krebs cycle are needed to deal with the eight fragments produced during β-oxidation.

The energy equivalent to the hydrolysis of two ATP molecules is *used* to combine the fatty acid with acetyl CoA.

Question

A Calculate the net gain of ATP for one molecule of palmitic acid, oxidised via β-oxidation and Krebs cycle.

Questions

1 Explain why a diet high in fat is also high in energy content.

2 Explain why palmitic acid, a large molecule, can pass into the matrix of the mitochondria.

3 Explain why children whose diet does not contain enough fat or carbohydrate can suffer from muscle wastage.

4 If a respiratory substrate contains more hydrogen atoms per mole, then it needs more oxygen to respire it and consequently it produces more metabolic water per mole. Camels' humps contain stored lipid.

  (a) Explain why lipid produces more metabolic water per mole than glucose.

  (b) Explain why the lipid in camels' humps is respired aerobically.

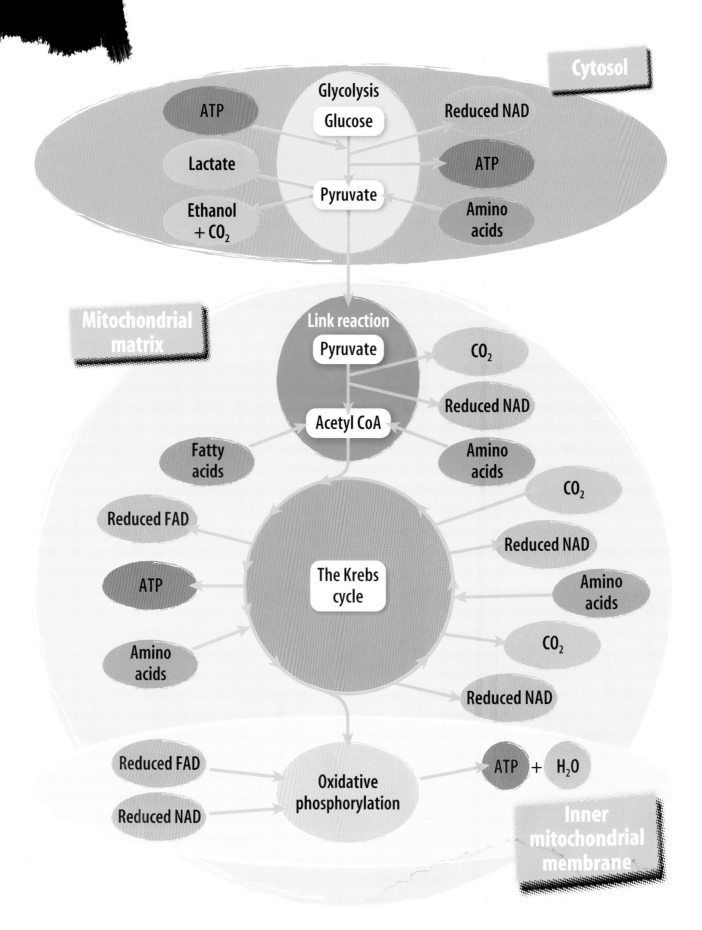

Practice questions

1. The main product from respiration is ATP. List four uses of ATP within cells. [4]

2. Where do the following stages of respiration take place in eukaryotic cells?
   (a) glycolysis
   (b) link reaction
   (c) Krebs cycle
   (d) oxidative phosphorylation. [4]

3. How many molecules of ATP are produced during the glycolysis of one molecule of glucose? [1]

4. How is pyruvate transported into the mitochondria in eukaryote cells? [1]

5. (a) Describe the structure of ATP. [3]
   (b) Explain the roles of each of the following during respiration:
   (i) ATP
   (ii) NAD
   (iii) electron carriers
   (iv) cristae
   (v) acetyl coenzyme A
   (vi) oxygen. [18]

6. Explain how the structure of a mitochondrion enables it to carry out its functions. [6]

7. Discuss how proton gradients and ATP synthase enzymes contribute to the formation of ATP. [10]

8. (a) What is meant by the term 'respiratory substrate'? [1]
   (b) Explain why more ATP is produced during the respiration of lipids than during the respiration of sugars. [3]

9. Describe three ways in which anaerobic respiration in yeast cells is different from anaerobic respiration in mammalian muscle cells. [3]

10. Under normal circumstances the brain cells can only use glucose as their respiratory substrate. However, during prolonged starvation, ketone bodies become the brain's major respiratory substrate. Liver cells convert stored fats to fatty acids and then to acetate. They then convert acetyl coenzyme A to the water-soluble ketone bodies. These are carried in the blood to the brain. There, they are converted back to acetyl coenzyme A, which can enter Krebs cycle.

    Heart and skeletal muscle use ketone bodies as respiratory substrates under normal circumstances.

    (a) Explain the significance of ketone bodies being water-soluble.
    (b) Explain why the oxygen supply to brain cells is crucial to survival.
    (c) Name the type of reaction used to change fats to fatty acids.
    (d) What molecules are produced when fats are broken down in this way?
    (e) Name the pathway used to convert fatty acids to acetate.
    (f) Explain why fats are used as a source of energy during prolonged starvation. [6]

1   Figure 1.1 shows the relationship between various metabolic processes.

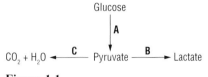

Figure 1.1

(a) **(i)**   Identify the three metabolic processes **A**, **B** and **C**.
                                                                                    [3]
(ii)   State the letter of the pathway in which acetyl coenzyme A is involved.
                                                                                    [1]
(iii) State the letter of the pathway in which ATP is utilised.
                                                                                    [1]

In an investigation, mammalian liver cells were homogenised (broken up) and the resulting homogenate centrifuged. Portions containing only nuclei, ribosomes, mitochondria and cytosol (residual cytoplasm) were each isolated. Samples of each portion, and of the complete homogenate, were incubated in four ways:

1   with glucose
2   with pyruvate
3   with glucose plus cyanide
4   with pyruvate plus cyanide.

Cyanide inhibits carriers in the electron transport chain. After incubation the presence or absence of carbon dioxide and lactate in each sample was determined. The results are summarised in Table 1.1.

(b) **(i)**   With reference to this investigation, name **two** organelles not involved in respiration.
                                                                                    [1]
(ii)   Explain why carbon dioxide is produced when mitochondria are incubated with pyruvate but **not** when incubated with glucose.
                                                                                    [3]
(iii) Explain why, in the presence of cyanide, lactate production does occur, but not carbon dioxide production.
                                                                                    [3]

This investigation may be repeated using yeast cells instead of liver cells.

(c)   State the products that would be formed by the incubation of glucose with cytosol from yeast.
                                                                                    [1]
                                                                            [Total: 13]
                                                                        (OCR 2804 Jan02)

2   Figure 2.1 is a diagram of a section through a mitochondrion.

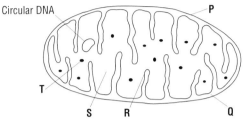

Figure 2.1

(a)   In each case, state the letter which indicates the site of:
the Krebs cycle
oxidative phosphorylation
decarboxylation.
                                                                                    [3]
(b)   Suggest **one** function of the loop of DNA shown in Figure 2.1.
                                                                                    [1]

Figure 2.2 is a diagrammatic representation of a section through the inner mitochondrial membrane showing the processes leading to ATP formation.

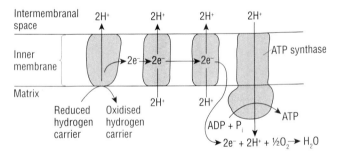

Figure 2.2

	Samples of homogenate									
	Complete		Nuclei only		Ribosomes only		Mitochondria only		Cytosol	
	Carbon dioxide	Lactate	Carbon dioxide	Lactate	Carbon dioxide	Lactate	Carbon dioxide	Lactate	Carbon dioxide	Lactate
1. glucose	✓	✓	✗	✗	✗	✗	✗	✗	✗	✓
2. pyruvate	✓	✓	✗	✗	✗	✗	✓	✗	✗	✓
3. glucose and cyanide	✗	✓	✗	✗	✗	✗	✗	✗	✗	✓
4. pyruvate and cyanide	✗	✓	✗	✗	✗	✗	✗	✗	✗	✓

✗ = absent ✓ = present

Table 1.1

Answers to examination questions will be found on the Exam Café CD.

(c) Name a hydrogen carrier that links the Krebs cycle to the electron transport chain. [1]

(d) Explain how oxidative phosphorylation results in the production of ATP. [5]

(e) The poison cyanide binds with one of the electron carriers. When this happens, the flow of electrons stops. Suggest how ingestion of cyanide by humans leads to death by muscle failure. [3]

[Total: 13]

(OCR 2804 Jun05)

3 Figure 3.1 is an outline of the glycolytic pathway.

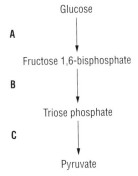

Figure 3.1

(a) With reference to Figure 3.1, state the letter, **A**, **B**, or **C**, in the glycolytic pathway where the following processes occur:
phosphorylation using ATP
dehydrogenation
formation of ATP
splitting of hexose. [4]

(b) State where glycolysis occurs in a cell. [1]

(c) State the **net gain** in ATP molecules when **one** molecule of glucose is broken down to pyruvate in glycolysis. [1]

(d) Describe what would happen to the pyruvate molecules formed under **anaerobic** conditions in mammalian muscle tissue. [3]

(e) Explain why, under **aerobic** conditions, lipids have a greater energy value per unit mass than carbohydrates or proteins. [2]

(f) Many chemicals will 'uncouple' oxidation from phosphorylation. In this situation, the energy released by oxidation of food materials is converted into heat instead of being used to form ATP. One such compound is dinitrophenol, which was used in munitions factories for the manufacture of explosives during the First World War. People working in these factories were exposed to high levels of dinitrophenol.
Suggest **and** explain why people working in munitions factories during the First World War became very thin regardless of how much they ate. [3]

[Total: 14]

(OCR 2804 Jan06)

4 **(a)** ATP is often described as the immediate source of energy for all living cells. Figure 4.1 is a diagram of the structure of an ATP molecule.

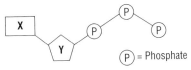

Figure 4.1

(i) Name the base labelled **X**.

(ii) Name the sugar labelled **Y**. [2]

(b) In a liver cell, ATP is formed during the respiratory pathway either directly (substrate level phosphorylation) or by oxidative phosphorylation. Indicate at which stage of respiration these two mechanisms occur by placing a tick (✓) or a cross (✗) in the appropriate box in Table 4.1. The glycolysis line has been completed for you.

Stage of respiratory pathway	Substrate level phosphorylation	Oxidative phosphorylation
Glycolysis	✓	✗
Link reaction		
Krebs cycle		
Electron transport chain		

Table 4.1 [3]

(c) A photosynthetic plant cell can also make ATP by photophosphorylation.

(i) Name the organelle in which photophosphorylation occurs. [1]

(ii) Describe the **similarities** between the mechanisms of photophosphorylation and oxidative phosphorylation. [5]

(d) ATP is used by nerve cells so that they are able to transmit nerve impulses.
Explain how ATP enables nerve cells to transmit impulses. [4]

[Total: 15]

(OCR 2804 Jan05)

Module 1
Cellular control

Introduction

In 1859 Charles Darwin published his book *On the Origin of Species by Natural Selection*. His theory of evolution drastically changed the way that humans perceived their place in nature and so was a highly significant scientific theory. Unfortunately Charles Darwin was not familiar with the work of Gregor Mendel, a monk now known as 'the father of modern genetics'. Mendel's discoveries revealed the mechanism behind evolution through natural selection.

During the next 100 years many scientists contributed to our knowledge of genetics and, when in 1953, Watson, Crick, Wilkins and Franklin discovered the structure of DNA, it was to radically change the face of biology. Since then scientists have learnt a lot more about the way in which genes work.

Genetic information flows from DNA, via RNA, to polypeptide (protein) construction. This concept is known as the 'central dogma of modern biology'. Much of the structure of living things consists of proteins. Through protein synthesis all the molecules needed for living organisms can be constructed, as enzymes catalyse reactions, some of which synthesise non-protein molecules. The endocrine and nervous systems rely on proteins and polypeptides to function. The way in which cells divide or undergo a programmed death and the way in which embryos develop are also under control of genes.

In the early 1980s, Professor Sir Alec Jeffreys developed DNA profiling (genetic fingerprinting) and in the early 1990s a collaborative project, the Human Genome Project, began in order to identify all the genes in the human genome and to sequence the base pairs in each length of DNA. Today there are many debates about stem cell research and genetic modification. In this module you will learn how genes control cells and hence whole organisms. You will also learn about Mendelian genetics (inheritance patterns), population genetics and how natural selection causes the evolution of new species. You will build on what you learn in this module for Module 2.

Test yourself

1 What is DNA profiling (fingerprinting) used for?
2 What is the *genetic code*?
3 What is meant by *mutation*?
4 Are all mutations harmful?
5 List the processes that cause genetic variation in living organisms.
6 Why aren't variations caused by the environment, such as a cat losing its tail in an accident, passed on to the offspring?
7 How did Lamarck's theory of evolution differ from Darwin's theory of evolution?
8 What is artificial selection?

Module contents

By the end of this spread, you should be able to ...

* State that genes code for polypeptides, including enzymes.
* Explain the meaning of the term *genetic code*.
* Describe, with the aid of diagrams, the way in which a nucleotide sequence codes for the amino acid sequence in a polypeptide.

Key definitions

A **gene** is a length of DNA that codes for one (or more) polypeptides.

A **polypeptide** is a polymer consisting of a chain of amino acid residues joined by peptide bonds.

The **genome** of an organism is the entire DNA sequence of that organism. The human genome consists of about 3 billion nucleotide base pairs.

A **protein** is a large polypeptide – usually 100 or more amino acids. Some proteins consist of one polypeptide chain and some consist of more than one polypeptide chain.

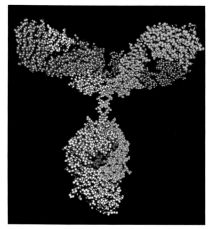

Figure 1 Immunoglobulin G antibody molecule

Examiner tip

DNA codes for the sequence of amino acids in a polypeptide, but remember there are *no* amino acids in the DNA molecule.

What is a gene?

A **gene** is a length of DNA, which means it is a sequence of nucleotide bases that codes for one (or more) **polypeptide**(s). It is a unit of heredity. In the human **genome** there are about 25 000 genes. A few are in the mitochondria. Most of them are situated on the linear chromosomes within the nucleus. Each gene occupies a specific place or **locus** on the chromosome. Remember that each chromosome consists of one molecule of DNA and each gene is just a part of a DNA molecule. The DNA in the chromosomes is associated with histone **proteins**.

Genes code for polypeptides such as:

* structural proteins including collagen and keratin
* haemoglobin
* immunoglobulins (antibodies)
* cell surface receptors
* antigens
* actin and myosin in muscle cells
* tubulin proteins in the cytoskeleton
* channel proteins
* electron carriers
* enzymes.

Since genes code for enzymes, they are involved in the control of all metabolic pathways and thus in the synthesis of all non-protein molecules found in cells.

The genetic code

The sequence of nucleotide bases on a gene (length of DNA) provides a code, with instructions for the construction of a polypeptide or protein. The genetic code has a number of characteristics.

* It is a triplet code. A sequence of three nucleotide bases codes for an amino acid. There are four bases arranged in groups of three so the number of different triplet sequences is 4^3 or 64. As there are only 20 amino acids used for protein synthesis, this is more than enough.
* It is a degenerate code. All amino acids except methionine have more than one code.
* Some codes don't correspond to an amino acid but indicate 'stop' – end of the polypeptide chain.
* It is widespread but not universal. For instance, the base sequence TCT codes for the amino acid serine in any organism. This has proved useful for genetic engineering as we can transfer a gene from one organism into another and it will usually still produce the same protein. However, there are some variations. In mammalian mitochondria there are two codes for methionine, and one of the standard stop codes codes for tryptophan. In ciliated protoctists, two of the standard stop codes code for glutamic acid.

How does the nucleotide sequence code for the amino acid sequence in a polypeptide?

Genes are on chromosomes in the cell nucleus but proteins are assembled in the cytoplasm, at ribosomes. A copy of the genetic code has to be made which can pass through a pore in the nuclear envelope to the cytoplasm. Messenger RNA (mRNA) is this copy.

Transcription

Transcription is the first stage of protein synthesis. A messenger RNA (mRNA) molecule is made. For this, one strand (the template strand) of the length of DNA is used as a template. There are free DNA nucleotides in the nucleoplasm and free RNA nucleotides in the nucleolus. The nucleotides are activated – they have two extra phosphoryl groups attached. There are four different activated RNA nucleotides: ATP, GTP, CTP and UTP.

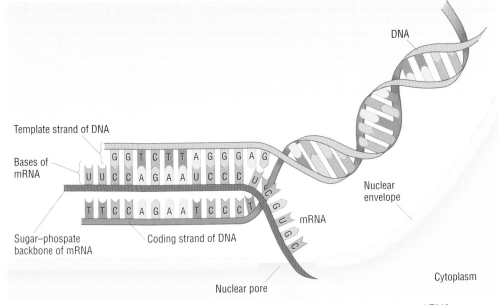

Figure 3 Transcription of a gene. The length of DNA unwinds and unzips. Free activated RNA nucleotides pair up and bind temporarily, with hydrogen bonds, to their complementary bases on the template strand of the unwound DNA. Their sugar–phosphate groups are bonded together to form the sugar–phosphate backbone. A single-stranded piece of mRNA, a copy of the coding strand of the DNA, leaves the nucleus through a pore in the nuclear envelope

Figure 2 Transcription of a gene: the backbone is DNA from the nucleus of an amphibian egg, with many strands of RNA extending from it (×8000)

Key definition

Transcription is the creation of a single-stranded mRNA copy of the DNA coding strand.

- A gene to be transcribed unwinds and unzips. To do this the length of DNA that makes up the gene dips into the nucleolus. Hydrogen bonds between complementary bases break.
- Activated RNA nucleotides bind, with hydrogen bonds, to their exposed complementary bases. U binds with A, G with C, and A with T on the *template* strand. This is catalysed by the enzyme RNA polymerase.
- The two extra phosphoryl groups (phosphates) are released. This releases energy for bonding adjacent nucleotides.
- The mRNA produced is complementary to the nucleotide base sequence on the template strand of the DNA and is therefore a *copy* of the base sequence on the *coding* strand of the length of DNA.
- The mRNA is released from the DNA and passes out of the nucleus, through a pore in the nuclear envelope, to a ribosome.

Questions

1 Explain why the genetic code would not work if nucleotide bases were arranged in pairs, rather than in triplets, to make the coding units.

2 Complete the table below.

DNA coding strand				GGA	
DNA template strand		TAC			
mRNA	UUA	AUG	CGU	GGA	UAA
Amino acid	Leucine	Methionine	Arginine	Glycine	Stop

3 What is meant by *degenerate code*?

The central dogma of molecular biology

By the late 1930s scientists knew that RNA was involved in protein synthesis. They also knew that protein synthesis occurred at ribosomes (organelles made of protein and RNA). By the early 1950s the structure of DNA, and how it could carry the genetic code, had been found. The puzzle was how DNA could direct protein synthesis, as DNA in eukaryotes never leaves the nucleus but proteins are assembled at ribosomes in the cytoplasm. In 1958 Francis Crick summarised the relationship between DNA, RNA and protein into what he described as the 'central dogma of molecular biology':

DNA directs its own replication and its **transcription** to RNA. The RNA (mRNA) in turn directs translation to proteins.

By the end of this spread, you should be able to . . .

✳ Describe, with the aid of diagrams, how the sequence of nucleotides within a gene is used to construct a polypeptide, including the roles of messenger RNA, transfer RNA and ribosomes.

✳ State that cyclic AMP activates proteins by altering their three-dimensional structure.

Key definition

Translation is the assembly of polypeptides (proteins) at ribosomes.

Translation is the second stage of protein synthesis, when the amino acids are assembled into a polypeptide. They are assembled into the sequence dictated by the sequence of **codons** (triplets of nucleotide bases) on the mRNA. The genetic code, copied from DNA into mRNA, is now translated into a sequence of amino acids. This chain of amino acids is a polypeptide. It happens at ribosomes, which may be free in the cytoplasm but many are bound to the rough endoplasmic reticulum.

Ribosomes

Ribosomes are assembled in the nucleolus of eukaryote cells, from ribosomal RNA (rRNA) and protein. Each is made up of two subunits and there is a groove into which the length of mRNA, with the code for the sequence of amino acids, can fit. The ribosome can then move along the mRNA, which can slide through the ribosomal groove, reading the code and assembling the amino acids in the correct order to make a functioning protein.

The sequence of amino acids in a protein is critical because:

* it forms the primary structure of a protein
* the primary structure determines the tertiary structure – how the protein folds up into its three-dimensional shape and is held in that 3D shape by hydrogen or ionic bonds and hydrophobic interactions forming between the R groups of amino acids
* the tertiary structure (shape) is what allows a protein to function. The tertiary structure of a protein is dependent upon its primary structure, and this primary sequence of amino acids in a polypeptide is determined by the genetic code, eventually leading to the protein having the correct shape so that it can function
* if the tertiary structure is altered, the protein can no longer function so effectively, if at all; for example, the active site of an enzyme may have an altered shape and the substrate molecules will no longer fit, or if a chloride ion channel protein in cell surface membranes has a different shape it won't allow the ions to pass through it.

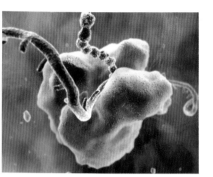

Figure 1 Artist's impression of a ribosome with mRNA (purple) synthesising a protein chain (yellow)

Transfer RNA

Another form of RNA, transfer RNA (tRNA), is made in the nucleus and passes into the cytoplasm. These are lengths of RNA that fold into hairpin shapes and have three exposed bases at one end where a particular amino acid can bind. At the other end of the molecule are three unpaired nucleotide bases, known as an **anticodon**. Each anticodon can bind temporarily with its complementary codon.

How the polypeptide is assembled

1 A molecule of mRNA binds to a ribosome. Two codons (six bases) are attached to the small subunit of the ribosome and exposed to the large subunit. The first exposed mRNA codon is always AUG. Using ATP energy and an enzyme, a tRNA with methionine and the anticodon UAC forms hydrogen bonds with this codon.

2 A second tRNA, bearing a different amino acid, binds to the second exposed codon with its complementary anticodon.

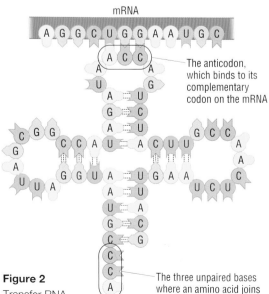

Figure 2
Transfer RNA

The anticodon, which binds to its complementary codon on the mRNA

The three unpaired bases where an amino acid joins

3 A peptide bond forms between the two adjacent amino acids. An enzyme, present in the small ribosomal subunit, catalyses the reaction.

4 The ribosome now moves along the mRNA, reading the next codon. A third tRNA brings another amino acid, and a peptide bond forms between it and the dipeptide. The first tRNA leaves and is able to collect and bring another of its amino acids.

5 The polypeptide chain grows until a stop codon is reached. There are no corresponding tRNAs for these three codons, UAA, UAC or UGA, so the polypeptide chain is now complete.

Some proteins have to be activated by a chemical, cyclic AMP (cyclic adenosine monophosphate or cAMP) that, like ATP, is a nucleotide derivative. It activates proteins by changing their 3D shape so that their shape is a better fit to their complementary molecules.

Protein synthesis in prokaryotes

In prokaryotes the DNA is not inside a nucleus; translation begins as soon as some mRNA has been made.

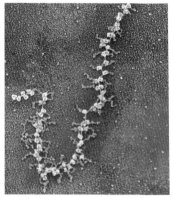

Figure 4 Scanning electron micrograph of translation: ribosomes (blue) move along the mRNA strand (pink) assembling proteins (green)

Figure 5 Protein synthesis in the bacterium *Escherichia coli*: DNA (pink) transcription yields mRNA strands (green), which are immediately translated by ribosomes (blue)

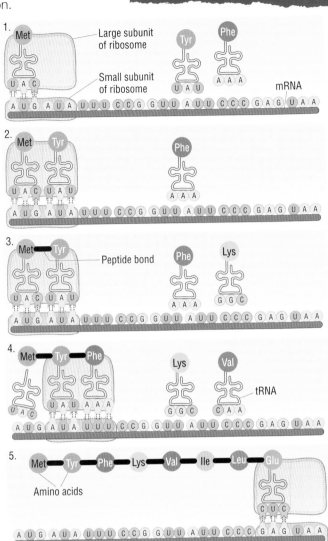

Figure 3 Translation of a length of mRNA at a ribosome and assembly of a polypeptide

Examiner tip

Remember that transcription is copying and then you will remember which stage is which in protein synthesis. Transcription is copying the DNA code onto a piece of mRNA. At ribosomes this code is translated into a protein.

STRETCH and CHALLENGE

Glycogen in muscle cells can be broken down by an enzyme, glycogen phosphorylase. Glycogen can be synthesised by the enzyme, glycogen synthase. If both were to be happening at the same time, it would waste the cell's energy, so there has to be a control mechanism to 'make or break' glycogen according to the cell's needs.

Glycogen phosphorylase is activated by cAMP but inhibited by ATP and by glucose 6-P. cAMP binds to an allosteric site (not the active site) of the enzyme and causes it to change its shape and bring its previously hidden active site to a more exposed position. If ATP or G-6-P bind, then the shape changes back and the active site becomes buried into the molecule.

Questions

A What do you think the effect of cAMP is on the activity of the enzyme glycogen synthase and how do you think the effect is brought about?

B What do you think the effect of G-6-P is on the activity of glycogen synthase?

Questions

1 In which type of cell, prokaryote or eukaryote, would you expect protein synthesis to be faster? Give reasons for your answer.

2 What is the minimum number of different tRNA molecules that are needed for protein synthesis?

(3) Mutations – 1

By the end of this spread, you should be able to ...

* State that mutations cause changes to the sequence of nucleotides in DNA molecules.
* Explain how the mutations can have beneficial, neutral or harmful effects on the way a protein functions.

Key definitions

A **mutation** is a change in the amount of, or arrangement of, the genetic material in a cell.

Chromosome mutations involve changes to parts of or whole chromosomes.

DNA mutations are changes to genes due to changes in nucleotide base sequences.

Mutation is a random change to the genetic material. This may be a change to DNA, e.g. by base deletion, addition or substitution; or by inversion or repeat of a triplet. **Chromosome mutation** involves a change to the structure of a chromosome, such as deletion, inversion or translocation (see spread 2.1.9). Mutations may occur during DNA replication. Certain substances (mutagens) may cause mutations. These include tar found in tobacco, UV light, X-rays and gamma rays.

In these two spreads we are only concerned with **DNA mutations**. Although the structure of DNA molecules makes them very stable and reduces the chances of corrupting the encoded genetic information that they hold, mistakes do occur. These may happen when DNA is replicating before nuclear division, by either mitosis or meiosis. Mutations associated with mitosis are somatic mutations and are not passed on to offspring. But they may contribute to the ageing process or may lead to cancer. Mutations associated with meiosis and gamete formation can be inherited (passed to offspring).

There are two main classes of DNA mutations:

* **Point mutations** in which one base pair replaces another. These are also called **substitutions**.
* **Insertion/deletion mutations** in which one or more nucleotide pairs are inserted or deleted from a length of DNA. These cause a **frameshift**.

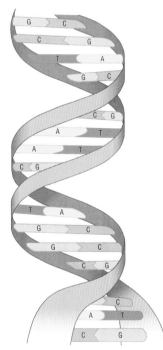

Figure 1 DNA molecule

Figure 2 DNA sequencing

First position	Second position				Third position
	T	C	A	G	
T	Phe	Ser	Tyr	Cys	T
	Phe	Ser	Tyr	Cys	C
	Leu	Ser	STOP	STOP	A
	Leu	Ser	STOP	Trp	G
C	Leu	Pro	His	Arg	T
	Leu	Pro	His	Arg	C
	Leu	Pro	Gln	Arg	A
	Leu	Pro	Gln	Arg	G
A	Ile	Thr	Asn	Ser	T
	Ile	Thr	Asn	Ser	C
	Ile	Thr	Lys	Arg	A
	Met	Thr	Lys	Arg	G
G	Val	Ala	Asp	Gly	T
	Val	Ala	Asp	Gly	C
	Val	Ala	Glu	Gly	A
	Val	Ala	Glu	Gly	G

Key:

- Asp Aspartic acid
- Lys Lysine
- Cys Cysteine
- Glu Glutamic acid
- Gly Glycine
- Phe Phenylalanine
- His Histidine
- Asn Asparagine
- Leu Leucine
- Ile Isoleucine
- Gln Glutamine
- Met Methionine
- Arg Arginine
- Trp Tryptophan
- Pro Proline
- Thr Threonine
- Tyr Tyrosine
- Val Valine
- Ser Serine
- Ala Alanine

Figure 3 The standard DNA triplet codes

The genetic code consists of three-letter words

You have seen that the genetic code consists of triplets of nucleotide bases within the DNA molecule, transcribed (copied) to three-letter codons on the mRNA molecule, which is a copy of the DNA coding strand.

Consider this sentence:

CAN THE BIG RED HEN LAY ONE EGG

It makes sense. What happens if we introduce a point mutation?

CAN THE BIG RED MEN LAY ONE EGG

Now the meaning is altered. However, only one word (triplet) is altered.

What happens if there is an insertion?

CAN ATH EBI GRE DHE NLA YON EEG G

Or a deletion?

CAN HEB IGR EDH ENL AYO NEE GG

You can see that the sense has been altered in both cases. Every triplet after the first one has changed, producing a frameshift. This is likely to have a greater effect on the resulting protein.

Below are some examples of possible mutations. A sequence of nucleotide bases is shown on the coding strand of part of a length of DNA.

Normal	ATG	CAG	CAG	CAG	TTT	TTA	CGC	AAT	CCC	DNA
	Met	Gln	Gln	Gln	Phe	Leu	Arg	Asn	Pro	Polypeptide
Point mutation	ATG	CAG	CAG	CAG	TTT	TCA	CGC	AAT	CCC	DNA
Missense	Met	Gln	Gln	Gln	Phe	Ser	Arg	Asn	Pro	Polypeptide
Point mutation	ATG	CAG	CAG	CAG	TTT	TAA	CGC	AAT	CCC	DNA
Nonsense	Met	Gln	Gln	Gln	Phe	Stop				Polypeptide
Point mutation	ATG	CAG	CAG	CAG	TTT	TTG	CGC	AAT	CCC	DNA
Silent mutation	Met	Gln	Gln	Gln	Phe	Leu	Arg	Asn	Pro	Polypeptide
Point mutation	ATG	CAG	CAG	CAG	TTT	TAC	GCA	ATC	CC	DNA
Frameshift	Met	Gln	Gln	Gln	Phe	Tyr	Val	Thr		Polypeptide

Many genetic diseases are the result of DNA mutations. Sickle-cell anaemia and cystic fibrosis both result from DNA mutations.

- In 70% of cases of cystic fibrosis, the mutation is the deletion of a triplet of base pairs, deleting an amino acid from the sequence of 1480 amino acids in the normal polypeptide.
- Sickle-cell anaemia results from a point mutation on codon 6 of the gene for the β-polypeptide chains of haemoglobin. This causes the amino acid valine to be inserted, at this position of the polypeptide chain, in place of glutamic acid.
- Growth-promoting genes are called protooncogenes. Some (such as the *RAS* gene in human bladder cancer) can be changed into oncogenes by a point mutation that alters the ability of the protooncogene to be switched off. They remain permanently switched on. Oncogenes promote unregulated cell division. Such cell division leads to a tumour.
- Huntington disease results from an expanded triple nucleotide repeat – a stutter. The normal gene for Huntington protein has repeating CAG sequences. If these expand to above a threshold number, the protein is altered sufficiently to cause Huntington disease, the symptoms of which manifest in later life and include dementia and loss of motor control.

Note

You do not need to learn the DNA base triplet sequences for amino acids. You will always be given this information in an exam question.

Examiner tip

Remember that the *coding* strand of the DNA has the sequence of bases that codes for the sequence of amino acids. mRNA is made *complementary* to the other strand, the *template* strand, and is a *copy* of the *coding* strand.

Questions

1 Describe the missense mutation shown in the table above.
2 Describe the nonsense mutation shown above.
3 Explain why the third mutation is called a silent mutation.
4 Is the frameshift shown above the result of an insertion or a deletion? What is its effect on the polypeptide?
5 Explain how the degenerate nature of the DNA code reduces the effects of point mutations.

By the end of this spread, you should be able to ...

* Explain how mutations can have beneficial, neutral or harmful effects on the way a protein functions.

Key definition

An **allele** is an alternative version of a gene. It is still at the same locus on the chromosome and codes for the same polypeptide but the alteration to the DNA base sequence may alter the protein's structure.

Mutations with neutral effects

If a gene is altered by a change to its base sequence, it becomes another version of the same gene. It is an **allele** of the gene. It may produce no change to the organism if:

* the mutation is in a non-coding region of the DNA
* it is a silent mutation. Although the base triplet has changed, it still codes for the same amino acid, so the protein is unchanged.

If the mutation does cause a change to the structure of the protein, and therefore a different characteristic, but the changed characteristic gives no particular advantage or disadvantage to the organism, then the effect is also neutral. For example, some people can smell honeysuckle flowers and some cannot. Whether or not we can smell honeysuckle does not seem to confer any selective advantage or disadvantage. However, it would be advantageous to be able to smell or taste something that could make us ill or kill us, so that we could avoid it.

Figure 1 Honeysuckle flowers

Try this

Clasp your hands together. Now look at your thumbs. Is left over right or right over left?

Now try to clasp your hands with your thumbs positioned in the opposite way (right over left if you did left over right). How does it feel?

About half the population do it one way and half do it the other and it feels very uncomfortable to do it the other way. You could survey a group in your school to see if it is about half and half.

Some people cannot taste a substance called PTC (phenylthiocarbamide). For those who can taste it, it tastes very bitter. In large quantities this substance is poisonous, so there is an advantage to being able to taste it. Some people can roll their tongue and some cannot. This is genetic. The ability to roll the tongue is dominant. However, scientists are not sure whether there is any advantage to being a tongue-roller. Some people have free ear lobes and some have attached lobes. There does not seem to be any advantage or disadvantage to these characteristics.

Mutations with harmful or beneficial effects

* Early humans in Africa almost certainly had dark skin.
* The pigment melanin protected them from the harmful effects of ultraviolet light. However, they could still synthesise vitamin D from the action of the intense sunlight on their skin. This is an important source of vitamin D, because much of the food that humans eat contains very little vitamin D. (Today it is added to some foods, but this is a very recent phenomenon.)

- Any humans with mutations to some of the genes determining skin colour, producing paler skin, would have burned and suffered from skin cancer.
- As humans migrated to more temperate climes, the sunlight was not intense enough to cause enough vitamin D to be made by those with dark skins.
- Humans with mutations producing paler skin (lack of pigment) would have an advantage over those with dark skin as they could synthesise more vitamin D. Lack of vitamin D leads to rickets and, in females, a narrow pelvis producing difficulties in childbirth, possibly leading to death of mother and/or child. Recent research suggests that vitamin D also helps to protect us from cancer and heart disease.
- The Inuit people have not lost all their skin pigments, although they do not live in an environment that has intense sunlight. However, they eat a lot of fish and seal meat, including the blubber, both rich sources of dietary vitamin D.

You can see that, depending on the environment, the same mutation for paler skin can be beneficial or harmful. The environment is never static. It often changes, and when it does, individuals within a population who have a certain characteristic may be better adapted to the new environment. The well-adapted organisms can out-compete those in the population that do not have the advantageous characteristic. This is natural selection, the mechanism for evolution. Without genetic mutations there would be no evolution.

Figure 2 Inuit hunter and seal

STRETCH and CHALLENGE

Some genes don't actually code for polypeptides. They code for RNA but it does not get translated. The RNA may switch other genes on or off.

Some genes can be switched on or off by the addition of methyl groups to the DNA, or through the action of histone proteins. Stem cells (such as those in an early embryo) are undifferentiated. They become specialised as they undergo epigenetic programming in which certain genes are turned off by methylation or association with histone proteins. However, it is possible to turn these genes back on again by epigenetic reprogramming. If the epigenetic modifications that turned an embryonic stem cell into a liver cell were removed, this cell could be persuaded to develop into different types of cells.

Question
A What useful applications could result from epigenetic reprogramming?

Examiner tip

Be aware that not all mutations are harmful. Some have no effect (they are neutral) and some are beneficial. We each have some mutations – we are all mutants!

Questions

1 Here is a sequence of DNA bases on a piece of DNA coding strand:

    ATG TTT CCT GTT AAA TAC CAT CGC

Below are four possible mutations of this piece of DNA:
- (a) Mutation 1 ATG TTT CCT GTT AAA TAA CAT CGC –
- (b) Mutation 2 ATG TTT CCT ATT AAA TAC CAT CGC –
- (c) Mutation 3 ATT TTT CCT GTT AAA TAC CAT CGC –
- (d) Mutation 4 ATG TTC CTG TTA AAT ACC ATC GC –

Explain each mutation and the effect it will have on the polypeptide.

2 For the sequence of bases shown in the normal piece of DNA above, write down:
- (a) the sequence of bases on the corresponding DNA template strand
- (b) the sequence of RNA bases on the mRNA
- (c) the tRNA anticodons for each mRNA codon
- (d) the sequence of amino acids in the polypeptide (you will need to refer to Figure 3 on spread 2.1.3).

By the end of this spread, you should be able to ...

✱ **Explain genetic control of protein production in a prokaryote using the *lac* operon.**

You already know that proteins are synthesised on ribosomes and are specified by mRNA. This idea arose from studies of the phenomenon known as enzyme induction, whereby bacteria vary the synthesis rates of specific enzymes in response to environmental changes, such as the type of food available.

Enzyme induction

The bacterium *Escherichia coli* (*E. coli*) can synthesise about 3000 different polypeptides. However, there is great variation in the numbers of different polypeptides within the cell. There may be 10 000 molecules of ribosomal polypeptides in each cell and just 10 molecules of some of the regulatory proteins.

Enzymes involved in basic cellular functions are synthesised at a fairly constant rate. Inducible enzymes are synthesised at varying rates, according to the cell's circumstances. Bacteria adapt to their environments by producing enzymes to metabolise certain nutrients only when those nutrients are present. *E. coli* normally respires glucose but it can also use lactose as a respiratory substrate.

E. coli grown in a culture medium with no lactose (disaccharide sugar) can be placed in a medium with lactose. (Lactose occurs in milk.) At first they cannot metabolise the lactose because they only have tiny amounts of the two enzymes needed to metabolise it. These enzymes are β-**galactosidase** (pronounced beta galactosidase), which catalyses the hydrolysis of lactose to glucose and galactose, and **lactose permease**, which transports lactose into the cell. A few minutes after lactose is added to the culture medium, *E. coli* bacteria increase the rate of synthesis of these two enzymes by about 1000 times. Lactose must trigger the production of the two enzymes, and is known as the **inducer**.

François Jacob and Jacques Lucien Monod

These French biologists, working at the Pasteur Institute, carried out investigations on the bacteria *E. coli*. They gave them glucose and lactose in the nutrient medium. They found that the bacteria used glucose first and then used lactose. The bacteria had two growth phases. In 1961 Jacob and Monod published a paper describing their theory of the *lac* operon. They received the Nobel Prize for this in 1965.

Lac system genes form an operon

Figure 1 The *E. coli lac* operon and its regulator gene. The operon consists of the structural genes encoding the enzymes needed for lactose metabolism (hence the name *lac* operon), and the control regions, O and P. O and P control the expression of the structural genes, Z and Y. Z codes for the enzyme β-galactosidase. Y codes for the enzyme lactose permease

The *lac* **operon** is a section of DNA within the bacterium's DNA. It consists of a number of parts:

- **The structural genes**: Z codes for the enzyme β-galactosidase, and Y codes for the enzyme lactose permease. Each consists of a sequence of base pairs that can be transcribed into a length of mRNA.
- **The operator region**, O, is a length of DNA next to the structural genes. It can switch them on and off.
- **The promoter region**, P, is a length of DNA to which the enzyme RNA polymerase binds to begin the transcription of the structural genes, Z and Y.

The regulator gene, I, is not part of the operon and is some distance from the *lac* operon.

Key definitions

An **operon** is a length of DNA, made up of structural genes and control sites. The structural genes code for proteins, such as enzymes. The control sites are the operator region and a promoter region.

The **operator** and **promoter** are both genes as they are lengths of DNA. However, they do not code for polypeptides.

How the *lac* operon works

When lactose is absent from the growth medium

1 The regulator gene is expressed (transcribed and translated) and the **repressor protein** is synthesised. It has two binding sites, one that binds to lactose and one that binds to the operator region.

2 The repressor protein binds to the operator region. In doing so it covers part of the promoter region, where **RNA polymerase normally attaches**.

3 RNA polymerase cannot bind to the promoter region so the structural genes cannot be transcribed into mRNA.

4 Without mRNA these genes cannot be translated and the enzymes β-galactosidase and lactose permease cannot be synthesised.

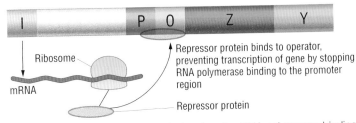

Figure 2 How the *lac* operon works by stopping RNA polymerase binding to the promoter region when lactose is absent from the growth medium

When lactose is added to the growth medium

1 Lactose (inducer) molecules bind to the other site on the repressor protein. This causes the molecules of repressor protein to change shape so that its other binding site cannot now bind to the operator region. The repressor dissociates (breaks away) from the operator region.

2 This leaves the promoter region unblocked. RNA polymerase can now bind to it and initiate the transcription of mRNA for genes Z and Y.

3 The operator–repressor–inducer system acts as a molecular switch. It allows transcription and subsequent translation of the structural genes, Z and Y, into the *lac* enzymes, β-galactosidase and lactose permease.

4 As a result, *E. coli* bacteria can use the lactose permease enzyme to take up lactose from the medium, into their cells. They can then convert the lactose to glucose and galactose using the β-galactosidase enzyme. These sugars can then be used for respiration, thus gaining energy from lactose.

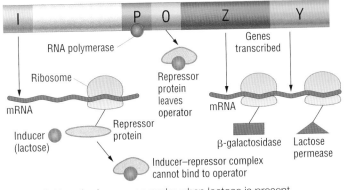

Figure 3 How the *lac* operon works when lactose is present

Escherichia coli

This bacterium, *E. coli*, sometimes gets a bad press. One strain, *E. coli* 157, may cause severe food poisoning. However, we all have harmless strains of *E. coli* living in our gut. *E. coli* has been used by many biologists for research and is used in schools and colleges for practical investigations. It was the first organism to have its genome mapped out.

Key definition

A **repressor protein** can bind to the operator region, and RNA polymerase binds to the promoter region to transcribe the structural genes.

Examiner tip

The regulator gene is not actually part of the operon, but its product, the repressor protein, plays an important part in the functioning of the operon. It is distinct from, although closely linked to, genes specifying the *lac* enzymes.

Questions

1 Match the components of the *lac* operon system and regulator gene with the correct functions.

Component	Function
A structural genes	I produces repressor protein
B regulator gene	II binds to repressor
C promoter	III codes for *lac* enzymes
D operator	IV binds to RNA polymerase

2 What is the function of the repressor protein?

3 What is the role of lactose?

4 Explain the advantage to *E. coli* bacteria of having a *lac* operon system to induce enzyme formation.

By the end of this spread, you should be able to . . .

∗ **Explain that the genes that control development of body plans are similar in plants, animals and fungi, with reference to homeobox sequences.**

Modern molecular genetics has allowed scientists to study the molecular basis of cell differentiation. Much of our knowledge of the processes involved comes from studies of the fruit fly, *Drosophila melanogaster*, an organism used by geneticists for just over a hundred years.

Drosophila development

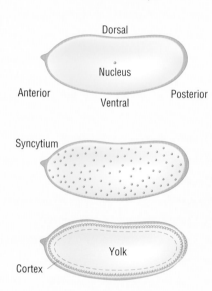

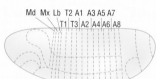

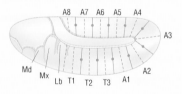

When the eggs are laid a series of mitotic divisions are triggered, at the rate of one every 6–10 minutes. (This rate of DNA replication is among the fastest known for eukaryotic organisms.)

- At first, no new cell membranes form and a multinucleate syncytium is formed.
- After the 8th division the 256 nuclei migrate to the outer part and by the 11th division the nuclei form an outer layer around a central, yolk-filled core.
- The division rate slows (the 14th division takes 60 minutes) and the nuclear genes switch from replicating to transcribing.
- The plasma membrane invaginates (folds inwards) around the 6000 nuclei and the resulting cells form a single outer layer.
- After another 2–3 hours the embryo divides into a series of segments. These correspond to the organism's organisation or body plan (which is the same as that of higher animals).
- Three segments (Md, Mx and Lb) merge to produce the head. There are three thoracic segments (T1–3) and eight abdominal segments (A1–8).
- At metamorphosis, when the larval form becomes the adult, legs, wings and antennae develop.

Genetic control of *Drosophila* development

The development is genetically mediated by **homeobox genes**.

- Some genes (maternal-effect genes) determine the embryo's polarity. Polarity refers to which end is head (anterior) and which end is tail (posterior).
- Other genes, called segmentation genes, specify the polarity of each segment.
- Homeotic selector genes specify the identity of each segment and direct the development of individual body segments. These are the master genes in the control networks of regulatory genes. There are two gene families:
  - the complex that regulates development of thorax and abdomen segments
  - the complex that regulates development of head and thorax segments.

Mutations of these genes can change one body part to another. This can be seen in the condition known as antennapedia – where the antennae of *Drosophila* look more like legs.

Key definition

Homeobox genes control the development of the body plan of an organism, including the polarity (head and tail ends) and positioning of the organs.

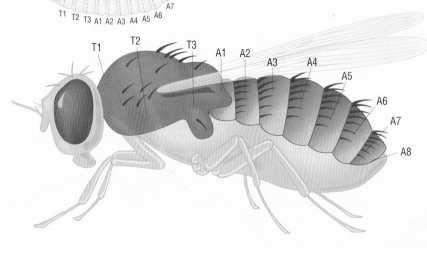

Figure 1 Development in *Drosophila*. The embryos and newly hatched larvae are the same size; the adult is much larger

Figure 2 Scanning electron micrograph of *Drosophilia melanogaster* with antennapedia mutation (×115)

Genetic control of development in other organisms

There are homeobox genes in the genomes of segmented animals from segmented worms (annelids) to vertebrates, including humans. The homeobox genes each contain a sequence of 180 base pairs (known as the homeobox) and this sequence produces polypeptides of about 60 amino acids. Some of these polypeptides are transcription factors and they bind to genes upstream (further along the DNA) and initiate transcription, so regulating the expression of other genes. Homeobox genes work in similar ways in most organisms including vertebrates, *Drosophila,* plants and fungi. When cuttings are taken from plants, the new roots and shoots develop in accordance with their polarity – roots at the end of the stem nearer the original roots.

The homeobox genes are arranged in clusters known as **Hox clusters**.
- Nematodes (roundworms) have one Hox cluster.
- *Drosophila* has two Hox clusters.
- Vertebrates have four clusters, of 9–11 genes, located on separate chromosomes.

The increase in the number of Hox clusters probably arose by duplication of a single complex present in segmented worms (annelids) and has allowed the more complex arthropods to evolve from the simpler annelids.

The homeobox genes are expressed in specific patterns in certain stages during the development of the embryo, in both vertebrates and invertebrates. They specify the identities and fates of embryonic cells and the development of the body plan. They are activated in the same order as they are expressed along the body of the organism, from anterior (head) to posterior (tail).

Retinoic acid and birth defects

Retinoic acid is a derivative of vitamin A. It activates homeobox genes in vertebrates in the same order that they are expressed in developing systems, such as the axial skeleton and the central nervous system. Both of these systems run head to tail. It is a **morphogen** (a substance that governs the pattern of tissue development). However, the amount of retinoic acid is crucial. Too much vitamin A (retinol) taken in by a pregnant woman, particularly during the first month of gestation, can interfere with the normal expression of these genes. This will cause birth defects, including cranial deformities.

Questions

1 Vitamin A is stored in the liver of all mammals. Suggest why pregnant women are advised not to eat liver.
2 Explain what is meant by 'a morphogen'.
3 What are transcription factors?
4 Explain the roles of
   (a) maternal-effect genes and
   (b) segmentation genes.

By the end of this spread, you should be able to ...

✴ Outline how apoptosis (programmed cell death) can act as a mechanism to change body plans.

Apoptosis (pronounced 'apo tosis') is programmed cell death that occurs in multicellular organisms. Cells should undergo about 50 mitotic divisions (the Hayflick constant) and then undergo a series of biochemical events that leads to an orderly and tidy cell death. This is in contrast to cell necrosis, an untidy and damaging cell death that occurs after trauma and releases hydrolytic enzymes.

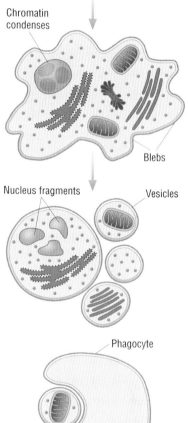

Nucleus

Chromatin condenses

Blebs

Nucleus fragments

Vesicles

Phagocyte

Figure 1 Series of events in apoptosis

Programmed cell death

Carl Vogt first described this phenomenon in 1842, and in 1885 Walter Flemming also described programmed cell death. In 1965, John Foxton Ross Kerr resurrected and researched the topic. He was studying tissues, using electron microscopy, and distinguished between necrosis and apoptosis. He was invited to Aberdeen and joined the research team of Professor Currie and his PhD student Andrew Wyllie. In 1972, the three of them published an article in the *British Journal of Cancer*. They chose the term *apoptosis* as it had been used by Hippocrates, to mean 'falling off bones'. It had also been used by Galen in the context of the dropping off of scabs. It is from two Greek words and means 'dropping off' as in flower petals dropping.

Hippocrates was an ancient Greek doctor and teacher of medicine. He is regarded as the father of medicine. He was born in 460 BC and died in 370 BC. Galen (AD 129–200) was a Roman doctor whose theories dominated Western medicine for over one thousand years.

Leonard Hayflick

Leonard Hayflick was born in 1928. He is professor of anatomy at the University of California and has studied ageing. Since the early 1900s it was thought that normal body cells were immortal. In 1962 Hayflick showed that normal body cells divide a limited number of times. He also showed that cancer cells are immortal. It took some time before the scientific establishment accepted this – at first a journal turned down his article – but he persevered in contradicting the current theory. His work is now accepted and has provided the basis for much of modern research into cancer. He has made many other important contributions to scientific knowledge. At the Wistar Institute, in Philadelphia, he developed a normal diploid strain of human cells, called WI38. These cells are used worldwide to produce many human viral vaccines, including polio, rubella, mumps, rabies and hepatitis A.

The sequence of events
- Enzymes break down the cell cytoskeleton.
- The cytoplasm becomes dense, with organelles tightly packed.
- The cell surface membrane changes and small bits called blebs form.
- Chromatin condenses and the nuclear envelope breaks. DNA breaks into fragments.
- The cell breaks into vesicles that are taken up by **phagocytosis**. The cellular debris is disposed of and does not damage any other cells or tissues.
- The whole process occurs very quickly.

Key definition

Phagocytosis is the endocytosis of large solid molecules into a cell.

How is it controlled?

The process is controlled by a diverse range of cell signals, some of which come from inside the cells and some from outside. The signals include cytokines made by cells of the immune system, hormones, growth factors and nitric oxide. Nitric oxide can induce apoptosis by making the inner mitochondrial membrane more permeable to hydrogen ions and dissipating the proton gradient.

Proteins are released into the cytosol. These proteins bind to apoptosis inhibitor proteins and allow the process to take place.

Apoptosis and development

Apoptosis is an integral part of plant and animal tissue development. There is extensive division and proliferation of a particular cell type followed by pruning through programmed cell death. The excess cells shrink, fragment and are phagocytosed so that the components are reused and no harmful hydrolytic enzymes are released into the surrounding tissue. Apoptosis is tightly regulated during development, and different tissues use different signals for inducing it. It weeds out ineffective or harmful T lymphocytes during the development of the immune system.

During limb development apoptosis causes the digits (fingers and toes) to separate from each other.

In children between the ages of 8 and 14 years, 20–30 billion cells per day undergo apoptosis. In 1 year this equates to a mass of cells equivalent to the total body mass. In adults, 50–70 million cells per day apoptose. The rate of cells dying should balance the rate of cells produced by mitosis.

If the rates are not balanced:
- not enough apoptosis leads to the formation of tumours
- too much leads to cell loss and degeneration.

Cell signalling plays a crucial role in maintaining the right balance.

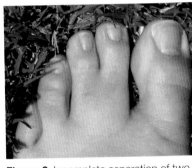

Figure 2 Incomplete separation of two toes (syndactyly) due to lack of apoptosi

STRETCH and CHALLENGE

If cells have damaged DNA, including a mutation to genes involved in regulating mitosis, they do not respond to signals from surrounding cells. They do not undergo apoptosis and keep on dividing, forming a tumour. If cells from the tumour break away, enter the blood or lymph and travel to another part of the body, they can set up secondary cancers. This is called metastasis and the cancerous tumour is malignant.

The Human papilloma virus (HPV) causes genital warts. This virus can be sexually transmitted and can affect cells of the cervix. It interferes with a protein called p53, prevents apoptosis and causes cervical cancer.

Question
A There is a vaccine available against the HPV virus. Think of pros and cons for vaccinating all boys and girls before they reach sexual maturity.

Questions

1 Explain the effect of making the inner mitochondrial membrane more permeable to hydrogen ions.
2 Explain why the process of apoptosis does not lead to damage of nearby cells and tissues.
3 Suggest how **(a)** ineffective and **(b)** harmful T lymphocytes would differ from effective T lymphocytes.

By the end of this spread, you should be able to ...

✱ Describe, with the aid of diagrams and photographs, the behaviour of chromosomes during meiosis, and the associated behaviour of the nuclear envelope, cell membrane and centrioles.

✱ Know the names of the main stages (but not sub-stages) of meiosis.

Key definition

Meiosis is a reduction division. The resulting daughter cells have half the original number of chromosomes. They are **haploid** and can be used for sexual reproduction.

All living organisms can reproduce. Reproduction can be either asexual or sexual. Asexual reproduction can be achieved by mitosis in eukaryotes or by binary fission in prokaryotes. The offspring produced by asexual reproduction are genetically identical to each other and to the parent. Genetic variation in this case is introduced only by random mutation.

In sexual reproduction, the offspring are genetically different from each other and from the parents. Each parent produces special reproductive cells, called **gametes**. Gametes (one from each parent) fuse together at fertilisation to produce a **zygote**.

The chromosome number must be halved

When two gametes fuse to make one cell the chromosomes are combined into one nucleus. Therefore the chromosome number in each gamete needs to be **haploid** (half that of the original cell). This ensures that, after fertilisation, the original chromosome number is restored. **Meiosis** is the type of nuclear division where the chromosome number is halved. It involves two separate divisions, referred to as meiosis I and meiosis II. Each division has four stages: prophase, metaphase, anaphase and telophase. In interphase, before meiosis I, the DNA replicates. As a result each chromosome consists of two identical sister chromatids, joined at the centromere. The cell now contains four, rather than the original two, copies of each chromosome.

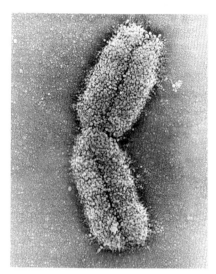

Figure 1 A chromosome consisting of two identical chromatids; (×9000)

Meiosis I

Prophase I

* The chromatin condenses and undergoes supercoiling so that chromosomes shorten and thicken. They can take up stains and be seen with a light microscope.
* The chromosomes come together in their **homologous pairs** (matching pairs) to form a **bivalent**. Each member of the pair has the same genes at the same loci. Each pair consists of one maternal and one paternal chromosome.
* The non-sister chromatids wrap around each other and attach at points called **chiasmata** (singular: chiasma).
* They may swap sections of chromatids with one another in a process called **crossing over**.
* The nucleolus disappears and the nuclear envelope disintegrates.
* A spindle forms. It is made of protein microtubules.
* Prophase I may last for days, months or even years, depending on the species and on the type of gamete (male or female) being formed.

Figure 2 Prophase I

Metaphase I

* Bivalents line up across the equator of the spindle, attached to spindle fibres at the centromeres. The chiasmata are still present.
* The bivalents are arranged randomly (random assortment) with each member of a homologous pair facing opposite poles.
* This allows the chromosomes to independently segregate when they are pulled apart in anaphase I.

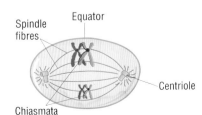

Figure 3 Metaphase I

Anaphase I

- The homologous chromosomes in each bivalent are pulled by the spindle fibres to opposite poles.
- The centromeres do not divide.
- The chiasmata separate and lengths of chromatid that have been crossed over remain with the chromatid to which they have become newly attached.

Telophase I

- In most animal cells two new nuclear envelopes form – one around each set of chromosomes at each pole – and the cell divides by cytokinesis. There is a brief interphase and the chromosomes uncoil.
- In most plant cells the cell goes straight from anaphase I into meiosis II.

Meiosis II

This division is in a plane at right angles to meiosis I.

Prophase II

- If a nuclear envelope has reformed, it breaks down again.
- The nucleolus disappears, chromosomes condense and spindles form.

Metaphase II

- The chromosomes arrange themselves on the equator of the spindle. They are attached to the spindle fibres at the centromeres.
- The chromatids of each chromosome are randomly assorted (arranged).

Anaphase II

- The centromeres divide and the chromatids are pulled to opposite poles by the spindle fibres. The chromatids randomly segregate.

Telophase II

- Nuclear envelopes reform around the haploid daughter nuclei.
- In animals, the two cells now divide to give four haploid cells.
- In plants, a tetrad of four haploid cells is formed.

Exchange between non-sister chromatids has occurred during crossover in prophase I

Homologous chromosomes are pulled to opposite poles

Figure 4 Anaphase I

Nuclear envelope forming

Cell dividing by cytokinesis

Figure 5 Telophase I

Centrioles replicate and move to poles

New spindle fibres form at right angles to previous spindle axis

Figure 6 Prophase II

Chromosomes lying on the equator of the cell

Figure 7 Metaphase II

Chromatid moving towards the pole

Figure 8 Anaphase II

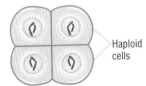

Haploid cells

Figure 9 Telophase II

Questions

1 What is a bivalent?
2 Some organisms have adult forms with haploid cells. They can reproduce sexually by producing diploid zygotes that then undergo meiosis. What sort of nuclear division is used to produce their gametes?
3 Complete the table to compare meiosis and mitosis.

	Mitosis	Meiosis
Number of divisions		
Products	Two genetically identical daughter cells	
Chromosome number	Maintained	
Do bivalents form?		
Does crossing over occur?		

Examiner tip

You must spell meiosis correctly so that it is absolutely clear which type of nuclear division you are talking about. The same goes for mitosis.

By the end of this spread, you should be able to ...

* **Explain how meiosis and fertilisation can lead to variation through the independent assortment of alleles.**
* **Explain the terms *allele*, *locus* and *crossing over*.**

Key definitions

An **allele** is a version of a gene. A gene is a length of DNA that codes for one or more polypeptides. An allele of the gene has a difference in the DNA base sequence that is expressed as (translates into) a slightly different polypeptide.

The **locus** is the position of a gene on a chromosome.

Crossing over is when lengths of DNA are swapped from one chromatid to another.

Maternal chromosomes are the set of chromosomes in an individual's cells that were contributed by the egg.

Paternal chromosomes are the set of chromosomes in an individual's cells that were contributed by the sperm.

The significance of meiosis

Sexual reproduction increases **genetic variation**, as genetic material from cells of two (usually unrelated) organisms combines. Genetic variation increases the chances of **evolution** as **natural selection** can favour the organisms that are best adapted to the ever-changing environment.

To maintain the original chromosome number, the chromosome number in **gametes** needs to be halved. Thus when two **haploid** gametes join, at **fertilisation**, the resulting **zygote** is **diploid**.

How meiosis and fertilisation lead to genetic variation

Meiosis increases genetic variation by:

* **crossing over** during prophase I, to shuffle **alleles**
* **genetic reassortment** due to the **random** distribution and subsequent **segregation** of the **maternal** and **paternal chromosomes** in the homologous pairs, during meiosis I
* **genetic reassortment** due to the **random** distribution and **segregation** of the sister **chromatids** at meiosis II
* **random mutation**.

Fertilisation increases genetic variation by randomly combining two sets of chromosomes, one from each of two genetically unrelated individuals.

Crossing over

This occurs during prophase I. The homologous chromosomes pair and come together to form bivalents. On average, between two and three cross-over events occur on each pair of human chromosomes.

* Non-sister chromatids wrap around each other very tightly and attach at points called **chiasmata**.
* The chromosomes may break at these points. The broken ends of the chromatids rejoin to the ends of non-sister chromatids in the same bivalent. This leads to similar sections of non-sister chromatids being swapped over. These sections contain the same genes but, often, different alleles.
* This is called crossing over.
* It produces new combinations of alleles on the chromatids (which will eventually become chromosomes in the daughter cells).
* The chiasmata remain in place during metaphase and they hold the maternal and paternal homologues together on the spindle, facing the way they will migrate.
* Holding the homologous pairs on the spindle equator ensures that when segregation occurs, at anaphase I, one member of each pair goes to each pole.

Reassortment of chromosomes

* This reassortment is the consequence of the random distribution of maternal and paternal chromosomes on the spindle equator at metaphase I, and the subsequent segregation into two daughter nuclei at anaphase I.
* Each gamete acquires a different mixture of maternal and paternal chromosomes.

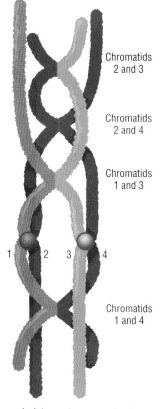

Chromatids 2 and 3

Chromatids 2 and 4

Chromatids 1 and 3

1　2　3　4

Chromatids 1 and 4

Figure 1 A homologous pair of chromosomes (maternal = red, paternal = blue) with four chiasmata. Each of the two chromatids on each chromosome can cross over with each of the two chromatids on the other chromosome in the bivalent. This example shows all possible chiasmata

From this process alone, theoretically one individual could produce 2^n genetically different gametes, where n = the haploid number of chromosomes. However, the actual number is very much greater than this because of crossing over, and the subsequent genetic recombination, during prophase I.

Reassortment of chromatids

- This is the result of the random distribution on the spindle equator, of the sister chromatids, at metaphase II.
- Because of crossing over, the sister chromatids are no longer genetically identical.
- How they align at metaphase II determines how they segregate at anaphase II.

Fertilisation

In humans only one ovum (actually it is a secondary oocyte and has not completed the second meiotic division) is (usually) released from an ovary at a time. There are about 300 million spermatozoa, all genetically different, and any one of them can fertilise the secondary oocyte. Whichever one fertilises the ovum, genetic material from two unrelated individuals is combined to make the zygote.

Mutation

DNA mutation may also occur during interphase when DNA replicates. This is not peculiar to meiosis as it can also occur in mitosis or binary fission. **Chromosome mutations** may also occur. However, mutation does increase genetic variation. If mutation occurs in the sperm or egg that are used in fertilisation then the mutated gene will be present in every cell of the offspring.

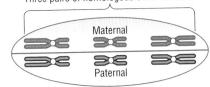
Three pairs of homologous chromosomes

Maternal

Paternal

Independent assortment of homologous chromosomes during meiotic division I

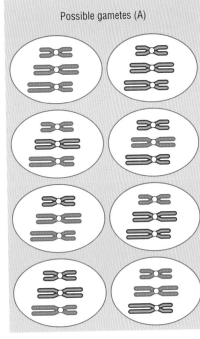
Possible gametes (A)

Figure 2 Possible gametes resulting from independent assortment of three pairs of chromosomes during meiosis I

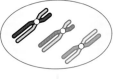
Chromatids after meiosis I. Alleles have been swapped

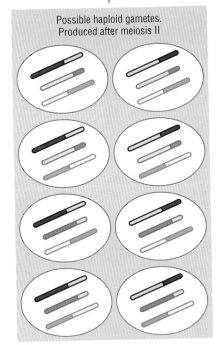
Possible haploid gametes. Produced after meiosis II

Figure 3 Possible gametes from random assortment of three sets of sister chromatids

STRETCH and CHALLENGE

Certain mutations can cause a reduced number of chiasmata. In some organisms this causes a higher rate of nondisjunction – the failure of chromosomes in a homologous pair to separate (segregate) at anaphase.

In mammals the male has sex chromosomes, X and Y, that do not fully match. However, there is a small region on each that matches the other.

Questions

A What is the effect of nondisjunction on the chromosome number in gametes?

B Why is a small homologous region on X and Y chromosomes important to make sure that each gamete contains either a Y or an X chromosome?

Key definition

Random change to the structure of a chromosome is called **chromosome mutation**. There are different types: inversion (a section of chromosome turns through 180°); deletion (a part is lost); translocation (a piece of one chromosome becomes attached to another); non-disjunction (homologous chromosomes fail to separate properly at meiosis 1 or chromatids fail to separate at meiosis 2; if this happens to a whole set of chromosomes, polyploidy results [spread 2.1.21]). The shuffling of alleles in prophase 1 is **not** an example of mutation.

Questions

1 What is the minimum number of genetically different gametes that can theoretically be produced by a human?

2 Explain why sexual reproduction has increased the diversity of living organisms on Earth.

Examiner tip

Don't use the word *gene* when you mean *allele*.

By the end of this spread, you should be able to . . .

✳ Explain the terms *genotype, phenotype, dominant, recessive, codominant, linkage.*

Genotype and phenotype

The **genotype** is the genetic makeup of an organism. It describes the organism in terms of the **alleles** it contains, usually in the context of a particular characteristic. An organism with two identical alleles for a particular gene is described as **homozygous**. Those with two different alleles of the same gene are heterozygotes (described as **heterozygous**).

Cystic fibrosis is caused by a mutation to a gene on one of the autosomes. (Autosomes are chromosomes *not* concerned with determining sex.) The mutation disrupts the transport of chloride ions and water across the membranes of cells lining the airways, gut and reproductive tracts. It changes the shape of the transmembrane chloride ion channels so that they no longer function correctly. In the lungs, the cilia are not properly hydrated and cannot shift the mucus. The mucus accumulates and dehydrates, allowing bacterial infections to occur. Recurrent infections may lead to lung failure.

- Heterozygotes have one normal allele (we'll represent this as CF) and one abnormal allele (cf). They have some abnormal chloride ion channels but enough normal channels for their lungs to function. They have no symptoms and are carriers. Their genotype is **CFcf**.
- People with cystic fibrosis are homozygous recessive. Their genotype is **cfcf**.
- People with the genotype **CFCF** are also homozygous and have two normal alleles. All their chloride ion channels function.

The **phenotype** refers to the characteristics that are expressed in the organism. It means those features that can be observed. These are determined by its genotype and the environment in which it has developed. Siamese cats have cream fur with dark fur at the face, ears, lower legs and tail. Their coat colouring is a partial form of albinism due to a mutation in the gene for the enzyme tyrosinase, involved in melanin production. The altered enzyme fails to work at normal body temperature but works at the cooler regions of the cat's body. (The front of the face is cooler because of air passing through the sinuses.)

Dominant and recessive

An allele is said to be **dominant** if it is always expressed in the phenotype, even if a different allele for the same gene is present in the genotype. The characteristic in question and its inheritance pattern are also described as dominant.

An allele is said to be **recessive** if it is only expressed, in the phenotype, in the presence of another identical allele, or in the absence of a dominant allele, for the same gene. The allele for cystic fibrosis is recessive, and only individuals with the genotype **cfcf** have symptoms of cystic fibrosis. Individuals with the genotype **CFcf** are symptomless carriers. Cystic fibrosis is a recessive characteristic and has a recessive inheritance pattern.

Codominant

Two alleles of the same gene are described as **codominant** if they are both expressed in the phenotype of a heterozygote. In cattle, one of the genes for coat colour has two alleles. C^R codes for red hairs and C^W codes for white hairs. Heterozygotes, C^RC^W, have both red and white hairs, and are described as roan. In humans, one way of classifying blood groups is the ABO system relating to antigens on the cell surface membranes of red blood cells. The antigens are determined by one gene that has three alleles, I^A, I^B, and I^o. I^o is recessive whilst I^A and I^B are codominant.

Examiner tip

There are no genes coding for genetic diseases. Genes code for proteins that are part of the organism's structure or for an enzyme to regulate a metabolic pathway. Genetic diseases may be the result of altered alleles that produce altered proteins.

Figure 1 Siamese cat with dark fur at head, ears, lower limbs, feet and tail regions

Key definition

Alleles are **codominant** if they both contribute to the phenotype.

Linkage

Linkage refers to two or more genes that are located on the same chromosome. The linked alleles (of these genes) are normally inherited together because they do not segregate independently at meiosis, unless chiasmata have been formed between them. At crossover, the alleles from one chromatid become linked to alleles on the other chromatid. Linkage reduces the number of phenotypes resulting from a cross.

In pea plants the gene for height (two alleles: tall and dwarf) and the gene for texture of mature seed pods (two alleles: smooth and wrinkled) are both on chromosome 4. (In peas $n = 7$. This means that each diploid pea cell has 14 chromosomes, two sets of 7.)

The gene for the colour of seeds (two alleles: yellow and green) and the gene for flower colour (two alleles: coloured and white) are both on chromosome 1.

We shall use genetic diagrams to solve problems about inheritance of linked characteristics in spread 2.1.16.

Sex linkage

A characteristic is sex-linked if the gene that codes for it is found on one of the sex (X and Y) chromosomes. In most animals the small Y chromosome has few genes so most sex-linked genes are likely to be found on the X chromosome. Sex-linked characteristics include haemophilia (A and B), red-green colour blindness, Duchenne muscular dystrophy, fragile X syndrome, and vitamin D-resistant rickets, but there are many others.

In mammals, females are the homogametic sex; they have XX sex chromosomes. However, in birds, butterflies and moths, males are homogametic and females are heterogametic.

Figure 2 In birds, females are heterogametic and males homogametic

Examiner tip

Note that when alleles are linked (on the same chromosome) they are shown together, PLpl, not PpLl as they would be in a normal dihybrid cross for unlinked alleles. If you use this convention it will be easier to solve any problems involving linkage.

Worked example

When homozygous purple-flowered sweet peas with long pollen grains (**PPLL**) were crossed with homozygous red-flowered sweet peas with short pollen grains (**ppll**), all the F_1 generation had purple flowers and long pollen grains. The genes for flower colour and length of pollen grain are on the same chromosome (linked). These plants were allowed to self-pollinate. The researchers expected a 3:1 ratio in the F_2 progeny. They obtained the following:

- 296 purple-flowered with long pollen grains
- 19 purple-flowered with short pollen grains
- 27 red-flowered with long pollen grains
- 86 red-flowered with short pollen grains.

Are these results significantly different from expectation? Can you suggest an explanation?

Answer

As the genes are linked, all gametes of purple-flowered, long-pollen plants are **PL**. All gametes of red-flowered, short-pollen plants are **pl**.
All F_1 plants have genotypes **PLpl** and are purple with long-pollen grains. Their gametes are **PL** and **pl**.
So expected F_2 ratio: **PLPL PLpl plPL plpl** (3 purple long to one red short). The observed ratios are significantly different.
An explanation is that in meiosis I crossing-over occurred with a chiasma between **P/p** and **L/l**.
Those with purple flowers and short pollen were **PlPl** or **Plpl**. Those with red flowers and long pollen were **pLpL** or **pLpl**.

Questions

1 What are the genotypes of people with the following blood groups? **(a)** Group AB; **(b)** Group O; **(c)** Group A; **(d)** Group B.

2 What is the genotype of the sex chromosomes for **(a)** a male cinnabar moth and **(b)** a female blackbird?

3 Explain why Siamese cats living in tropical countries have paler extremities than those living in cold climates.

4 Suggest why Siamese kittens are cream all over when first born and do not develop the darker extremities until they are a few weeks old.

By the end of this spread, you should be able to . . .

✳ **Use genetic diagrams to solve problems involving sex linkage.**

When constructing genetic diagrams there are certain conventions:
- Start by showing the parental phenotypes.
- The gene is represented by a single letter, with upper case for the dominant allele and lower case for the recessive allele.
- Ideally the letter chosen should be one where it is easy to distinguish between its lower and upper case forms. **Rr** is better than **Cc**. However, if you use letters such as **C**, make the upper case version large and the lower case version very small or use **c′**.
- Where the gene in question has more than two alleles (any individual will only have two alleles but there may be more than two in the population), the gene has an upper case letter and the alleles are denoted in superscript. This is the case for the alleles for blood groups, I^A, I^B, and I^o.
- Where there is codominance (both alleles contribute to the phenotype) the gene is represented by an upper case letter and the alleles in superscript. In some flowers the colours red and white are codominant and written as C^R and C^W. In blood group inheritance, I^A and I^B are codominant. I^o is recessive.

Examples of genetic diagrams

Haemophilia A

In the second century AD a rabbi exempted a male infant from circumcision because his mother's sister had three sons who died of bleeding following circumcision. This is one of the earliest recognitions of human inheritable disease. Several factors are needed for blood to clot following a wound. It is a complex series of events. One of these factors is a protein, factor VIII, that is coded for by a gene on the X chromosome. The recessive allele expresses an altered protein that does not function. This leads to an increase in blood-clotting time. Internal bleeding, resulting from knocks, and bleeding into joints are particularly harmful.

Males have only one X chromosome and if it has the allele for haemophilia A they will suffer from haemophilia. Such males are hemizygous – they have only one allele for a particular characteristic. Haemophilia A shows a recessive inheritance pattern.

Figure 1 shows the possible inheritance of haemophilia in a family where the father is normal and the mother is a carrier.

H = allele for normal factor VIII
h = allele for non-functioning factor VIII

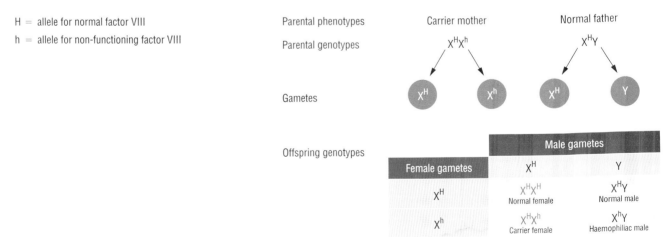

Figure 1 Inheritance of haemophilia from a carrier mother and a normal father

You can see that the possible offspring phenotypes are 25% carrier female, 25% normal female, 25% affected (haemophiliac) male and 25% normal male.

Duchenne muscular dystrophy (DMD)

The DMD gene for a muscle protein, dystrophin, is on the X chromosome in humans. Dystrophin is a large protein, involved in structures needed for muscle contraction. Mutations of the gene usually result in a severely truncated dystrophin protein or no dystrophin. Boys with the disease develop muscle weakness in early childhood and are usually wheelchair-bound by the age of 10. Death often occurs due to complications of muscle degeneration (skeletal and heart muscle) by the early 20s.

Figure 2 shows a pedigree of a family where some members are affected by muscular dystrophy.

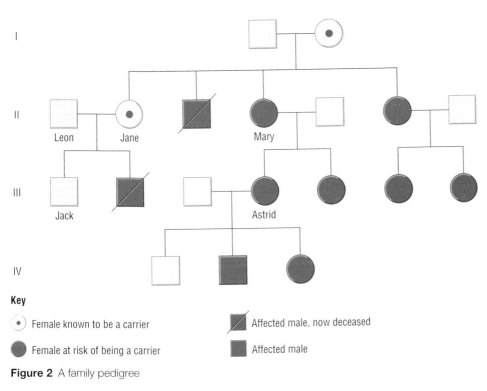

Examiner tip

Solving genetics problems, such as this one, involves trial and error. This is a good method. You need to scribble some possible genetic diagrams, in pencil, and then cross them through before presenting the one you think is correct. Remember also that the clues are often in the individual showing the recessive characteristic – you know what the genotype is of a recessive character. Always link the phenotypes of the offspring to their genotypes.

Key

⊙ Female known to be a carrier

⬤ Female at risk of being a carrier

◨ (diagonal) Affected male, now deceased

◼ Affected male

Figure 2 A family pedigree

Questions

Use genetic diagrams to solve the following problems:

1 Female haemophiliacs are rare. What would be the genotypes of parents who produced a haemophiliac girl?

2 Red-green colour blindness is caused by a sex-linked allele that is recessive to the allele for normal colour vision. In a family a boy and his grandfather both have red-green colour blindness. Is this grandfather the father of his mother or of his father?

3 In fruit flies, *Drosophila melanogaster,* the gene for eye colour is on the X chromosome. Red is dominant to white. Females are homogametic. If white-eyed females are crossed with red-eyed males, all the female progeny in the F_1 (first filial [first offspring]) generation have red eyes and all the males have white eyes. If the F_1 males and females are allowed to interbreed, in the F_2 generation half the females and half the males have white eyes, whilst the rest have red eyes. Draw genetic diagrams to explain these crosses.

4 What are the genotypes of Mary, Astrid, Jack and Jane, shown in the pedigree in Figure 2?

5 If Leon and Jane have another child, what are the chances that it will have Duchenne muscular dystrophy? Use a genetic diagram to explain your answer.

By the end of this spread, you should be able to . . .

✱ **Use genetic diagrams to solve problems involving codominance.**

Key definition

Alleles are **codominant** if they both contribute to the phenotype.

Sickle-cell anaemia

Sickle-cell anaemia was the first human disease to be understood at the molecular level.

- All individuals with the disease have the same mutation.
- The β-strands of haemoglobin differ by one amino acid at position 6.
- In normal haemoglobin, glutamic acid is at position 6, but in sickle-cell haemoglobin, valine is present instead.
- When this abnormal haemoglobin is deoxygenated it is not soluble and becomes crystalline and aggregates into more linear and less globular structures. This deforms the red blood cells, making them inflexible (and often sickle-shaped) and unable to squeeze through capillaries.
- After many cycles of oxygenation and deoxygenation, some cells become irreversibly sickled. Some are destroyed.
- If enough sickle-cells become lodged in capillaries blood flow is impeded. Organs, particularly bones, do not receive enough oxygen, leading to a painful crisis.
- Eventually organs, especially heart, lungs and kidneys, become damaged.
- The genotype of people with normal haemoglobin can be denoted as H^AH^A.
- The genotype of people with sickle-cell anaemia is H^SH^S.
- The genotype of symptomless heterozygotes is H^AH^S.

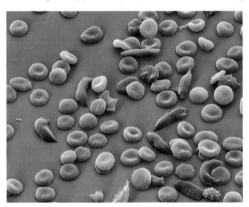

Figure 1 Sickled and normal red blood cells (×1000)

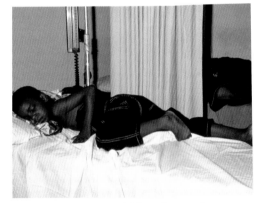

Figure 2 Patient with sickle-cell anaemia experiencing a painful crisis

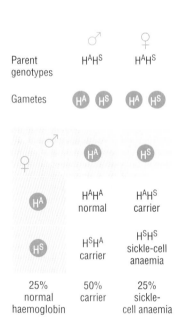

| | ♂ H^AH^S | ♀ H^AH^S |
| Parent genotypes | | |

Gametes: H^A H^S H^A H^S

♀ \ ♂	H^A	H^S
H^A	H^AH^A normal	H^AH^S carrier
H^S	H^SH^A carrier	H^SH^S sickle-cell anaemia
	25% normal haemoglobin	50% carrier	25% sickle-cell anaemia

Key:

H^A allele for normal haemoglobin
H^S allele for abnormal haemoglobin

Figure 3 Genotypes and phenotypes of offspring of two parents, both heterozygous for sickle-cell disease

In heterozygotes, red blood cells are made in the bone marrow with half their haemoglobin normal and half sickled. The presence of normal haemoglobin prevents sickling in the red cells when they are in the circulation and deoxygenated. Thus, heterozygotes are symptomless carriers and at whole-organism level this condition could be considered to be recessive. However, at the molecular and cellular level, because both alleles contribute to the phenotype as observed in red blood cells, it is codominant.

What are the possible genotypes and phenotypes of offspring of two parents, both carriers of sickle-cell disease?

We can see from Figure 3 that we expect 25% to have normal haemoglobin, 50% to be heterozygous carriers and 25% to have sickle-cell anaemia.

Roan cattle

One of the genes for coat colour in shorthorn cattle has two alleles: C^R codes for red hairs and C^W codes for white hairs.

- Homozygous individuals with genotype C^RC^R have red (chestnut) coats.

- Homozygous individuals with genotype $C^W C^W$ have white coats.
- Heterozygotes, genotype $C^R C^W$, have red and white hairs and the coat is roan.

If red and white shorthorn cattle are interbred then all the offspring are roan – a mixture of red and white.

If two roan shorthorn cattle are interbred then the probable phenotypes are 25% red, 50% roan and 25% white (Figure 5).

STRETCH and CHALLENGE

Patients with sickle-cell anaemia can be treated by blood transfusions about every three months. Some patients have been treated with chemotherapy. They are given hydroxyurea that causes an increase in fetal haemoglobin formation and produces an improvement in red blood cell function.

Some patients have been treated with bone marrow transplants, although this is reserved for severely affected patients as there is a risk of graft-versus-host reaction.

Questions

A Explain why an increase in fetal haemoglobin improves red cell function in sickle-cell patients. (*Hint: revisit AS work on transport in animals.*)

B Suggest what is meant by 'graft-versus-host reaction'. (*Hint: revisit AS work on the immune system.*)

C Suggest why blood transfusions are given every three months. (*Hint: revisit AS work on transport in animals.*)

There are various ways to diagnose sickle-cell disease. A blood smear can be exposed to very low oxygen concentrations to see if red cells become sickled. However, this can give false positives for carriers, as under these laboratory conditions, red blood cells from heterozygotes may also sickle. A more accurate diagnosis involves electrophoresis of haemoglobin. A sample of extracted haemoglobin is placed in a well in an electrophoresis gel and a buffer solution added. Electrodes are placed at either end and the protein migrates towards the positive electrode, according to the overall electric charge on the protein. Glutamic acid has a negative charge and valine is neutral.

Question

D Predict how the migration patterns of normal and sickle-cell haemoglobins will differ from each other.

An analysis of the gene encoding the β-globulin from the individual's DNA can diagnose the presence of the mutation. This can also be done on fetal cells obtained prenatally by chorionic villus sampling or amniocentesis.

Question

E Explain why cells such as skin cells, and not red blood cells, are used for DNA analysis for the presence of a mutated β-globulin gene.

Figure 4 Red shorthorn bull

Parental phenotypes	Cow (♀) Red	Bull (♂) White
Genotypes	$C^R C^R$	$C^W C^W$
Gametes	C^R	C^W
Offspring genotypes	All $C^R C^W$	
Offspring phenotypes	All roan	

Figure 5 Genetic diagram of cross between red shorthorn cow and white shorthorn bull

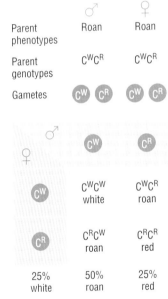

	♂ Roan	♀ Roan
Parent phenotypes		
Parent genotypes	$C^W C^R$	$C^W C^R$
Gametes	C^W C^R	C^W C^R

♀ \ ♂	C^W	C^R
C^W	$C^W C^W$ white	$C^W C^R$ roan
C^R	$C^R C^W$ roan	$C^R C^R$ red
	25% white / 50% roan	25% red

Figure 6 Genetic diagram showing probable outcome from interbreeding roan shorthorn cattle

Questions

1 A shorthorn bull is run with a herd of red shorthorn cows. Of the calves produced in the first year, 11 are red and 10 are roan. What is the genotype and phenotype of the bull?

2 Ducks of the breed called Swedish blue have grey plumage. If two Swedish blue ducks mate, some offspring are grey, some are white and some are black. Explain these observations and describe how a poultry breeder could make sure he produces all Swedish blue ducks.

3 **(a)** Suggest what sort of mutation causes sickle-cell disease.
  (b) Explain why one of the symptoms of sickle-cell disease is severe anaemia.

Examiner tip

Always provide a key with your genetic diagrams. Indicate which symbol relates to which allele.

You may be told the genotype of the female parent and of the male parent. Don't get confused and assume that the gene in question is sex-linked.

By the end of this spread, you should be able to . . .

✳ Describe the interactions between loci (epistasis).
✳ Predict phenotypic ratios in problems involving epistasis.

Key definition

Epistasis is the interaction of different gene loci so that one gene locus masks or suppresses the expression of another gene locus.

We have already seen how interactions between alleles (whether dominant/recessive or codominant) at the same locus (i.e. of the same gene) may affect the phenotype.

There are cases where different genes on different loci interact to affect one phenotypic characteristic. Where one gene masks or suppresses the expression of another gene this is called **epistasis** (from a Greek word meaning 'stoppage'). The genes involved may control the expression of one phenotypic characteristic in one of the following ways:

• They may work against each other (antagonistically) resulting in masking.
• They may work together in a complementary fashion.

Working antagonistically

The homozygous presence of a recessive allele may prevent the expression of another allele at a second locus. The alleles at the first locus are **epistatic** to the alleles at the second locus, which are described as **hypostatic**.

Epistasis is not inherited, it is an interaction between two gene loci. It reduces phenotypic variation.

Recessive epistasis

An example of **recessive epistasis** is the inheritance of flower colour in *Salvia*.
• Two gene loci, A/a and B/b, on different chromosomes are involved.
• A pure-breeding pink-flowered variety of *Salvia*, genotype **AAbb**, was crossed with a pure-breeding white-flowered variety, genotype **aaBB**. All the F_1 generation, genotype **AaBb**, had purple flowers.
• Interbreeding the F_1 to give the F_2 generation resulted in purple, pink and white flowers in the ratio of 9:3:4.
• The homozygous **aa** is epistatic to both alleles of the gene B/b. Neither the allele, **B**, for purple nor the allele, **b**, for pink can be expressed if there is no dominant allele, **A**, present.

Dominant epistasis

A second type of epistasis, **dominant epistasis**, occurs when a dominant allele at one gene locus masks the expression of the alleles at a second gene locus. An example is fruit colour in summer squash.
• Two gene loci, D/d and E/e, are involved. The presence of one **D** allele results in white fruits, regardless of the alleles present at the second locus (E/e).
• In homozygous **dd** individuals, the presence of one **E** allele produces yellow fruits and the presence of two **e** alleles produces green fruits.
• If two white-coloured, double heterozygotes (**DeEe**) are crossed, the offspring show the following phenotype ratio: 12 white (**D-E-** or **D-ee**): 3 yellow (**ddE-**): 1 green (**ddee**). (- indicates that either allele of the gene may be present.)

Another example of dominant epistasis is the inheritance of feather colour in chickens.
• There is an interaction between two gene loci, I/i and C/c.
• Individuals carrying the dominant allele, **I**, have white feathers, even if they also have the dominant allele, **C**, for coloured feathers.
• Birds that are homozygous for **c** (genotype **IIcc**, **Iicc** or **iicc**) are also white.

Figure 1 Summer squash

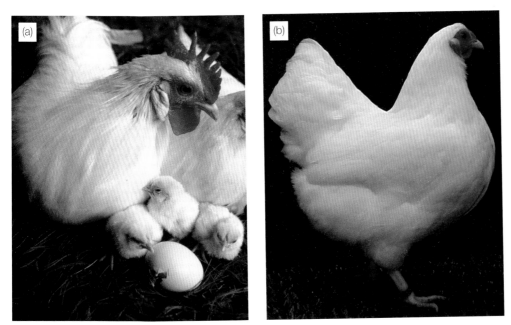

Figure 2 (a) White Leghorn chicken; **(b)** White Wyandotte chicken

Pure-breeding White Leghorn chickens have the genotype **IICC**. Pure-breeding white Wyandotte chickens have the genotype **iicc**.
- If two such birds are mated, all the progeny are white, with genotype **IiCc**.
- If the progeny interbreed, they produce white-feathered chickens and coloured-feathered chickens, in a ratio of 13:3 respectively.

Working in a complementary fashion

William Bateson and Reginald Punnett (inventor of the Punnett square) crossed two strains of white-flowered sweet peas (ccRR × CCrr).
- Unexpectedly, all the F_1 plants had purple flowers.
- They allowed the F_1 to interbreed and the F_2 had purple flowers and white flowers in a ratio of 9:7.
- The explanation suggests that at least one dominant allele for both gene loci (**C-R-**) has to be present for flowers to be purple.
- All other genotype combinations, such as **ccR-** or **C-rr**, produce white flowers.
- This is because the homozygous recessive condition at *either* locus masks the expression of the dominant allele at the other locus.

The way the two gene loci may produce these results is if they complement each other – if one gene codes for an intermediate colourless pigment and the second locus codes for an enzyme that converts the intermediate compound to the final purple pigment.

Figure 3 Sweet peas

$$\text{gene C/c} \qquad\qquad \text{gene R/r}$$
$$\downarrow \qquad\qquad\qquad \downarrow$$

precursor substance ⟶ intermediate compound ⟶ final pigment
(colourless)   (colourless)   (purple)

Questions

1 Write down all the genotypes that produce **(a)** purple *Salvia* flowers, **(b)** pink *Salvia* flowers and **(c)** white *Salvia* flowers.
2 Write down all the genotypes that produce **(a)** white chickens and **(b)** coloured chickens.
3 Write down all the genotypes that produce **(a)** white-flowered sweet peas and **(b)** purple-flowered sweet peas.
4 Explain why epistasis reduces phenotypic variation.

Examiner tip

You are not expected to construct genetic diagrams to explain the results of these crosses.

All of these epistatic ratios are variations on the normal 9:3:3:1 ratio from a dihybrid inheritance of two unlinked genes (see spreads 2.1.15 and 2.1.16).

Remember:
- A 9:3:4 ratio suggests recessive epistasis.
- A 12:3:1 ratio or a 13:3 ratio suggests dominant epistasis.
- A 9:7 ratio suggests epistasis by complementary action.

What determines sex?

We have seen that in humans, males are heterogametic (having XY sex chromosomes) and females are homogametic (having XX sex chromosomes).

This isn't the case throughout the animal kingdom. In birds, butterflies and moths, females are heterogametic and males are homogametic.

In some animals, such as turtles, sex of offspring is determined by the temperature at which eggs are incubated. Those incubated at a higher temperature become females. Some molluscs, such as oysters, may change sex once or more during their lifetime.

Plymouth rock chickens

Figure 1 A Plymouth rock chicken

Newly hatched chicks have to be sexed so that farmers can keep the females as they are the egg-layers. This can be difficult as male birds do not have a penis. In the Plymouth rock breed of chickens, there is a dominant sex-linked allele, **B**, which causes the normally black feathers to have a white bar. Newly hatched chicks with the allele **B** have barred feathers, seen as a white spot on the otherwise black head. Those with only the allele **b** have non-barred feathers and have a black head.

A poultry breeder wants to know which genotype birds to mate in order to produce phenotypes that can be easily sexed.

- In birds males are homogametic, XX, and females are heterogametic, XY.
- A female with a white spot must have the genotype X^BY.
- A male with a black head must have the genotype X^bX^b (as this characteristic is recessive).

We can use a genetic diagram to show the cross between these two and predict the phenotype ratios in the offspring.

B = allele for barred (white) feathers
b = allele for non-barred (black) feathers

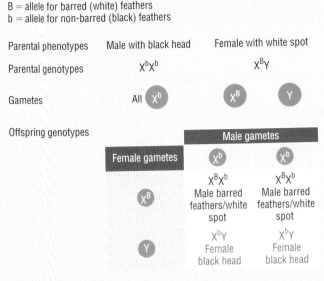

Figure 2 Crossing Plymouth rock chickens

From Figure 2 we can see that the offspring phenotype ratios are 50% females with a black head and 50% males with white spots. Therefore the ratio is 1:1.

The chance of producing birds with white spots is 50% or 0.5. But, as all the females have black heads and all the males have a white spot, this enables the breeder to easily sex the progeny.

Question

A   Suggest why the cross shown in Figure 2 is more reliable than crossing males with white spots with black-headed females.

Pale brindled moths

There is a rare black form of the pale brindled moth. They are all female. When such a female is mated with a normal male, all her female offspring are black while all her male offspring are normal.

Question

B   Use genetic diagrams to suggest an explanation for the observations about the rare form of pale brindled moth. Remember that female moths are heterogametic.

Cinnamon budgerigars

The wing markings of budgerigars are normally black. The sex-linked allele, **b**, for cinnamon, is recessive to the allele, **B**, for black. The allele, **b**, in the absence of the allele, **B**, produces brown wings. The gene is on the X chromosome.

Question

C   A breeder has several budgerigars, all black winged. Budgerigars can live for many years and breed each year. He buys a cinnamon female. Use genetic diagrams to work out the breeding programme he should carry out to obtain birds that will produce only progeny with brown wings.

By the end of this spread, you should be able to ...

* Use predicted phenotypic ratios to solve problems involving epistasis.

Coat colour in mice

Coat colour in mice may be agouti (alternating bands of pigment on each hair so that fur looks grey), black or albino. The gene for agouti has two alleles, A/a. Allele **a** is a mutation; when found in homozygous individuals it produces a black coat. A gene B/b at a separate locus controls the formation of pigment. Individuals of genotype **B-** can produce pigment, but those of genotype **bb** cannot, and are albino.

The production of the agouti pigmentation occurs in two steps:

$$\text{precursor substance} \xrightarrow{\text{gene B/b}} \text{black pigment} \xrightarrow{\text{gene A/a}} \text{agouti pattern}$$
(colourless)

When several pairs of agouti individuals of genotype **AaBb** are crossed, the total offspring are 28 agouti, 10 black and 12 albino, giving a ratio very close to 9:3:4.

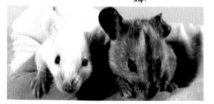

Figure 1 Coat colour in mice

Combs of domestic chickens

Four comb shapes in domestic chickens are shown in Figure 2. Two gene loci, P/p and R/r, interact to affect comb shape. The effect of the P/p alleles depends on which of the R/r alleles are present in the bird's genotype.

When true-breeding pea-combed chickens, genotype **PPrr**, are bred with true-breeding rose-combed chickens, genotype **ppRR**, the progeny have walnut combs, **PpRr**.

When walnut-combed chickens are interbred, the progeny show four phenotypes – walnut comb, rose comb, pea comb, and single comb in the classic Mendelian 9:3:3:1 ratio for dihybrid F_2 progeny.

Genotype	Phenotype
P-R-	Walnut comb
ppR-	Rose comb
P-rr	Pea comb
pprr	Single comb

Table 1 Genotypes and chicken comb phenotypes

Although you don't have to be able to construct genetic diagrams to show the results of this cross, it helps to see one (Figure 3).

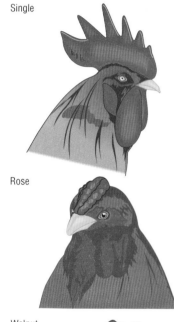
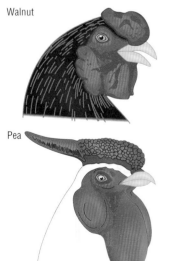

Figure 2 Comb shapes in domestic chickens

Gametes

♀ \ ♂	PR	Pr	pR	pr
PR	PPRR Walnut	PPRr Walnut	PpRR Walnut	PpRr Walnut
Pr	PPRr Walnut	PPrr Pea	PpRr Walnut	Pprr Pea
pR	PpRR Walnut	PpRr Walnut	ppRR Rose	ppRr Rose
pr	PpRr Walnut	Pprr Pea	ppRr Rose	pprr Single

Figure 3 Cross between walnut-combed chickens

STRETCH and CHALLENGE

Eye colour in humans

The colour of the iris in human eyes is determined by at least two genes, D/d and E/e, that are found at separate loci but interact with one another. Both have dominant alleles that cause the pigment melanin to be produced. The recessive alleles do not express melanin. There are two layers of pigment in the iris, one in front of the other. No melanin in either gives the albino condition. The iris looks pink as blood vessels within it can be seen. When the rear layer contains melanin and the front layer does not the iris is blue. As more and more melanin is produced in the front layer, the iris looks darker blue, green, brown and almost black.

Table 2 shows the iris colour phenotypes and their genotypes.

Examples of genotypes	Number of D and E alleles	Colour of iris
DDEE	4	Dark brown/black
DDEe or DdEE	3	Medium brown
DdEe, ddEE, DDee	2	Light brown
Ddee or ddEe	1	Dark blue
ddee	0	Pale blue

Table 2 Human eye colour, phenotypes and genotypes

Questions

A Show the genotypes and phenotypes, and phenotype ratio, of the possible offspring from two parents who are heterozygous at both gene loci and have light brown eyes.

B Is it possible for two blue-eyed parents to produce a child with brown eyes?

C Some humans have green eyes and some have violet eyes. What does this observation suggest about the genes involved in determining eye colour?

Questions

1 Write down all the genotypes that produce albino mice.

2 Write down all the genotypes that produce agouti mice.

3 Write down all the genotypes that produce black mice.

4 Is the above an example of recessive epistasis or complementary gene action, or could it be described as both? Give reasons for your answer.

5 Which two genotypes should a poultry breeder cross to produce chickens, so that 25% of the progeny have walnut combs, 25% have rose combs, 25% have pea combs and 25% have single combs?

6 A possible explanation for crossing pea-combed hens with rose-combed cockerels, and getting progeny that have walnut combs, is that comb shape is determined by one gene with two codominant alleles: C^R for rose comb and C^P for pea comb. Explain how the F_2 progeny negate this suggestion. (*Hint: you will need to predict the F_2 phenotypes using this model.*)

7 A black cat and a white cat interbreed and produce several litters of kittens. Eight are white, four are black and four are brown. A pure-breeding brown cat of the same breed and a pure-breeding black cat together produce several litters of brown kittens. There are two gene loci involved in determining their coat colour. Suggest an explanation for these observations. (*Hint: make some deductions and do some trial and error scribbling.*)

By the end of this spread, you should be able to ...

✳ **Use the chi-squared test to test the significance of the difference between observed and expected results. (The formula for χ^2 will be provided.)**

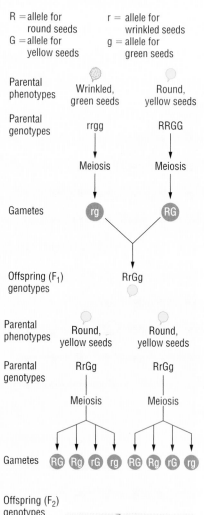

R = allele for round seeds
r = allele for wrinkled seeds
G = allele for yellow seeds
g = allele for green seeds

Parental phenotypes — Wrinkled, green seeds — Round, yellow seeds

Parental genotypes — rrgg — RRGG

Meiosis — Meiosis

Gametes — rg — RG

Offspring (F₁) genotypes — RrGg

Parental phenotypes — Round, yellow seeds — Round, yellow seeds

Parental genotypes — RrGg — RrGg

Meiosis — Meiosis

Gametes — RG Rg rG rg — RG Rg rG rg

Offspring (F₂) genotypes

♂ Gametes			
♀ Gametes	RG	Rg	rG	rg
RG	RRGG	RRGg	RrGG	RrGg
Rg	RRGg	RRgg	RrGg	Rrgg
rG	RrGG	RrGg	rrGG	rrGg
rg	RrGg	Rrgg	rrGg	rrgg

Offspring (F₂) phenotypes
9 round, yellow seeds
3 round, green seeds
3 wrinkled, yellow seeds
1 wrinkled, green seed

Figure 1 A genetic explanation for one of Mendel's dihybrid crosses

On the previous spread (2.1.15) you have seen the ratio we expect to get from a dihybrid (two genes/characteristics) cross, where the two genes in question are not linked on the same chromosome and are therefore independently assorted at meiosis.

In one of his experiments, Gregor Mendel crossed true-breeding pea plants that produced round, yellow seeds with true-breeding pea plants that produced green, wrinkled seeds. All the F₁ progeny grew into plants that produced round, yellow seeds. He allowed these plants to self-pollinate. He collected all the seeds, a large number, and grew them all. He found in the F₂ generation, 9/16 were yellow and round, 3/16 were yellow and wrinkled, 3/16 were green and round and 1/16 green and wrinkled.

We can see the genetic explanation for Mendel's results in Figure 1.

When scientists have developed a theory to explain a phenomenon, they can test it by using it to make predictions and then carrying out an investigation to see if the predictions are correct. If we think that two characteristics are determined by two unlinked genes, we *expect* a 9:3:3:1 ratio in the F₂ generation. However, even with a large sample size, it is unlikely that the *observed* offspring phenotypes will appear in exactly this ratio.

We can use a statistical test to analyse the results and test their statistical similarity. We can see if the observed numbers of offspring of each phenotype are close to the expected numbers (calculated using the ratio 9:3:3:1) or if the observed and expected numbers are very different. If the observed and expected numbers are similar we can assume any differences are due to chance and our prediction is correct. However, if the difference between observed and expected numbers is significant we must assume our prediction, and the model upon which it is based, are wrong.

The **chi** (pronounced 'ki'; symbol χ)-**squared** test tests the *null hypothesis*. The null hypothesis is a useful starting point in examining the results of a scientific investigation. It is based on the assumption that 'there is no (statistically) significant difference between the observed and expected numbers, and any difference is due to chance'. If there is no significant difference between observed and expected then we accept the null hypothesis.

In an investigation like the one shown in Figure 1, 288 plants were grown in the F₂ generation and they produced the following observed phenotypes:

169 yellow and round; 54 green and round;
51 green and round; 14 green and wrinkled.

This appears to be quite close to a 9:3:3:1 ratio. Out of 288 individuals, the expected phenotypes would be:

yellow and round: 9/16 × 288 = 162
green and round: 3/16 × 288 = 54

yellow and wrinkled: 3/16 × 288 = 54
green and wrinkled: 1/16 × 288 = 18

Key definition

The χ^2 test is a statistical test to find out if the difference between observed categorical data (data in categories) and expected data is small enough to be due to chance.

The χ^2 test can be used:
- for categorical data (data in categories)
- where there is a strong biological theory that we can use to predict expected values.

Other criteria must also be met:
- The sample size must be relatively large.
- Only raw counts, not percentages or ratios, can be used.
- There are no zero scores.

The formula for calculating a value of χ^2 is:

$$\chi^2 = \frac{\text{the sum of (observed numbers (O) – expected numbers (E))}^2}{\text{expected numbers (E)}}$$

This is usually written as:

$$\chi^2 = \sum \frac{(O - E)^2}{E}$$

This looks complex but it has a simple basis. Three factors are taken into account:

- The differences may be positive or negative so they are squared to stop any negative values cancelling out the positive values.
- Dividing by E takes into account the size of the numbers.
- The sum sign (Σ) takes into account the number of comparisons being made.

It is easier to calculate the value of χ^2 using a table.

- The bigger the value of χ^2 the more certain we are that there is a significant difference between observed and expected values. Or, conversely, the smaller the value of χ^2 the more certain we are that the difference between observed and expected is due to chance and is not a significant difference.
- We look up the calculated value of χ^2 in a distribution table showing values of χ^2. The critical value of χ^2 with three degrees of freedom (see below) and for a p (probability) value of 0.05 is 7.82. (The probability value of 0.05 means this is the value which could occur by chance just 1 time in 20, or 5 in 100).
- If our calculated value of χ^2 is smaller than that critical value then we say that the difference is due to chance and is not statistically significant. (If it is greater than the critical value, then the difference is probably not due to chance and we have to think again about our genetic explanation for the results.)

In this example there are four categories and three degrees of freedom (number of categories –1). Using the χ^2 table we see that our calculated value of χ^2 is smaller than the critical value. Therefore the difference between our observed and expected results is not significant. Further to this we can see our calculated value of χ^2 falls between 1.21 and 2.37 on the row for 3 degrees of freedom. These values correspond to between a 0.5 (50%) and 0.75 (75%) probability that the difference between observed and expected phenotypes is due to chance alone. This is more than a 5% chance, so we accept the null hypothesis, that is, the results are consistent with our genetic explanation.

Class/category	Observed (O)	Expected (E)	O – E	(O – E)²	$\frac{(O - E)^2}{E}$
Yellow round	169	162	7	49	0.30
Green round	54	54	0	0	0.00
Yellow wrinkled	51	54	–3	9	0.17
Green wrinkled	14	18	–4	16	0.88
					$\chi^2 = 1.35$

Number of classes	Degrees of freedom	χ^2							
2	1	0.00	0.10	0.45	1.32	2.71	3.84	5.41	6.64
3	2	0.02	0.58	1.39	2.77	4.61	5.99	7.82	9.21
4	3	0.12	1.21	2.37	4.11	6.25	7.82	9.84	11.34
5	4	0.30	1.92	3.36	5.39	7.78	9.49	11.67	13.28
6	5	0.55	2.67	4.35	6.63	9.24	11.07	13.39	15.09
Probability that deviation is due to chance alone		0.99 (99%)	0.75 (75%)	0.50 (50%)	0.25 (25%)	0.10 (10%)	0.05 (5%)	0.02 (2%)	0.01 (1%)

Accept null hypothesis (any difference is due to chance and not significant)

Reject null hypothesis; accept experimental hypothesis (difference is significant, not due to chance)

Critical value of χ^2 0.05 p level; this is the level at which we are 95% certain the result is not due to chance, agreed on by statisticians as a cut-off point

Figure 2 Part of a χ^2 table

Examiner tip

You will always be given the formula for χ^2. When you have worked out its value, compare it with the critical value – the one on the row of the correct degrees of freedom (number of categories –1) and at the $p = 0.05$ level.

Question

1 In mice one of the genes for coat colour has alleles Y/y. **Y** produces yellow fur and is dominant to **y** for grey fur. When two heterozygous yellow-furred mice are crossed repeatedly, they produce many litters. 140 pups are yellow and 68 are grey.

(a) What is the genotype of grey mice?

(b) What ratio of phenotypes would you expect if this is a normal Mendelian monohybrid cross?

(c) Carry out a χ^2 test to find out if the difference between observed and expected data is significant.

(d) Suggest an explanation for the results.

(e) To find out the genotype of an individual showing a dominant characteristic we do a test cross. The individual is crossed with one with the recessive characteristic. If a yellow mouse, genotype **Yy**, is crossed with a grey mouse, what ratio of phenotypes would you expect in the progeny?

By the end of this spread, you should be able to ...

* Describe the differences between continuous and discontinuous variation.
* Explain the basis of continuous and discontinuous variation with reference to the number of genes that influence the variation.
* Explain that both genotype and environment contribute to phenotypic variation.
* Explain why variation is essential in selection.

The rediscovery of Mendel's work

Mendel's use of maths to analyse his observed data was an unusual approach. His conclusions from such analyses did not fit well with the existing theories about the causes of variation among organisms. Charles Darwin and Alfred Russel Wallace believed that offspring were a blend of their parents' phenotypes. They said that the variation was of a continuous nature. Mendel theorised that the variation was due to discrete or particulate units and was therefore discontinuous.

In 1879 Walter Flemming discovered chromosomes. He observed them in the nuclei of cells from a salamander and described how they behaved during cell division. This discovery supported Mendel's theory of discrete units of inheritance. In the early twentieth century, Hugo de Vries reached conclusions about plant genetics, and then read Mendel's work and realised that Mendel had already reached those conclusions 50 years earlier. Carl Correns and Erich Tschermak also independently reached the same conclusions as Mendel.

In 1902, Sutton and Boveri published papers linking their observations of chromosome behaviour at meiosis with Mendel's principles of independent assortment and segregation. This initiated the chromosomal theory of inheritance, developed over the following 20 years.

Figure 1 Maize cobs

Discontinuous variation

This describes *qualitative* differences between phenotypes. Qualitative differences fall into clearly distinguishable categories. There are no intermediate categories – you are either male or female; you have blood group O, A, B or AB; pea plants may be tall or dwarf; vegetable squash shape is spherical, disc-shaped or elongated.

Continuous variation

This describes *quantitative* differences between phenotypes. These are phenotypic differences where there is a wide range of variation within the population, with no distinct categories. Examples are height and mass in humans, cob length in maize varieties, grain colour in wheat, seed length in broad beans, milk yield in cattle and egg size in poultry.

The genetic basis of discontinuous and continuous variation

Both types of variation may be the result of more than one gene. However, in discontinuous variation, if there is more than one gene involved, the genes interact in an epistatic way (see spread 2.1.13) where one gene masks or influences the expression of another gene. In many examples of discontinuous variation there may only be one gene involved. Such examples are described as *monogenic*. The condition cystic fibrosis, for example, occurs when a person has two faulty alleles of the *CFTR* gene. This gene codes for the cystic fibrosis transmembrane regulatory protein – a chloride ion channel in the membranes of epithelial cells in lungs, gut and reproductive tracts.

In discontinuous (qualitative) variation:
- different alleles at a single gene locus have large effects on the phenotype
- different gene loci have quite different effects on the phenotype
- examples include codominance, dominance and recessive patterns of inheritance.

In continuous (quantitative) variation:
- traits exhibiting continuous variation are controlled by two or more genes
- each gene provides an additive component to the phenotype
- different alleles at each gene locus have a small effect on the phenotype
- a large number of different genes may have a combined effect on the phenotype. These are known as **polygenes** and the characteristic they control is described as **polygenic**. The genes are unlinked – they are on different chromosomes.

At first, scientists disagreed as to whether Mendelian principles of inheritance could be applied to polygenic characteristics. After much research by various scientists, it was discovered that the many genes involved can each contribute to the phenotype. There may be many genes involved in controlling length of corn cobs. The alleles of each gene may add 1 cm (for recessive allele) or 2 cm (for dominant allele) to the cob length. Hence, the effect of each gene is additive. If three genes, A/a, B/b and C/c, each contribute 2 or 1 cm to the length of an ear of grain, then individual plants with the genotype **AABBCC** will have an ear length of 12 cm. Individuals with genotype **aabbcc** will have an ear length of 6 cm.

Genotype and environment contribute to the phenotype

Although a plant with genotype **AABBCC** has the genetic potential to produce ears of grain of length 12 cm, some plants may not produce such long ears. This could be because they are short of water, light or minerals such as nitrates and phosphates . Any of these environmental factors may limit the expression of the genes.

In humans, intelligence is partly determined by genes and partly by environment. Children inherit many genes, with alleles from each parent, giving a genetic potential. However, that potential is only realised with the help of a stimulating learning environment both at home and at school. It is also aided by good nutrition for growth and development of organs, including the brain and nervous system.

The expression of polygenic traits is influenced more by the environment than is the expression of monogenic traits.

Variation and selection

Whether the environment (as in natural selection) or humans (as in artificial selection) are doing the selecting, genetic (inheritable) variety within the population is necessary.

When the environment changes, those individuals that are well adapted will survive and reproduce, passing on their advantageous alleles to their offspring. This is the basis for evolution by natural selection. Over the last 10 000 years, since settled agriculture began, humans have selected plants and animals with desired characteristics, and bred from them. Many of the traits required in agriculture are polygenic and so the study of quantitative genetics is particularly useful to crop plant and farm animal breeders.

Examiner tip

No calculations of heritability will be required.

Questions

1 In a plant, four genes contribute to stem height. The alleles are A/a, B/b, C/c and D/d. Each contributes 4 or 2 cm to the stem height. What is the genetic potential height of the plants with the following genotypes?

AABBCCDD; aabbccdd; AaBbCcDd; AAbbCcDd; AABBccDD

2 Two plants with the same genotype for seed size are grown, one at 15 °C and one at 20 °C. The one grown at 20 °C has larger seeds. Suggest why this is so.

3 Two maize plants with the same genotype for cob length are grown in different soils, at the same temperature and with the same light intensity. Suggest how the differences in soil types may influence the expression of the genes for cob length.

By the end of this spread, you should be able to ...

* Use the Hardy–Weinberg principle to calculate allele frequencies in populations.

A group of individuals carries a larger number of different alleles than an individual. This gives rise to a pool of genetic diversity that can be measured using the **Hardy–Weinberg equation**. Factors such as migration, selection, genetic drift and mutation can alter the amount of genetic variation within a population.

You have seen in AS biology how Darwin deduced that:
* there is a competitive struggle for survival
* there is variation between individuals
* those best adapted are more likely to survive and breed.

The birth of population genetics

Although Mendel was familiar with Darwin's work, Darwin had not read Mendel's and did not know of a mechanism to produce the variation between individuals in a population. However, as biologists developed the concept of genes and alleles, they began to understand the genetic basis of inherited variation. As they studied evolution they realised that populations, rather than individuals, are the functional units in this process. Scientists realised that they needed to consider the frequency of alleles in the population and not just offspring from individual matings.

Early in the twentieth century, many scientists, including Hardy and Weinberg, developed the basic principles of **population genetics**.

In population genetics, biologists focus on the genetic structure of populations. They measure changes in alleles and in genotype frequency from generation to generation.

Measurement of allele and genotype frequencies

We observe the phenotype (and not the genotype) of individuals. To measure the frequency of an allele, we need to know:
* the mechanism of inheritance of a particular trait
* how many different alleles of the gene for that trait are in the population.

For traits that show codominance, the frequency of the heterozygous phenotype is the same as the frequency for the heterozygous genotype.
* In the MN blood group, the gene L has two alleles, L^M and L^N.
* Each allele controls the production of a specific antigen on the surface of red blood cells.
* An individual may be (pheno)type M (genotype: L^ML^M, or **MM**), type N (L^NL^N, or **NN**) or type MN (L^ML^N, or **MN**).
* Because these alleles are codominant, we can determine the frequency of the alleles in a population.

However, if one allele (of a gene for a particular trait) is recessive, the heterozygotes show the same phenotype as the homozygous dominant individuals. This means that the frequency of the alleles cannot be directly determined.

Key definitions

A **population** is a group of individuals of the same species that can interbreed. Populations are dynamic – they can expand or contract due to changes in birth or death rates or migration.

The set of genetic information carried by a population is the **gene pool**.

Worked example

In a population of 100 individuals, 36 are **MM**, 48 are **MN** and 16 are **NN**.

The 36 **MM** individuals represent 72 M alleles.

The 48 **MN** individuals hold another 48 M alleles.

So out of a total of 200 alleles (for this trait) in this population, 120 are M.

So the frequency of M in the population is 120/200 = 0.6.

As the frequency of M + N = 1, the frequency of N alleles is 0.4 (1 − 0.6).

The Hardy–Weinberg principle

The British mathematician Godfrey Hardy and the German doctor Wilhelm Weinberg developed a mathematical model to calculate the allele frequencies, in populations, for traits with dominant and recessive alleles. It is one of the fundamental concepts in population genetics.

It makes the following assumptions:

- The population is very large (this eliminates sampling error).
- The mating within the population is random.
- There is no selective advantage for any genotype.
- There is no mutation, migration or genetic drift.

Worked example: cystic fibrosis

Heterozygotes (**CFcf**) are symptomless carriers. Those affected with cystic fibrosis have the genotype **cfcf**.

In a population of 2000, one person suffers from cystic fibrosis. We wish to know how many of the population are carriers.

p represents the frequency of the dominant allele, **CF**
q represents the frequency of the recessive allele, **cf**
q^2 is the frequency of the genotype **cFcf**
p^2 is the frequency of the genotype **CFCF**
$2pq$ is the frequency of the genotype **CFcf** assuming random mating within the population, where any two individuals of **CFcf** (pq) mate with each other, the resulting genotypes of the offspring could be **CFCF**, **CFcf**, **CFcf**, **cfcf**; these translate as $p^2 + 2pq + q^2$

Within the population the frequency of the alleles, $p + q$, adds up to 1, or 100%

Within the population, the frequency of the genotypes, $p^2 + 2pq + q^2$, adds up to 1, or 100%

We know that q^2 is 1 in 2000, so q^2 is 0.0005
So q is the square root of 0.0005, which is 0.022
If $p + q = 1$, then $p = 1 - 0.022$, which is 0.978
The frequency of carriers is given by $2pq$
So $2pq = 2 \times 0.978 \times 0.022$
So $2pq = 0.043$

This means that 4.3 people in 100 are carriers.

To find out how many are carriers in our population of 2000:

$2000 \times 4.3/100 = 86$

Examiner tip

Remember that individuals have genomes, whereas populations have gene pools.

Questions

1 What factors can alter the amount of genetic variation within a population?
2 Explain why the Hardy–Weinberg principle does not need to be used to calculate the frequency of codominant alleles.
3 The incidence of the disease cystic fibrosis within the population has been reduced by genetic screening during early pregnancy, followed, in some cases, by termination of the pregnancy.
   (a) Has the frequency of the cfcf alleles in the population also dropped?
   (b) What are the implications for using the Hardy–Weinberg principle, today, to calculate allele frequencies within populations?

The roles of genes and environment in evolution

By the end of this spread, you should be able to ...

* Explain, with examples, how environmental factors can act as stabilising or evolutionary forces of natural selection.
* Explain how genetic drift can cause large changes in small populations.
* Explain the role of isolating mechanisms in the evolution of new species, with reference to ecological (geographic), seasonal (temporal) and reproductive mechanisms.

Environmental factors can act as stabilising or evolutionary forces of natural selection

All organisms can reproduce and therefore have the potential to increase their population size. Many populations reach their carrying capacity (the maximum size that the environment can sustain) and then remain stable. Therefore, not all the young produced survive to adulthood. If they did, and then they also produced young, the population would continue to expand.

The environmental factors that limit the growth of a population may include space (for plants to grow, or for animals to defend a feeding territory and to rear young), availability of food, light, minerals or water, predation and infection by pathogens. These factors offer **environmental resistance**. Some are abiotic (caused by non-living components of the environment) and some are biotic (caused by other living organisms).

Over a period of time, the population size will fluctuate around a mean level. If environmental resistance is great enough, the population size will shrink. This reduces competition and the population will grow. As it increases, there will be more intraspecific (within the population) competition for resources such as food, shelter and mates, so the population size falls again.

What determines which individuals in the population will survive?

Because of variation within a population, some members will be better adapted and be able to out-compete other members. The predator that can run faster, has sharper teeth or claws, or is camouflaged and can surprise its prey, will have an advantage in the struggle for existence. It will have a greater chance of surviving to adulthood and producing young, many of which will carry the alleles giving the same advantageous characteristics.

Selection pressures

Prey animals, such as rabbits, that are well camouflaged will be more likely to escape predation and live to reproductive age. Here, predation is the **selection pressure**. It increases the chances of alleles for an agouti coat being passed on to the young. It reduces the chances of alleles for a white or black coat being passed on to the next generation. Here, natural selection (the environment selects the individuals that will survive and reproduce) keeps things the way they are. This is **stabilising selection**. If a new phenotype does arise it is unlikely to confer an advantage and will not be selected.

However, if the environment changes, the selection pressure changes – for instance should the climate change and the ground be covered in snow, then those rabbits with white fur would now have a selective advantage. They would be more likely to survive and breed to pass on the alleles for white fur. The frequency of these alleles in the gene pool would change, as would the frequency of individuals with white coats within the population. This is **directional selection** and leads to evolutionary change. It is an **evolutionary force** of natural selection.

Key definition

Selection pressure – an environmental factor that confers greater chances of survival to reproductive age on some members of the population.

Figure 1 The slender-snouted crocodile, *Crocodylus cataphractus*, has been around for 200 million years – natural selection has kept it the same

What prevents a population from freely interbreeding

A large population of organisms may be split into sub-groups by various **isolating mechanisms**:

- geographic (ecological) barriers, such as a river or mountain range
- seasonal (temporal) barriers such as climate change throughout a year
- reproductive mechanisms (members may no longer be able to physically mate – their genitals may be incompatible or their breeding seasons or courtship behaviours may vary).

This leaves two sub-populations, isolated from each other. In each case different alleles will be eliminated or increased within each sub-population. Eventually the sub-populations will not be able to interbreed and will be different species.

How does genetic drift lead to large changes in small populations?

Let's look at an extreme example. If a population is formed from just one set of heterozygous parents (**Aa** × **Aa**) and they can produce only two offspring, the allele frequency can drastically change. The probability of the occurrence of allele frequencies and genotypes in the subsequent generation (the F_2) is shown in Table 1. In the parental generation, the frequency of both alleles, **A** and **a**, is 0.5.

Figure 2 The red campion, *Silene dioica*, grows in shade and mainly flowers in late spring; the white campion, *Silene latifolia*, prefers sun and mainly flowers in early summer. They are both ecologically and temporally separated. However, a few individuals of both species continue to flower through the summer. In habitats with a mix of light and shade, especially hedgerows, they do interbreed to form a pale pink intermediate (above)

Possible genotypes of two offspring	Probability	Allele frequency	
		A	a
AA and AA	(1/4)(1/4) = 1/16	1.00	0.00
Aa and aa	2(1/4)(1/4) = 2/16	0.25	0.75
AA and aa	2(1/4)(1/4) = 2/16	0.50	0.50
Aa and Aa	(2/4)(2/4) = 4/16	0.50	0.50
AA and Aa	2(2/4)(1/4) = 4/16	0.75	0.25
aa and aa	(1/4)(2/4) = 1/16	0.00	1.00

Table 1 All possible pairs of offspring produced by two heterozygous parents

You can see from the table that in 10 out of 16 possible genotypes from the F_2 generation (offspring of matings from the first offspring) the new allele frequencies are drastically different from those of the original parents. In two cases, either the **A** or **a** allele would be eliminated from the population in a single generation.

This is an extreme example. At the other end of the spectrum is a large population, with small changes in allele frequency. As the size of populations decreases, the degree of fluctuation will increase. Such fluctuations or changes in allele frequency are called **genetic drift**. In extreme cases, genetic drift may lead to the chance elimination of one allele from the population. It reduces genetic variation and may reduce the ability of a population to survive in a new environment. It could contribute to the extinction of a population or species or it could lead to the production of a new species.

How do small populations occur in nature?

A natural disaster, such as a large volcanic eruption or a disease pandemic, may cause a population bottleneck (shrinkage to a small size). In 1775, in the Pingelap atoll in the western Pacific ocean, a storm and a famine reduced the population to 30 people. Today's 2000 inhabitants are all descended from those 30 survivors. About 5% of them have a form of eye defect (achromatopsia) caused by a recessive allele. This disorder is extremely rare in other human populations. Pedigree studies show that one of the original survivors was a chief who was heterozygous for this condition. If he was the only carrier in the original population of 30, the allele frequency was 1/60 or 0.016. The 5% who suffer this disorder are recessive homozygotes and the frequency of the allele in the present population is calculated (using the Hardy–Weinberg equation) as 0.23.

Examiner tip

When answering questions about selection, always refer to increasing the chances of favourable *alleles* (don't say 'genes') being passed on to offspring.

Questions

1 List three biotic factors and three abiotic factors that limit population growth.

2 Explain why inbreeding amongst some animals, such as dogs and race horses, may be harmful.

3 Explain why marriages between relatives may not be desirable.

4 Explain why, in small populations where marriage with outsiders is discouraged, there may be high incidences of genetic disorders. (For example in the Amish community in the USA, there are higher than normal frequencies of achondroplasia (short limbs and trunk) and hexadactyly (having 6 digits).)

By the end of this spread, you should be able to...

✱ Explain the significance of the various concepts of the species, with reference to the biological species concept and the phylogenetic (cladistic/evolutionary) species concept.

What is a species?

The concept of a species may be regarded as the starting point in the classification of living organisms.

The biological species concept

One definition of a species is 'a group of similar organisms that can interbreed and produce fertile offspring and is reproductively isolated from other such groups'. This is the **biological species concept**. This concept is problematic when biologists want to classify living organisms that do not reproduce sexually. Also some members of the same species may look very different from each other. In some species the males look very different from the females and some species, such as ants, have castes which look very different from one another. Some isolated populations may appear to be very different from each other.

The phylogenetic species concept

All living organisms have DNA, RNA and proteins. Closely related organisms have similar molecular structures for these substances. With improved methods of DNA sequencing, biologists have used systematic molecular analysis to compare particular base sequences (haplotypes) on chromosomes of specific organisms. Analysis of the base pair sequences is carried out and the differences, caused by base substitutions, are expressed as % divergence.

$$\% \text{ divergence} = \frac{\text{number of substitutions}}{\text{number of base pairs analysed}} \times 100$$

Any group of organisms with haplotypes that are more similar to each other than to those in any other group is called a **clade**. Hence the use of molecular systematics (analysis) is a **cladistic approach** to classification. It assumes that classification of living organisms corresponds to their **phylogenetic** descent and that all valid **taxa** (groups) must be **monophyletic**.

A clade is a taxonomic group comprising a single ancestral organism and all its descendants. For this reason it is described as a **monophyletic** group.

Cladistics is the hierarchical classification of species, based on their evolutionary ancestry. It is different from taxonomic classification systems as:

- it focuses on evolution (phylogenetic relationships), rather than on similarities between species
- it places great importance on using objective and quantitative (molecular) analysis
- it uses DNA and RNA sequencing
- it uses computer programmes and the data obtained from nucleic acid sequencing to generate dendrograms or cladograms that represent the evolutionary tree of life, because manual creation of such diagrams would be extremely difficult when dealing with large numbers of species
- it makes no distinction between extinct and extant species and both may be included in cladograms.

Key definition

The **biological species concept** is a group of similar organisms that can interbreed and produce fertile offspring.

Key definitions

The **phylogenetic species concept** is 'a group of organisms that have similar morphology (shape), physiology (biochemistry), embryology (stages of development) and behaviour, and occupy the same ecological niche'.

A **monophyletic** group is one that includes an ancestral organism and all its descendent species.

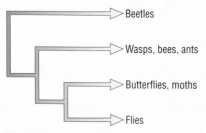

Beetles

Wasps, bees, ants

Butterflies, moths

Flies

Figure 1 A cladogram showing the evolutionary relationship between seven insect groups

The German entomologist (studier of insects) Willi Hennig called it phylogenetic systematics. The term 'cladistics' comes from an ancient Greek word *Klados*, meaning branch.

It is different from Linnaean classification because it does not use the groups kingdom, phylum or class as it regards the evolutionary tree as very complex. As such, it is not helpful to use a fixed number of levels in the classification of living organisms.

However, the Linnaean system of classification also reflects the phylogenies (evolutionary relationships) between the different species of organisms. But, unlike cladistics, it shows both monophyletic and **paraphyletic** groups as taxa.

The cladistic approach has often confirmed the Linnaean classification of organisms but has sometimes led to organisms being reclassified. It has helped biologists to understand the evolutionary relationships between species.

Key definition

A **paraphyletic** group includes the most recent ancestor but not all its descendants. It is a monophyletic group with one or more clades excluded. For example, the grouping of reptiles is paraphyletic as it excludes birds, which are descendants of reptiles. The group Prokaryotes is paraphyletic as it comprises bacteria (eubacteria) and Archaea (Archaebacteria) but excludes the eukaryotes.

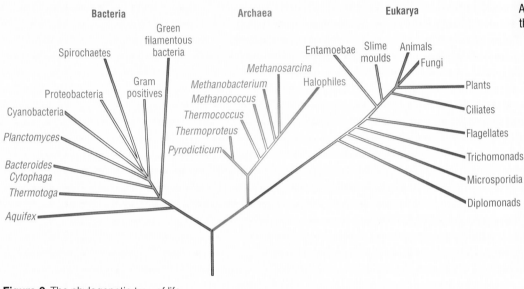

Figure 2 The phylogenetic tree of life

Question

1 (a) Suggest why the phylogenetic species concept should be used to classify bacteria.
   (b) Explain which type of classification system should be used to classify a newly discovered fossil organism.

By the end of this spread, you should be able to ...

✳ **Compare and contrast natural selection and artificial selection.**
✳ **Describe how artificial selection has been used to produce the modern dairy cow and to produce bread wheat, *Triticum aestivum*.**

Natural and artificial selection

Natural selection

We have seen that natural selection (spread 2.1.19) is a mechanism for evolution. Those organisms best adapted to their environments are more likely to survive to reproductive age and to pass on the favourable characteristics (via alleles) to their offspring. In this case, the environment (nature) is doing the selecting.

Artificial selection

Charles Darwin first used this term.

About 10 000 years ago, in the area now called the Near and Middle East, humans (who had previously roamed and been hunter-gatherers) began to practise settled agriculture. Animals such as goats, sheep and dogs were domesticated for meat, milk, wool, leather and protection. Grasses such as wild wheat were cultivated for grain. Since then, they and other animals and plants have been artificially selected. In artificial selection:

- humans select the organisms with useful characteristics
- humans allow those with useful characteristics to breed and prevent the ones without the characteristics from breeding
- thus, humans have a significant effect upon the evolution of these populations or species.

However, the genetic processes underlying both types of selection are the same.

Figure 1 (a) Modern Chillingham White cattle – the ancestors of domestic cattle are thought to have looked like this; **(b)** A Holstein–Friesian dairy cow – note the large udder

How artificial selection has produced the modern dairy cow

Cattle have been domesticated for several thousand years. Humans selected animals for docility, meat and milk production, and to survive in the environment. There are now several breeds. Some have thick coats and can live in the Scottish Highlands. Some can survive in arid areas. The main breeds of dairy cattle, with high milk yields, are Holstein–Friesian, Brown Swiss, Guernsey, Ayrshire, Jersey and Milky Shorthorn.

The original wild cattle which were first domesticated are thought to have looked like modern Chillingham White cattle. By repeatedly selecting cows with high milk yields and allowing them to breed over many generations, humans have artificially selected improved breeds with higher milk production. Today, breeders still practise artificial selection.

- Each cow's milk yield is measured and recorded.
- The progeny of bulls is tested to find which bulls have produced daughters with high milk yields.
- Only a few good-quality bulls need be kept as the semen from one bull can be collected and used to artificially inseminate many cows.
- Some elite cows are given hormones so they produce many eggs.
- The eggs are fertilised *in vitro* (in glass) and the embryos are implanted into surrogate mothers.
- These embryos could also be cloned and divided into many more identical embryos.

In this way a few elite cows can produce more offspring than they would naturally.

How artificial selection has produced bread wheat, *Triticum aestivum*

The genus *Triticum* includes wild and domestic species of wheat. The genus *Aegilops*, or wild goat grass, has contributed its genome to the modern bread wheat. Most wild species of wheat are diploid with 14 chromosomes; n = 7. Grasses, like many other domesticated plants, are able to undergo **polyploidy** – their nuclei can contain more than one diploid set of chromosomes. Modern bread wheat is hexaploid, 6n, having 42 chromosomes in the nucleus of each cell. Because the nuclei need to be larger to contain the extra chromosomes their cells are also bigger. Genetic analysis of modern species of domesticated wheat has shown that it is a hybrid containing three distinct genomes, A^UA^UBBDD (see Figure 2).

- The genome A^UA^U has come from a wild wheat species, such as *T. urartu*.
- The genome **BB** has come from wild emmer wheat, *T. turgidum* (a tetraploid, 4n, species thought to be a hybrid of *T. uratu* and a wild goat grass like *Ae. speltoides*).
- The genome **DD** has come from a wild goat grass such as *Ae. tauschii* or *Ae. squarrosa*.

Using the Linnaean system of classification, all wheats that can interbreed are classified as being in the same species. However, more recently a genetic classification has been used. Both classifications are valid but should not be confused.

Winter wheat can be grown in the UK as the climate is not extreme. This wheat produces soft grains with low protein content suitable for biscuits but not for bread. Spring wheat is grown where winters are colder (Europe and North America). The grains are harder and have a higher protein content, making them suitable for bread.

Artificial selection to produce improved varieties of *Triticum* wheat is still going on

Wheat is a very important crop. It can grow in large areas of the world and makes up 33% of all cereal crops. Breeders continue to carry out selection programmes to produce improved varieties. Characteristics that they focus on include:

- resistance to fungal infections
- high protein content
- straw (stem) stiffness
- resistance to lodging (stems bending over in wind and rain)
- increased yield.

Each year, in the UK, the Camden and Chorleywood Food Research Association (C&CFRA), formerly The Flour Milling and Baking Research Association, surveys the wheat varieties grown in the UK and classifies them according to their suitability for making bread or biscuits or for use as animal feed.

Questions

1 Make a table to compare natural selection and artificial selection. (Include similarities as well as differences.)
2 In the example above, has the biological species concept or the cladistic species concept been used to classify *T. aestivum* and einkorn wheat?
3 The protein in wheat grains has a specific quality. When hydrated and kneaded it forms a sticky, rubbery mass that holds bubbles of gas produced from yeast, giving the characteristic light texture of bread. The only other cereal grain that produces a protein able to do this is rye.
  (a) Describe the nature and origin of the gas produced by yeast.
  (b) Explain why wheat flour used for bread-making needs to have a higher protein content than wheat flour used for making biscuits.

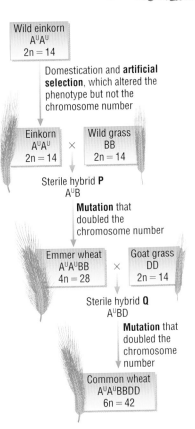

Figure 2 How artificial selection has produced modern bread wheat from wild ancestors. The letters A^U, B and D denote sets of chromosomes (genomes)

Figure 3 (a) Einkorn wheat – a diploid species, still grown in some parts of the world for animal feed; **(b)** modern bread wheat, *Triticum sp.*, a hexaploid species

145

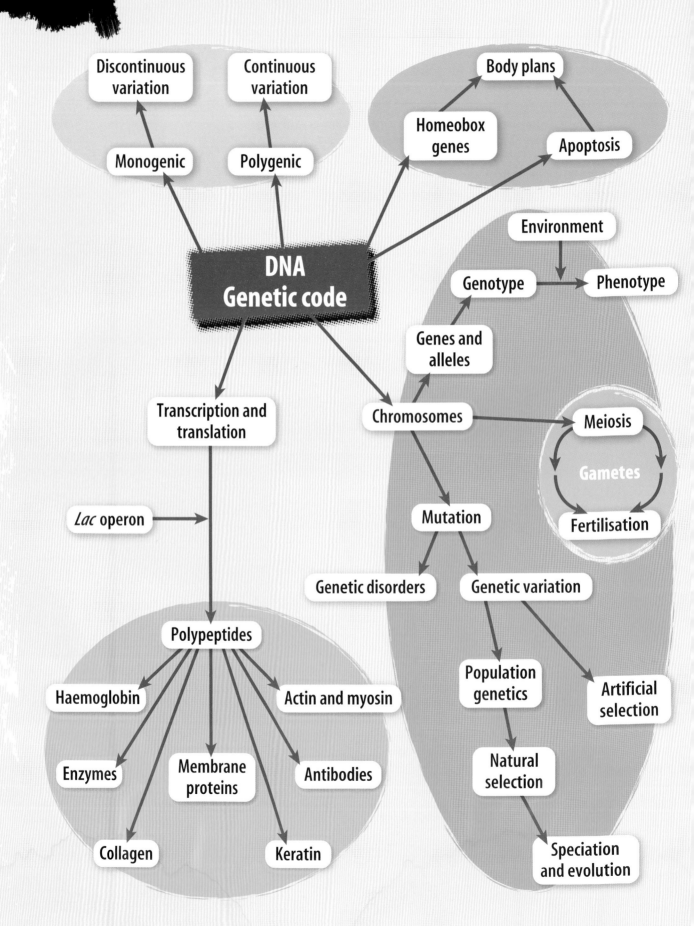

Practice questions

(1) What is meant by the *genetic code*? [3]

(2) **(a)** Why is the genetic code described as degenerate? [1]

(b) Explain the significance of it being a degenerate code. [2]

(3) Copy and complete the table below. [6]

	DNA replication	DNA transcription
Enzyme		
Does DNA unwind?		
Amount of DNA involved		
When in the cell cycle does it occur?		
Where in the cell does it occur?		
What are the products?		

(4) Explain how meiosis can lead to genetic variation. [4]

(5) Explain how genetic variation is essential for the processes of natural and artificial selection. [5]

(6) Explain the differences between continuous and discontinuous variation. [4]

(7) Suggest how artificial selection has been used to produce modern race horses. [5]

(8) Fragile X syndrome is a genetic condition, characterised by severe learning difficulties and narrow face with large forehead, jaw and ears. It affects about 1 in 1250 males and 1 in 2000 females and all ethnic groups. It is the most common inherited form of mental retardation in humans. The condition is X-linked. It is due to an expanding triple nucleotide repeat, in a gene that encodes an RNA-binding protein. The triplet varies normally from 6 to 52 repeats. If the repeat expands to more than 230 (known as a full mutation), the 5′ end of the gene becomes methylated (CH$_3$ groups added) and transcription stops. The gene in question is towards the end of the long arm of the X chromosome and makes the end of the chromosome kinked and fragile.

Most female carriers have between 52 and 230 repeats of this nucleotide triplet. This situation is known as a premutation. During meiosis the triplet expands more. The number of increases is greater when the parent carrier is female. The risk of expansion from premutation to full mutation (of 230+ repeats) increases as the premutation triplet increases from 52–70 to over 90. No full mutations arise from normal length repeats, only from premutations. Affected daughters are the daughters of premutation mothers and not premutation fathers. The penetrance (appearance of the symptoms) increases as it is transmitted through the generations of an affected family.

The codon triplet also expands during mitosis, and affected individuals have a mosaic body tissue pattern.

(a) What is meant by saying that the gene is X-linked? [1]

(b) What will be the immediate effect on transcription if the gene is shut off? [1]

(c) Explain what a premutation is. [2]

(d) Suggest what *mosaic body tissue pattern* means. [1]

(e) Explain why the penetrance increases down the generations. [4]

(9) Male-pattern baldness is caused by a gene on one of the autosomes (chromosomes other than X or Y). It is an autosomal dominant trait. It is transmitted from male to male. It is expressed more in males than in females.

(a) Explain the term *autosomal dominant trait*. [3]

(b) How does the fact that this trait (condition) is transmitted from father to sons support the hypothesis that the gene is not on the X chromosome? [1]

(c) Suggest why this gene is expressed more in males than in females. [2]

(10) In many regions of the world, such as African and Pacific Rim countries, most adult humans are lactose-intolerant. They lack the enzyme lactase and cannot digest lactose, the sugar found in milk and other dairy products, such as cheese and chocolate. These foods give them serious intestinal discomfort, flatulence or vomiting.

The gene coding for lactase enzyme is switched on in baby mammals but switched off in adults.

In human societies where dairy produce has long been an important part of the diet, the frequency of lactose-intolerant individuals in the population is low.

(a) Suggest how natural selection has led to a higher frequency of lactose-tolerant individuals in European countries. [3]

(b) Lactose-intolerant individuals can eat yoghurt. Explain why this is so. [2]

(c) Suggest why adult cats should not be given milk. [2]

(d) There is special milk for adult cats available. What treatment has this milk undergone? [2]

1 **(a)** The following are different stages in meiosis. Each stage has been given a letter.

anaphase II   metaphase II   anaphase I
M   **N**   **P**
prophase I   telophase II   metaphase I
Q   **R**   **S**

(i) Using **only** the letters, arrange these stages in the correct sequence. [1]

(ii) State the letter of the stage when each of the following processes occur.
pairing of chromosomes
centromeres divide
crossing over
bivalents align on equator
nuclear membrane reforms [5]

(iii) State **two** processes that occur in a cell during interphase to prepare for a meiotic division. [2]

(b) Haemophilia A is a sex-linked genetic disease which results in the blood failing to clot properly. It is caused by a recessive allele on the X chromosome. Figure 1.1 shows the occurrence of haemophilia in one family.

☐ = Male   ◯ = Female   ■ = Male haemophiliac

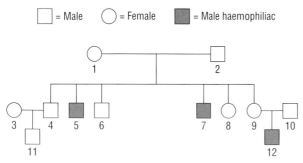

Figure 1.1

(i) Using the following symbols:

H = dominant allele **h** = recessive allele

state the genotypes of the following individuals. The first one has been completed for you.
1                              $X^H X^h$
2; 5; 6; 9. [4]

(ii) State the probability of individual 8 being a carrier of haemophilia. [1]

(iii) Explain why only females can be carriers of haemophilia. [2]

[Total: 15]
(OCR 2804 Jun06)

2 The tiger, *Panthera tigris*, is the largest and most distinctive cat in the world.

(a) Complete the following table to show the classification of the tiger.

kingdom	
	chordata
	mammalia
order	carnivora
family	felidae
genus	
	tigris

[5]

Tigers are further classified into a number of sub-species (races) based on marked phenotypic differences, such as body size and colour. Figure 2.1 shows the distribution of the different sub-species 100 years ago and in 2004. The names of the sub-species are shown on the map.

Key:

☐ 100 years ago   ■ 2004

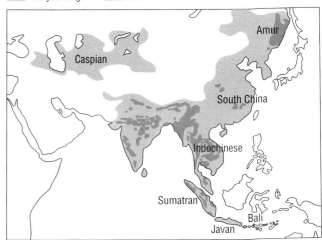

Figure 2.1

(b) **(i)** Describe the changes shown in Figure 2.1. [2]

(ii) Explain how the distinct phenotypic differences between the sub-species may have arisen. [4]

(iii) Suggest why these populations of tigers are classified as different sub-species rather than as different species. [2]

[Total: 13]
(OCR 2804 Jan06)

3 Pollen from a pure-bred tomato plant with white flowers and yellow fruit was transferred to the stigmas of a pure-bred plant with yellow flowers and red fruit. All the F_1 generation had yellow flowers and red fruit.

(a) State **three** practical precautions that must be taken by a plant breeder to ensure that the offspring produced are **only** from the desired cross. [3]

Answers to examination questions will be found on the Exam Café CD.

(b) In a test cross, pollen from the F_1 generation was transferred to pure-bred plants with white flowers and yellow fruit. The ratio of phenotypes **expected** among the offspring of a dihybrid test cross such as this is 1:1:1:1.

Seeds from the test cross were collected and grown, giving plants with the following phenotypes:

yellow flowers and red fruit	87
yellow flowers and yellow fruit	13
white flowers and red fruit	17
white flowers and yellow fruit	83
	$\underline{200}$

A chi-squared (χ^2) test can be carried out to check whether the numbers of each phenotype of offspring resulting from the test cross are in agreement with a 1:1:1:1 ratio.

$$\chi^2 = \sum \frac{(O - E)^2}{E}$$

$\sum$ = 'sum of ...'

O = observed value; E = expected value

Table 3.1

Phenotypes	Yellow flowers red fruit	Yellow flowers yellow fruit	White flowers red fruit	White flowers yellow fruit
Observed number (O)	87	13	17	83
Expected ratio	1	1	1	1
Expected number (E)	50	50	50	50
$O–E$	37			33
$(O–E)^2$	1369			1089
$(O–E)^2/E$	27.38			21.78
$\sum (O–E)^2/E = \chi^2$				

(i) Complete the boxes in Table 3.1 to calculate χ^2 for these results. [3]

(ii) State the number of degrees of freedom applicable to these results. [1]

(iii) Use the calculated value of χ^2 and the table of probabilities provided in Table 3.2 to find the probability of the results of the test cross differing by chance from the expected ratio. [1]

(iv) State what statistical conclusion may be drawn from the probability found in **(b) (iii)** about the difference between expected and actual results. [2]

(c) Explain the discrepancy between the expected and the actual results of the test cross. [5]

Table 3.2

Distribution of χ^2					
Degrees of freedom	Probability, p				
	0.10	0.05	0.02	0.01	0.001
1	2.71	3.84	5.41	6.64	10.83
2	4.61	5.99	7.82	9.21	13.82
3	6.25	7.82	9.84	11.35	16.27
4	7.78	9.49	11.67	13.28	18.47

[Total: 15]

(OCR 2805/02 Jun03)

4 The bacterium *Escherichia coli* can use either glucose or lactose as a respiratory substrate. When *E. coli* is grown in a medium containing lactose, but no glucose, the genes coding for the enzymes required to utilise lactose are switched on. These genes are located together in the *lac* operon as shown in Figure 4.1.

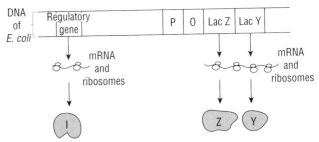

Figure 4.1

(a) Complete the table below stating the functions of the parts of the *lac* operon.

Part of the *lac* operon	Function
Regulatory gene	Controls production of repressor protein
O – operator	
P – promoter	
I – repressor molecule	
Z – beta galactosidase	
Y – lactose permease	

[5]

(b) Explain why beta galactosidase and lactose permease are not produced when lactose is absent. [3]

(c) Describe the events that occur within *E. coli* when lactose is the only respiratory substrate available. [5]

[Total: 13]

(OCR 2804 Jan04 Q5)

Module 2
Biotechnology and gene technologies

Introduction

Biotechnology and gene technology are advancing rapidly. They have, and will continue to have, far-reaching consequences for our lives, our health and our understanding of the world around us. All of us use the products of biotechnology – in our food manufacture and washing powders, for example – and many of us are healthy thanks to biotechnology, such as insulin production and gene therapy.

In this unit you will learn how most plants naturally possess the ability to reproduce asexually. This ability has been exploited in the production of genetically uniform crops. Animals, too, can be cloned; Dolly the sheep was big news.

Developments of the Human Genome Project and the sequencing of the genomes of other species have led to a new field in biology, that of genomics. You will learn how genomes are sequenced and what information is being gained from comparing the DNA sequences of different organisms. This, and other progress in gene technology, has led to the harnessing of many more natural biological processes in order to make products that are useful to humans. You will also learn how genetic engineering is carried out and some of the current and potential human benefits of gene technologies, for example the engineering of human insulin and Golden Rice™, and gene therapy.

The implications of genetic techniques are subject to much public debate. We will give some consideration to the ethical dimension of gene technology.

Test yourself

1 Biotechnology includes processes that humans have been using for centuries. Name some of these processes.
2 What is the disease associated with a lack of insulin in humans?
3 Before human insulin was available through biotechnological processes, where did the insulin used in treatments come from?
4 Why do some plants reproduce asexually?
5 What are the biotechnological ingredients of washing powders?
6 What have you heard about the debate on gene technology? Do you have a personal opinion on this?
7 Why would you expect to be able to gain useful scientific information by comparing the genomes of different organisms? What have all living organisms got in common that allows such comparisons?

Module contents

By the end of this spread, you should be able to . . .

✳ Describe the production of natural clones in plants using the example of vegetative propagation in elm trees.

What exactly are clones?

By definition, a **clone** is an exact copy, but the term is used in biology to describe genes, cells or whole organisms that carry identical genetic material because they are derived from the same original DNA.

Identical twins are produced when a zygote splits into two. These twins are natural clones. When plants reproduce asexually by producing runners, the new plants are clones. When bacteria (single-celled organisms) divide asexually by binary fission, all the resulting bacteria are clones of the original bacterium. In all of these processes identical copies of the 'original' DNA generate new organisms with the same *cloned* DNA.

It is important to draw a distinction between cloned *genes*, cloned *cells* and cloned *organisms*. It is also essential to remember that the production of cloned DNA, cells and organisms is a natural process for growth and reproduction that can also be achieved by artificial means.

The advantages and disadvantages of asexual reproduction

Prokaryotes divide by binary fission. Their DNA replicates and the cell divides into two. Provided there are no mutations, the two resulting cells are genetically identical to each other and to the parent cell.

The basis of asexual reproduction in nearly all eukaryotes is mitosis. The genetic material replicates and separates to form two new nuclei, each containing an exact copy of the original DNA. In single-celled eukaryotic organisms, the cell splits to produce two daughter cells that are cloned offspring.

In multicellular organisms, particularly plants, some of the cells produced by mitosis can grow into new, separate organisms with DNA that is identical to the parent plant, so they are clones of the parent plant.

The advantages of asexual reproduction are that:
- it is quick, allowing organisms to reproduce rapidly and so take advantage of resources in the environment
- it can also be completed if sexual reproduction fails or is not possible
- all offspring have the genetic information to enable them to survive in their environment.

The disadvantage of asexual reproduction is that:
- it does not produce any genetic variety, so any genetic parental weakness will be in all the offspring. If the environment changes, for example with the introduction of a new disease-causing organism, then all genetically identical organisms will be equally susceptible.

Natural vegetative propagation in plants

Asexual reproduction in plants takes place naturally in a variety of different ways. For example, a number of plant species, including the English elm (*Ulmus procera*), are adapted to reproduce asexually following damage to the parent plant. This allows the species to survive catastrophes such as disease or burning. New growth in the form of root suckers, or **basal sprouts**, appears within 2 months of the destruction of the main trunk. These suckers grow from meristem tissue in the trunk close to the ground, where least damage is likely to have occurred.

Examiner tip

Mitosis is the basis of asexual reproduction in eukaryotes. Binary fission occurs in prokaryotes. So the replication of DNA is a vital first step in the process. The only *genetic* differences that can occur between asexually reproduced organisms are those resulting from mutations.

Figure 1 A row of English elms in a hedgerow. The smaller ones have grown from root suckers sent out by the larger ones

Key definition

Vegetative propagation refers to the production of structures in an organism that can grow into new individual organisms. These offspring contain the same genetic information as the parent and so are clones of the parent.

Module 2
Biotechnology and gene technologies
Clones in nature

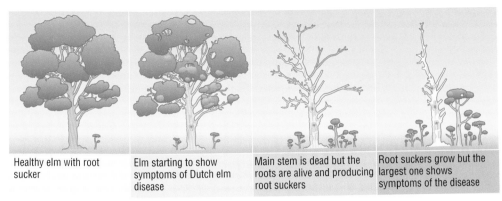

Healthy elm with root sucker

Elm starting to show symptoms of Dutch elm disease

Main stem is dead but the roots are alive and producing root suckers

Root suckers grow but the largest one shows symptoms of the disease

Figure 2 Vegetative propagation of elms by root suckers helps them survive damage and spread

Advantages and disadvantages to the elm of vegetative propagation

Root suckers help the elm spread, because they can grow all around the original trunk. When the tree is stressed or the trunk dies – for example when the tree is felled as part of the coppice cycle – the suckers grow into a circle of new elms called a **clonal patch**. This, in turn, puts out new suckers so that the patch keeps expanding as far as resources permit.

In the twentieth century, Dutch elm disease spread through Europe's elms: the leaves withered, followed by death of the branches and trunks, as a result of a fungal disease carried by a beetle. The English elm responds to the destruction of the main trunk by growing root suckers, as described above. However, once the new trees get to about 10 cm in diameter, they become infected and die in turn. Because the new trunks are clones of the old one, they do not have any resistance to the fungal attack so they remain just as vulnerable as the original tree. There is no genetic variation within the cloned population, so natural selection cannot occur.

Clonal patch of elm in a woodland with other trees competing. The root suckers spread out but the other trees shade them out

The trees are cleared (felled for timber)

Trees start to regrow but the elm root suckers have a head start; in this way the clonal patch spreads

Figure 3 Rapid repopulation of woodland by elm suckers after felling

---

STRETCH and CHALLENGE

Root suckers are not the only type of vegetative propagation. Look on your plate at your next meal and you might find others – see Table 1.

Question

A What are the advantages and disadvantages of each method to the plant and to humans?

Vegetative propagation type	Examples
Basal sprouts (**root suckers**) form from meristem tissue close to ground	English elm, raspberry
Specialised underground stems become swollen with nutrient molecules, forming **tubers**, from which new plants grow	Potato
Condensed shoots with very short stems and fleshy leaf bases (**bulbs**) form, containing nutrients; buds at the sides develop into new bulbs	Onion, daffodil
Specialised stems (**runners**) grow along the ground from the parent plant; at the tips they form roots and shoots	Strawberry

Table 1 Examples of natural vegetative propagation

Questions

1 Explain why asexual reproduction is quicker than sexual reproduction.

2 What advantages are there for elm trees of the capacity to reproduce *both* sexually and asexually?

3 Explain why asexual reproduction in plants is described as vegetative propagation.

4 Hybrid varieties of elm that are resistant to Dutch elm disease have been cultivated. Rooted cuttings of several varieties are grown. When they are 7 years old, they are inoculated with spores of the fungi, *Ophiostoma novo-ulmi* and *O. ulmi*, that cause Dutch elm disease. Suggest how scientists at horticultural research stations then produce large numbers of elms that are resistant to Dutch elm disease.

By the end of this spread, you should be able to . . .

✳ Describe the production of artificial clones of plants from tissue culture.
✳ Discuss the advantages and disadvantages of plant cloning in agriculture.

Artificial vegetative propagation

We saw in the previous spread how many plants can produce vegetative structures which can then form new and separate individuals. For many years, farmers and growers have also been able to artificially propagate valuable plants. The two main methods are:

* Taking cuttings – a section of stem is cut between leaf joints (nodes). The cut end of the stem is then often treated with plant hormones to encourage root growth, and planted. The cutting forms a new plant which is a clone of the original parent plant. Large numbers of plants, such as geraniums, can be produced quickly this way.
* Grafting – a shoot section of a woody plant (often of a fruit tree or a rosebush) is joined to an already growing root and stem (known as a rootstock). The graft grows and is genetically identical to the parent plant, but the rootstock is genetically different.

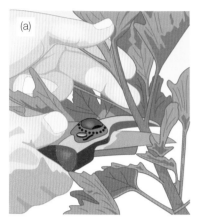

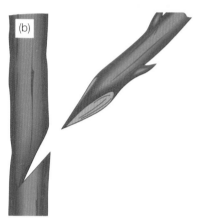

Figure 1 (a) Taking cuttings: the cut stem is trimmed of lower leaves, dipped in hormone rooting formula, placed in compost and covered in clear plastic to reduce water loss; **(b)** Side grafting, one of several methods of grafting: the rootstock is cut to match the wedge-shaped stem to be grafted (scion). The vascular tissue is lined up then bindings are wrapped around the graft area to hold it in place until growth supports the grafted section

Figure 2 Britain's favourite cooking apple is the Bramley. Apple seeds have a genetic mix and will not breed true, so every Bramley apple tree in the world is a genetically identical graft. The first graft was taken in 1856 from a tree in Nottinghamshire. The tree in this picture still has a visible graft scar at its base

Artificial propagation using tissue culture: large-scale cloning

Although useful, cuttings and grafts cannot produce huge numbers of cloned plants very easily. Also some plants do not reproduce well from either cuttings or grafts. More modern methods of artificial propagation use plant **tissue culture** in order to generate huge numbers of genetically identical plants from a very small amount of plant material.

Tissue culture can be used to generate large stocks of a particularly valuable plant very quickly, with the added advantage that these stocks are known to be disease-free.

Micropropagation by callus tissue culture

The most common method used in the large-scale cloning of plants is micropropagation. Many houseplants are produced in this way, notably orchids.

* A small piece of tissue is taken from the plant to be cloned, usually from the shoot tip. This is called an **explant**.
* The explant is placed on a nutrient growth medium.

Key definition

Tissue culture refers to the separation of cells of any tissue type and their growth in or on a nutrient medium. In plants, the undifferentiated callus tissue is grown in nutrient medium containing plant hormones that stimulate development of the complete plant.

Module 2
Biotechnology and gene technologies
Artificial clones and agriculture

- Cells in the tissue divide, but they do not differentiate. Instead they form a mass of undifferentiated cells called a **callus**.
- After a few weeks, single callus cells can be removed from the mass and placed on a growing medium containing plant hormones that encourage shoot growth.
- After a further few weeks, the growing shoots are transferred onto a different growing medium containing different hormone concentrations that encourage root growth.
- The growing plants are then transferred to a greenhouse to be acclimatised and grown further before they are planted outside.

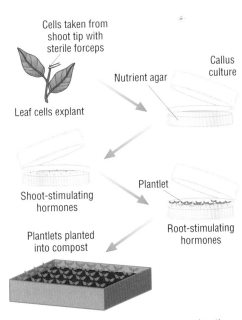

Cells taken from shoot tip with sterile forceps

Leaf cells explant

Nutrient agar

Callus culture

Shoot-stimulating hormones

Plantlet

Root-stimulating hormones

Plantlets planted into compost

Figure 3 Micropropagation of plants using tissue culture

Figure 4 Many orchids are produced by micropropagation for sale as houseplants

The advantages and disadvantages of plant cloning in agriculture

Over thousands of years agriculture has sought to provide high-quality crops in terms of yield and resistance to environmental conditions such as drought, pests or weeds. Selective breeding over generations has resulted in crop plants having reduced genetic variation. This is because farmers have identified and grown only those crops with useful features. Some crops such as fruit trees cannot be grown from seed because the new tree will have a combination of genes that will not give the correct fruit. Bananas have to be grown by cloning because all cultivated bananas are sterile.

Propagation using callus culture means:
- farmers know what the crop plant produced will be like because it is cloned from plants with known features such as high yield, taste, colour and disease-resistance
- farmers' costs are reduced because all the crop is ready for harvest at the same time.

This is essentially a 'refinement' of selective breeding.

Micropropagation is much faster than selective breeding, because huge numbers of genetically identical plants can be generated from a small number of plants or a single valuable plant.

The disadvantages of using cloned plants in agriculture are the same as those described for asexually reproducing organisms. Genetic uniformity means that all plants are equally susceptible to any new pest, disease or environmental change. The most striking example of problems associated with genetic monoculture was seen in Ireland between 1845 and 1851 in the potato famines. Around one million people died of starvation and disease as much of the potato crop across the country was lost due to infection by the fungus-like protoctist, *Phytophthora infestans*. Although the potato crop at that time was not grown using plant culture, the plants used were genetically uniform and so were all susceptible to the disease.

Farming methods are now more regulated and, although genetically uniform crops are grown, the areas given to a specific crop and the distance between areas of the same crop are controlled in order to limit the effects of the arrival of new pathogens.

Examiner tip

Cloning plants in this way is *not* genetic engineering; it is a method of generating many genetically identical individuals. Whilst it is possible at the callus stage to introduce genes, and so genetically engineer the plant, you must consider the engineering part as a separate process.

Questions

1 Why is it essential that meristem cells are present in the plant tissues used for any artificial vegetative reproduction method?

2 Suggest an advantage of placing a graft from a tree that produces a particularly desirable fruit onto a rootstock of a different tree species.

3 Explain how limiting the area given over to a particular crop, and the distance between areas of the same crop, help in limiting the effects of a newly arrived pathogenic organism.

By the end of this spread, you should be able to . . .

* Describe how artificial clones of animals can be produced.
* Discuss the advantages and disadvantages of cloning animals.
* Outline the differences between reproductive and non-reproductive cloning.

Examiner tip

You must remember that clones are described as *genetically* identical. You will lose marks if you simply say they are identical. Questions on cloning can easily be linked to those on evolution and natural selection. You should make links in your answers on genetic information between the different parts of the specification that look at how life processes are influenced by the genes of organisms.

Key definition

A **cloned** animal is one that has been produced using the same genetic information as another animal. Such an animal has the same genotype as the donor organism.

Artificial cloning in animals – two possible ways

In animals, only embryonic cells are naturally capable of going through the stages of development in order to generate a new individual. As you learned in AS biology, we describe these cells as being **totipotent stem cells**, i.e. capable of differentiating into any type of adult cell found in the organism. These cells are able to 'switch on' any of the genes present on the genome.

There are two methods of artificially cloning animals:

Splitting embryos – 'artificial identical twins'

Cells from a developing embryo can be separated out, with each one then going on to produce a separate, genetically identical organism. This method was developed in 1979 and has been used to clone sheep, cattle, rabbits and toads. In 2000, the first primate – a rhesus monkey called Tetra – was cloned in this way.

Nuclear transfer – using enucleated eggs

A differentiated cell from an adult can be taken, and its nucleus placed in an egg cell which has had its own nucleus removed. Such a cell is described as enucleated. The egg then goes through the stages of development using genetic information from the inserted nucleus. The first animal cloned by this method was the sheep, named Dolly in honour of Dolly Parton, in 1996. The cell was taken from the mammary gland of a 6-year-old ewe (female sheep), its nucleus transplanted into a cell from a second sheep and then inserted into the uterus of a third sheep, and then a fourth, to develop. This was the only success from 277 attempts.

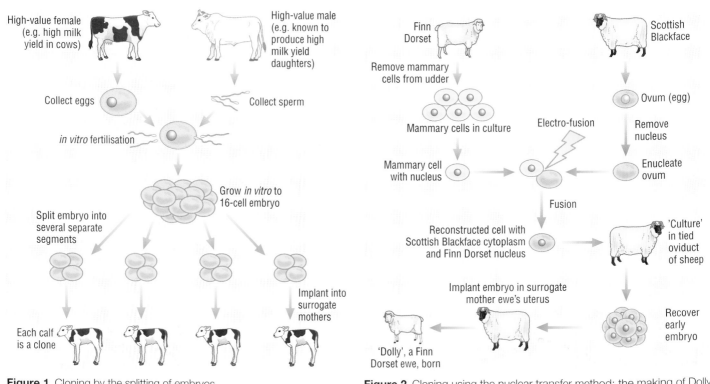

Figure 1 Cloning by the splitting of embryos

Figure 2 Cloning using the nuclear transfer method; the making of Dolly

Advantages	Disadvantages
High-value animals, for example cows giving high milk yield, can be cloned in large numbers.	High-value animals are not necessarily produced with animal welfare in mind. Some strains of meat-producing chickens have been developed that are unable to walk.
Rare animals can be cloned to preserve the species.	As with plants – excessive genetic uniformity in a species makes it unlikely to be able to cope with, or adapt to, changes in the environment.
Genetically modified animals – for example sheep that produce pharmaceutical chemicals in their milk – can be quickly reproduced.	It is still unclear whether animals cloned using the nuclear material of adult cells will remain healthy in the long term. Dolly the sheep was put down at age 6 in 2003, suffering from a form of lung cancer caused by a virus. The post mortem revealed nothing unusual for an animal of her age and weight, although at the time her death was wrongly reported as being due to premature ageing caused by cloning.

Table 1 Advantages and disadvantages of cloning animals

What is non-reproductive cloning?

The cloning of plants and animals as described so far has been done to generate new organisms. One of the most significant potential developments in cloning is the possibility of using cloned cells to generate cells, tissues and organs to replace those damaged by diseases or accidents. There are numerous advantages of using cloned cells:

- Being genetically identical to the individual's own cells means that they will not be 'rejected' because the immune system will not recognise them as foreign.
- Cloning and cell culture techniques could mean an end to the current problems of waiting for donor organs to become available for transplant.
- Cloned cells can be used to generate any cell type because they are totipotent. Damage caused by some diseases and accidents cannot currently be repaired by transplantation or other treatments.
- Using cloned cells is likely to be less dangerous than a major operation such as a heart transplant.

There are many possibilities for non-reproductive cloning. These include:

- the regeneration of heart muscle cells following a heart attack
- the repair of nervous tissue destroyed by diseases such as multiple sclerosis
- repairing the spinal cord of those paralysed by an accident that resulted in a broken back or neck.

The techniques are often referred to as *therapeutic cloning*. It is important to note that some people object to its use in humans. There are ethical objections to the use of human embryonic material and some scientific concerns about a lack of understanding of how cloned cells will behave over time.

A third way – turning the clock back on differentiated cells

In 2008, researchers in Kyoto (Japan) and Wisconsin and California (USA) reported that they had successfully reprogrammed human skin cells to become pluripotent, almost identical to embryonic stem cells. The researchers identified four essential regulator genes in this process and have called the cells *induced pluripotent stem cells (iPS cells)*.

If such cells can be safely developed from any individual's own skin cells and used to generate the 200+ different cell types found in a human, the technique could replace the more controversial nuclear transfer method used by scientists working on therapeutic cloning.

STRETCH and CHALLENGE

Transferring a nucleus from an adult cell into an egg cell carries most but not all of the DNA from the donor individual. Some DNA is found in mitochondria and this DNA codes for around 60 different polypeptides. So a cloned individual has nuclear DNA from the donor and mitochondrial DNA from the organism that provided the enucleated egg.

Question

A In normal sexual reproduction, which parent's mitochondrial DNA is found in the offspring?

B Suggest the functions of polypeptides produced from mitochondrial genes.

Questions

1 What are the differences between cloning by the splitting of embryos and cloning by nuclear transfer?

2 Give reasons why repairing a damaged heart using cloned cells could potentially be less dangerous than receiving a heart transplant.

By the end of this spread, you should be able to . . .

✳ **State that biotechnology is the industrial use of living organisms (or parts of them) to produce food, drugs or other products.**

✳ **Explain why microorganisms are often used in biotechnological processes.**

What is biotechnology?

Key definition

Biotechnology is technology based on biology and involves the exploitation of living organisms or biological processes, to improve agriculture, animal husbandry, food science, medicine and industry.

The term **biotechnology** was first used in 1919 by Karl Ereky, a Hungarian agricultural engineer. In its widest sense, it refers to all technological processes that make use of living organisms or parts of living organisms in order to manufacture useful products or provide useful services for human exploitation. In Ereky's terms, this would include the mechanisation of farming and the selective breeding of plants and animals over generations.

Ancient biotechnology methods include yoghurt-making, cheese-making, baking and brewing, which have been carried out for thousands of years.

In 1917, Chaim Weizmann first used a pure culture of the bacterium *Clostridium acetobutylicum* to produce acetone, needed to make explosives during World War I.

Modern biotechnology is characterised by recombinant DNA technology, and the US Supreme Court ruling in 1980 that a genetically modified *Pseudomonas* bacterium, developed to digest crude oil in oil spills, could be patented was a landmark.

In recent years, our growing understanding of genetics and ability to manipulate living organisms in a variety of ways has led to a huge expansion in biotechnological processes. Biotechnology has applications in four major areas that affect our lives:
* healthcare and medical processes – this includes the production of drugs by microorganisms and gene therapy to treat some genetic disorders
* agriculture – this includes micropropagation of plants and the development of genetically modified plants
* industry – this includes genetically modifying organisms to produce enzymes
* food science – this includes developing foods with improved nutrition or better taste, texture and appearance.

There is also blue biotechnology applied to marine and aquatic environments.

Figure 1 Alexander Fleming, a British scientist who, by serendipity, discovered the antibiotic penicillin in 1928. The Australian Howard Florey and the German refugee Ernst Chain developed a method for producing it and this led to huge strides forward in pharmaceutical biotechnology. The three were awarded the Nobel Prize for medicine in 1945.

The genetic engineering of crops to be herbicide-resistant, plus other manipulations of organisms, use biotechnological techniques, as we shall see later. We shall focus here on biotechnological processes that have a commercial and large-scale application.

Table 1 Some biotechnological processes and the organisms involved

Purpose of biotechnological process	Examples	Organisms involved
The production of foods	Cheese and yoghurt-making	Bacterial (*Lactobacillus*) growth in milk changes the flavour and texture of the milk to generate a different food. These bacteria prevent the growth of other bacteria that would cause spoilage, and so preserve the food.
	Mycoprotein (Quorn meat alternative)	Growth of a specific fungus (*Fusarium*) in culture. The fungal mycelium produced is separated and processed as food.
	Naturally brewed soya sauce	Roasted soya beans are fermented with yeast or fungi such as *Aspergillus*.
The production of drugs or other pharmaceutical chemicals	Penicillin, an antibiotic medicine	The fungus *Penicillium* grown in culture produces the antibiotic as a by-product of its metabolism.
	Insulin, a hormone used by diabetic patients whose own production of hormone is not sufficient	Bacteria (*E. coli*) are genetically modified to carry the human insulin gene. Organisms secrete the insulin protein as they grow.
The production of enzymes or other chemicals for commercial use	Pectinase, used in fruit juice extraction	The fungus *A. niger* grown in certain conditions produces and secretes pectinase enzyme.
	Calcium citrate (used in detergents)	The fungus *A. niger* produces citric acid as a by-product of its normal metabolism.
	Bio-gas fuel production	Methanogenic bacteria, grown on concentrated sewage, respire anaerobically and generate gases that can be used as fuel.
The bioremediation of waste products	Waste water treatment	A variety of bacteria and fungi use organic waste in the water as nutrients and make the waste harmless; for example *Fusarium* grown on corn steep liquor, a waste product of the corn milling industry.

The use of microorganisms in biotechnology

Many biotechnological processes make use of microorganisms. The use of both bacteria and fungi in various processes is widespread because microorganisms:

- grow rapidly in favourable conditions, with a generation time (time taken for numbers to double) of as little as 30 minutes
- often produce proteins or chemicals that are given out into the surrounding medium and can be harvested
- can be genetically engineered to produce specific products
- grow well at relatively low temperatures, much lower than those required in chemical engineering of similar processes
- can be grown anywhere in the world and are not dependent on climate
- tend to generate products that are in a more pure form than those generated via chemical processes
- can often be grown using nutrient materials that would be otherwise useless or even toxic to humans.

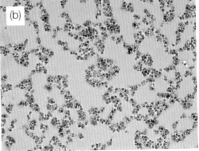

Figure 2 (a) Researchers operating a reactor for converting crop waste into biofuels; **(b)** Yeast cells (*Saccharomyces cerevisiae*) are commonly used in biotechnological processes including the brewing of wine and beer (×500)

Examiner tip

The specification requires you to be able to explain why microorganisms are used widely in biotechnological processes. You should state what it is about the microorganisms *and why* that feature is important as part of a biotechnological process. Do not make vague statements about them being 'cheap' or 'easy to grow'.

Questions

1 Explain why microorganisms are particularly useful in waste water treatments.

2 Suggest the advantage to an organism, such as *Penicillium*, of naturally producing an antibiotic chemical as a by-product of its metabolism.

3 **(a)** Find out what the following branches of biotechnology are: **(i)** pharmacogenomics and **(ii)** nutrigenomics.

(b) How do these link to the Human Genome Project?

By the end of this spread, you should be able to ...

* Describe and explain, with the aid of diagrams, the standard growth curve of a population of microorganisms in a closed culture.
* Describe the differences between primary and secondary metabolites.

The standard growth curve

A small number of organisms placed in a fresh 'closed **culture**' environment will undergo population growth in a very predictable, standard way. Plotting the growth in population over time gives the standard growth curve shown in Figure 1. A closed culture refers to the growth of microorganisms in an environment where all conditions are fixed and contained. No new materials are added and no waste products or organisms removed.

Key definition

A **culture** is a growth of microorganisms. This may be a single species (which would be called a pure culture) or a mixture of species (called a mixed culture). Microorganisms can be cultured in a liquid such as nutrient broth, or on a solid surface such as nutrient agar gel.

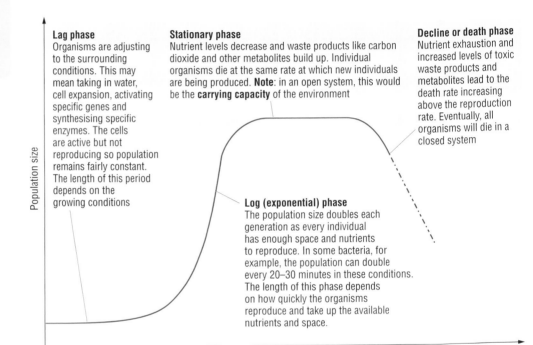

Lag phase
Organisms are adjusting to the surrounding conditions. This may mean taking in water, cell expansion, activating specific genes and synthesising specific enzymes. The cells are active but not reproducing so population remains fairly constant. The length of this period depends on the growing conditions

Stationary phase
Nutrient levels decrease and waste products like carbon dioxide and other metabolites build up. Individual organisms die at the same rate at which new individuals are being produced. **Note**: in an open system, this would be the **carrying capacity** of the environment

Decline or death phase
Nutrient exhaustion and increased levels of toxic waste products and metabolites lead to the death rate increasing above the reproduction rate. Eventually, all organisms will die in a closed system

Log (exponential) phase
The population size doubles each generation as every individual has enough space and nutrients to reproduce. In some bacteria, for example, the population can double every 20–30 minutes in these conditions. The length of this phase depends on how quickly the organisms reproduce and take up the available nutrients and space.

Figure 1 The standard growth curve for a culture of microorganisms

Figure 2 The tanks used in the brewing industry are known as fermenters or fermentation tanks

Fermentation and fermenters

The term fermentation was originally applied only to the use of anaerobic respiration to produce substances, in particular, the production of ethanol (alcohol) through anaerobic respiration of yeast. These naturally produced fermentation products are by-products of anaerobic respiration pathways, as we saw in spread 1.4.8.

Fermentation now also refers to the culturing of microorganisms both aerobically and anaerobically in fermentation tanks. The substances generated by growth of the microorganism culture are separated and treated to produce the final useful product.

Module 2
Biotechnology and gene technologies
The growth curve

Metabolism is a process, metabolites are the products

As you learned in AS biology, metabolism refers to the sum total of all of the chemical reactions that go on in an organism. These processes produce:

- new cells and cellular components
- chemicals such as hormones and enzymes
- waste products. The waste products produced vary depending on the type of organism and metabolic process involved, ranging from gases such as carbon dioxide and oxygen to soluble molecules like urea, ammonia and nitrates.

The waste products of some organisms' metabolic processes are the vital nutrients required by other organisms.

Primary and secondary metabolites

The terms **primary** and **secondary metabolite** are often used when referring to the metabolic processes of microorganisms.

- **Primary metabolites** – are substances produced by an organism as part of its normal growth; they include amino acids, proteins, enzymes, nucleic acids, ethanol and lactate. The production of primary metabolites matches the growth in population of the organism.
- **Secondary metabolites** – are substances produced by an organism that are not part of its normal growth. The antibiotic chemicals produced by a number of microorganisms are almost all secondary metabolites. The production of secondary metabolites usually begins after the main growth period of the organisms and so does not match the growth in population of the organism.

It is important to remember that whilst all microorganisms produce primary metabolites (they need to in order to grow) only a relatively small number of microorganisms produce secondary metabolites.

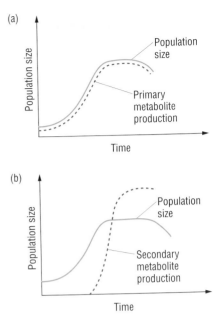

Figure 3 Growth and product curves showing the production of **(a)** a primary metabolite and **(b)** a secondary metabolite

Examiner tip

Examiners often ask questions relating to the standard growth curve. You should not only be able to describe the curve but also to explain what is happening at each stage. You should also be able to explain how changes, such as adding nutrients or removing waste products, would affect the shape of the curve.

STRETCH and CHALLENGE

The standard growth curve above shows how the population of a species of microorganism increases over time in a closed culture. In fact, any type of organism can be placed in a new environment. If a small number of organisms, of a particular species, are placed in a particular location, then provided they can grow and reproduce in that location, the population growth will follow the same phases and be subject to similar constraints.

Consider for example the arrival of rabbits or cane toads in Australia, the arrival of Rhododendron plants into gardens of Great Britain and their subsequent escape into the surrounding countryside, or the arrival of a single finch species onto a Galapagos Island.

Questions

A What factors would govern whether a newly arrived species could survive and reproduce in a specific location?

B What impact might the arrival of a new species have on existing species in that location?

C Would the population eventually go into decline as shown by the standard growth curve? Explain your answer.

D Do some research to find out why cane toads were introduced into Australia.

E Explain the term 'biological control'.

Questions

1 Describe how adding nutrients to a growing culture of microorganisms might affect the population growth if they were added in:
   (a) the lag phase
   (b) the stationary phase.

2 Explain why the production of primary metabolites in a microorganism matches the overall growth of the microorganism population.

3 Suggest why secondary metabolites such as antibiotics are only produced after the main growth phase of a microorganism.

(6) Commercial applications of biotechnology

By the end of this spread, you should be able to . . .

* Compare and contrast the processes of continuous and batch culture.
* Explain the importance of manipulating the growing conditions in a fermentation vessel in order to maximise the yield of product required.
* Explain the importance of asepsis in the manipulation of microorganisms.

Industrial-scale fermenters and 'scaling up'

Commercial applications of biotechnology often require the growth of a particular microorganism on an enormous scale. An industrial-scale fermenter is essentially a huge tank which may have a capacity of tens of thousands of litres. The growing conditions in it can be manipulated and controlled in order to ensure the best possible yield of the product.

The precise growing conditions depend on the microorganisms being cultured, and on whether the process is designed to produce a primary or secondary metabolite. They are:

* temperature – too hot and enzymes will be denatured; too cool and growth will be slowed
* type and time of addition of nutrient – growth of microorganisms requires a nutrient supply, including sources of carbon, nitrogen and any essential vitamins and minerals. The timing of nutrient addition can be manipulated, depending on whether the process is designed to produce a primary or secondary metabolite
* oxygen concentration – most commercial applications use the growth of organisms under aerobic conditions, so sufficient oxygen must be made available. A lack of oxygen will lead to the unwanted products of anaerobic respiration and a reduction in growth rate
* pH – changes in pH within the fermentation tank can reduce the activity of enzymes and so reduce growth rates.

Such large cultures need large 'starter' populations of the microorganism. These are obtained by taking a pure culture and growing it in sterile nutrient broth.

Examiner tip

There are many commercial applications of biotechnology which use fermentation tanks like that shown in Figure 1. Do not try to learn details of lots of different examples; it is your understanding of the general principles of how growing conditions are manipulated that is required.

Figure 2 'Starter' cultures are often grown in a flask placed on a shaking or 'tilting' base, which helps mix the organisms and nutrients

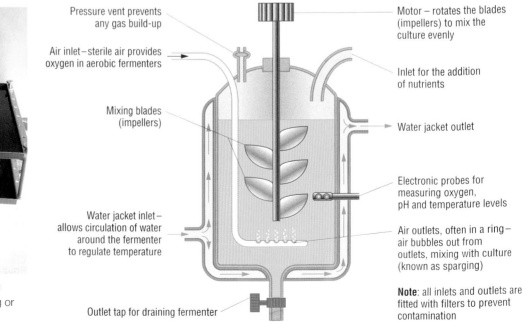

Pressure vent prevents any gas build-up

Air inlet – sterile air provides oxygen in aerobic fermenters

Mixing blades (impellers)

Water jacket inlet – allows circulation of water around the fermenter to regulate temperature

Outlet tap for draining fermenter

Motor – rotates the blades (impellers) to mix the culture evenly

Inlet for the addition of nutrients

Water jacket outlet

Electronic probes for measuring oxygen, pH and temperature levels

Air outlets, often in a ring – air bubbles out from outlets, mixing with culture (known as sparging)

Note: all inlets and outlets are fitted with filters to prevent contamination

Figure 1 Generalised diagram showing the features of a large-scale industrial fermenter

Module 2
Biotechnology and gene technologies
Commercial applications of biotechnology

Batch and continuous culture

Industrial-scale fermentations can be operated in two ways:

- A **batch** culture, where the microorganism starter population is mixed with a specific quantity of nutrient solution, then allowed to grow for a fixed period with no further nutrient added. At the end of the period, the products are removed and the fermentation tank is emptied. Penicillin is produced using batch culture of *Penicillium* fungus.
- A **continuous** culture, where nutrients are added to the fermentation tank and products removed from the fermentation tank at regular intervals – or even, as the name suggests, continuously. Human hormones such as insulin are produced from continuous culture of genetically modified *Escherichia coli* bacteria.

Asepsis is vital in biotechnological processes involving microorganisms

The nutrient medium in which the microorganisms grow could also support the growth of many unwanted microorganisms. Any unwanted microorganism is called a **contaminant**. Unwanted microorganisms:

- compete with the culture microorganisms for nutrients and space
- reduce the yield of useful products from the culture microorganisms
- may cause spoilage of the product
- may produce toxic chemicals
- may destroy the culture microorganism and their products.

In processes where foods or medicinal chemicals are being produced, contamination means that all products must be considered unsafe and so must be discarded.

The term **aseptic technique** refers to the measures taken to ensure **asepsis**, that is, that contamination of the culture does not occur at any point from isolation of the initial culture, through scaling up, fermentation and product harvesting.

Key definitions

Aseptic technique refers to any measure taken at any point in a biotechnological process to ensure that unwanted microorganisms do not contaminate the culture that is being grown or the products that are extracted.

Asepsis is the absence of unwanted microorganisms.

Batch culture	Continuous culture
Growth rate is slower because nutrient level declines with time	Growth rate is higher as nutrients are continuously added to the fermentation tank
Easy to set up and maintain	Set up is more difficult, maintenance of required growing conditions can be difficult to achieve
If contamination occurs, only one batch is lost	If contamination occurs, huge volumes of product may be lost
Less efficient, fermenter is not in operation all of the time	More efficient, fermenter operates continuously
Very useful for processes involving the production of secondary metabolites	Very useful for processes involving the production of primary metabolites

Table 1 Advantages and disadvantages of batch and continuous culture

Aseptic techniques and measures at laboratory and starter culture level	Aseptic techniques and measures at large-scale culture level
• All apparatus for carrying/moving microorganisms is sterilised before and after use, for example by heating in a flame until glowing or by UV light. Some equipment is steam-sterilised at 121 °C for 15 minutes in an autoclave (large pressure cooker). • Work can be carried out in a fume cupboard or a laminar flow cabinet where air circulation carries any airborne contaminants away from the bench space. • Cultures of microorganisms are kept closed where possible and away from the bench surface when open and in use.	• Washing, disinfecting and steam-cleaning the fermenter and associated pipes when not in use removes excess nutrient medium and kills microorganisms. • Fermenter surfaces made of polished stainless steel prevent microbes and medium sticking to surfaces. • Sterilising all nutrient media before adding to the fermenter prevents introduction of contaminants. • Fine filters on inlet and outlet pipes avoid microorganisms entering or leaving the fermentation vessel.

Table 2 Summary of aseptic techniques

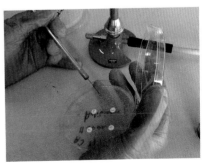

Figure 3 A wire loop is used to transfer small quantities of microorganisms. The loop is sterilised by holding in a Bunsen flame until glowing, then allowed to cool before reuse

Questions

1 Yeast cells can be cultured both aerobically and anaerobically. Explain why the fermentation products differ in these two conditions.

2 Microorganisms that do not require oxygen for growth but grow better in its presence are known as 'facultative anaerobes'. Suggest what is meant by the terms *strict (obligate) anaerobe* and *obligate aerobe*.

3 Explain why a failure in aseptic conditions is more problematic:
  (a) in continuous culture than it is in batch culture
  (b) in large-scale production than in bench level or starter culture.

By the end of this spread, you should be able to ...

* **Describe how enzymes can be immobilised.**
* **Explain why immobilised enzymes are used in large-scale production.**

Enzymes are powerful catalysts

You learned in AS biology how enzymes act as catalysts in metabolic reactions. The features of enzymes that make them so useful in industrial processes are:

* specificity – enzymes can catalyse reactions between specific chemicals, even in mixtures of many different chemicals. This means that fewer by-products are formed and less purification of products is necessary
* temperature of enzyme action – most enzymes function well at relatively low temperatures, much lower than those needed for many industrial chemical processes. This saves a great deal of money on fuel costs. However, enzymes from thermophilic bacteria (bacteria that thrive at high temperatures) have been extracted and used in reactions that need a high temperature.

In the biotechnological processes previously described, whole organisms are cultured on a large scale to generate particular products. In many areas of clinical research and diagnosis and in some industrial processes, the product of a single chemical reaction is required. It is often more efficient to use isolated enzymes to carry out the reaction rather than growing the whole organism or using an inorganic catalyst.

Isolated enzymes can be produced in large quantities in commercial biotechnological processes. The extraction of enzyme from the fermentation mixture is known as **downstream processing**, a term used to describe the processes involved in the separation and purification of any product of large-scale fermentations.

Immobilising enzymes

In order for the product of an enzyme-controlled reaction to be generated, enzyme and substrate must be able to collide and form enzyme–substrate complexes. This is most easily achieved by mixing quantities of substrate and isolated enzyme together under suitable conditions for the enzyme to work. The product generated then needs to be extracted from the mixture. This can be a costly process.

It is possible to **immobilise** enzymes so that they can continue to catalyse the enzyme-controlled reaction but do not mix freely with the substrate as they would normally in a cell or isolated system. There are advantages and disadvantages to the use of immobilised enzymes in large-scale production as shown in Table 1.

Figure 1 Some examples of the industrial and commercial use of enzymes: biological detergents containing protein-digesting enzymes; biofuel production; fruit juice extracted using pectinase; and cheesemaking

Key definition

Immobilisation of enzymes refers to any technique where enzyme molecules are held, separated from the reaction mixture. Substrate molecules can bind to the enzyme molecules and the products formed go back into the reaction mixture leaving the enzyme molecules in place.

Advantages	Disadvantages
Enzymes are not present with products so purifications/downstream processing costs are low.	Immobilisation requires additional time, equipment and materials and so is more expensive to set up.
Enzymes are immediately available for reuse. This is particularly useful in allowing for continuous processes.	Immobilised enzymes can be less active because they do not mix freely with substrate.
Immobilised enzymes are more stable because the immobilising matrix protects the enzyme molecules.	Any contamination is costly to deal with because the whole system would need to be stopped.

Table 1 Advantages and disadvantages of using immobilised enzymes

Methods for immobilising enzymes

There are four possible methods for immobilising enzymes. The precise method used for a particular process depends on a range of factors including ease of preparation, cost, relative importance of enzyme 'leakage' and efficiency of the particular enzyme that is immobilised.

Adsorption and covalent bonding involve binding enzymes to a support, whereas entrapment and membrane separation hold them in place without binding.

Adsorption

Enzyme molecules are mixed with the immobilising support and bind to it due to a combination of hydrophobic interactions and ionic links. Adsorbing agents used include porous carbon, glass beads, clays and resins.

Because the bonding forces are not particularly strong, enzymes can become detached (known as leakage). However, provided the enzyme molecules are held so that their active site is not changed and is displayed, adsorption can give very high reaction rates.

Covalent bonding

Enzyme molecules are covalently bonded to a support, often by covalently linking enzymes together and to an insoluble material (for example clay particles) using a cross-linking agent like gluteraldehyde or sepharose. This method does not immobilise a large quantity of enzyme but binding is very strong so there is very little leakage of enzyme from the support.

Entrapment

Enzymes may be trapped, for example in a gel bead or a network of cellulose fibres. The enzymes are trapped in their natural state (i.e. not bound to another molecule so their active site will not be affected). However, reaction rates can be reduced because substrate molecules need to get through the trapping barrier. This means the active site is less easily available than with adsorbed or covalently bonded enzymes.

Membrane separation

Enzymes may be physically separated from the substrate mixture by a partially permeable membrane. Most simply, the enzyme solution is held at one side of a membrane whilst substrate solution is passed along the other side. Substrate molecules are small enough to pass through the membrane so that the reaction can take place. Product molecules are small enough to pass back through the membrane.

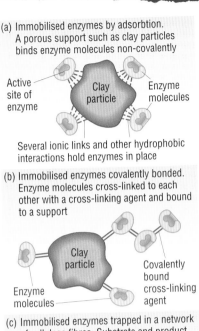

(a) Immobilised enzymes by adsorbtion. A porous support such as clay particles binds enzyme molecules non-covalently

(b) Immobilised enzymes covalently bonded. Enzyme molecules cross-linked to each other with a cross-linking agent and bound to a support

(c) Immobilised enzymes trapped in a network of cellulose fibres. Substrate and product molecules can pass through the cellulose fibres.

Figure 2 Methods of immobilising enzymes

Examiner tip

Remember that the rate of enzyme catalysis is dependent on the shape of the active site and on the access of substrate molecules to the active site. Immobilisation of enzymes can affect both of these features. You will need to be able to explain why.

Producing new antibiotics

There has been an increase in the number of antibiotic-resistant strains of pathogenic bacteria in recent years. Scientific research into developing new antibiotics often focuses on changing the structure of available antibiotics so that the target microorganisms are no longer resistant.

Immobilised penicillin acyclase enzyme reactors have been used to convert the antibiotic penicillin into amino penicillanic acid on a large scale. The amino penicillanic acid is then used as a base molecule to produce a range of different penicillin-type antibiotics.

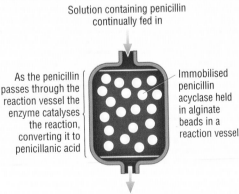

As the penicillin passes through the reaction vessel the enzyme catalyses the reaction, converting it to penicillanic acid

Solution containing penicillin continually fed in

Immobilised penicillin acyclase held in alginate beads in a reaction vessel

Solution containing 6-amino penicillanic acid leaves the reaction vessel

Figure 3 The production of penicillanic acid using an immobilised enzyme reactor

Questions

1 Compare and contrast the different methods of immobilising enzymes.

2 Explain how covalently bonding enzymes to a support could completely remove the enzymes' activity.

3 Suggest and explain the features of bacterial cells that are most likely to be the target of the antibiotic chemicals that would be of most use to human medicine.

By the end of this spread, you should be able to ...

* ✳ Outline the steps involved in sequencing the genome of an organism.
* ✳ Outline how gene sequencing allows for genome-wide comparisons between individuals and species.

Understanding and manipulating DNA

Since the discovery, in 1953, of the structure of DNA there have been a number of advances in science that have used our understanding of the structure and role of DNA in organisms. These include:

* **DNA profiling (genetic fingerprinting)** used in forensic crime scene analysis and paternity and maternity testing
* **genomic sequencing** and **comparative genome mapping** used in research into the function of genes and regulatory DNA sequences
* **genetic engineering** used in the production of pharmaceutical chemicals, genetically modified organisms and xenotransplantation
* **gene therapy** used to treat conditions such as cystic fibrosis.

Gene technology is advancing rapidly, but it is also important to note that many of the techniques involved have their basis in natural processes.

* DNA strands can be cut up into smaller fragments using restriction endonuclease enzymes (see spread 2.2.11).
* The fragments can be separated by size using *electrophoresis* and replicated many times to produce multiple copies using a process called the *polymerase chain reaction* (see spreads 2.2.9 and 2.2.10).
* DNA fragments can be analysed to give their specific base sequence (see spread 2.2.9).
* DNA fragments can be sealed together using ligase enzyme (see spread 2.2.11).
* DNA probes can be used to locate specific sequences on DNA fragments (see spread 2.2.9).

Such techniques mean that sections of DNA, including whole genes, can be identified and manipulated.

The genomic age

The DNA of all organisms contains sections known as **genes** – these code for the production of polypeptides and proteins. However, this **coding DNA** forms only a small part of the DNA found in an organism. In fact, only 1.5% of the genome of humans actually codes directly for polypeptides and proteins. Much DNA is **non-coding DNA** and has been referred to as junk DNA. The use of the term 'junk' is misleading, since this non-coding DNA carries out a number of regulatory functions, many of which are still to be discovered. A great deal of research is going on into trying to work out how genomes work as a whole. **Genomics** – the study of genomes – is seeking to map the whole genome of an increasing number of organisms. Comparing genes and regulatory sequences of different organisms will help us to understand the role of genetic information in a range of areas including health, behaviour and evolutionary relationships between organisms.

Sequencing the genome of an organism – an outline

The sequencing reaction can only operate on a length of DNA of about 750 base pairs. This means that the genome must be broken up and sequenced in sections. In order to ensure that the assembled code is accurate, sequencing is carried out a number of times on overlapping fragments, with the overlapping regions analysed and put back together to form the completed code. The stages involved are as follows:

Key definition

Genomics refers to the study of the whole set of genetic information in the form of the DNA base sequences that occur in the cells of organisms of a particular species. The sequenced genomes of organisms are placed on public access databases.

An example of a chromosome map. The markers include positions of microsatellites and stains which bind to nine nucleotide clusters

The chromosome is sheared into sections of around 100 000 base pairs

Each section is inserted into a BAC and transferred into an *E. coli* cell

Figure 1 Genome/chromosome maps allow identification of the location that the bacterial artificial chromosome (BAC) sample has come from

Module 2
Biotechnology and gene technologies
Studying whole genomes

- Genomes are first **mapped** to identify which part of the genome (i.e. which chromosome or section of chromosome) they have come from. Information that is already known is used – for example using the location of **microsatellites** (short runs of repetitive sequences of 3–4 base pairs found in several thousand locations on the genome).

- Samples of the genome are sheared (mechanically broken) into smaller sections of around 100 000 base pairs. This is sometimes referred to as a 'shotgun' approach.

- These sections are placed into separate **bacterial artificial chromosomes** (**BACs**) and transferred to *E. coli* (bacterial) cells. As the cells grow in culture, many copies (clones) of the sections are produced. These cells are referred to as **clone libraries**.

In order to sequence a BAC section:

1 Cells containing specific BACs are taken and cultured. The DNA is extracted from the cells and restriction enzymes used to cut it into smaller fragments. The use of different restriction enzymes on a number of samples gives different fragment types.

2 The fragments are separated using a process known as electrophoresis.

3 Each fragment is sequenced using an automated process.

4 Computer programmes then compare overlapping regions from the cuts made by different restriction enzymes in order to reassemble the whole BAC segment sequence.

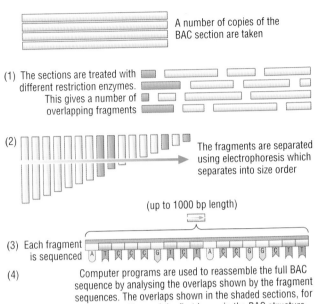

A number of copies of the BAC section are taken

(1) The sections are treated with different restriction enzymes. This gives a number of overlapping fragments

(2) The fragments are separated using electrophoresis which separates into size order

(up to 1000 bp length)

(3) Each fragment is sequenced

(4) Computer programs are used to reassemble the full BAC sequence by analysing the overlaps shown by the fragment sequences. The overlaps shown in the shaded sections, for example, will reveal the first bases in the BAC structure

Figure 2 Sequencing a BAC section using overlapping fragment analysis

Comparing genomes

A wide variety of organism genomes have now been sequenced. Knowing the sequence of bases in a gene of one organism and being able to compare genes for the same (or similar) proteins across a range of organisms is known as *comparative gene mapping*. This has a wide range of applications:

- The identification of genes for proteins found in all or many living organisms gives clues to the relative importance of such genes to life.

- Comparing the DNA/genes of different species shows evolutionary relationships. The more DNA sequences organisms share, the more closely related they are likely to be.

- Modelling the effects of changes to DNA/genes can be carried out. For example, a number of studies have tested the effects of mutations on genes obtained from yeast that are also found in the human genome. Yeast is a haploid organism, so a mutation to a gene is always shown in the phenotype.

- Comparing genomes from pathogenic and similar but non-pathogenic organisms can be used to identify the genes or base-pair sequences that are most important in causing the disease. This can lead to identification of targets for developing more effective drug treatments and vaccines.

- The DNA of individuals can be analysed. This analysis can reveal mutant alleles, or the presence of alleles associated with increased risk of particular diseases, such as heart disease or cancer.

Figure 3 The genomes of many organisms have been sequenced recently. Examples of some undertaken during 2007 are shown here: *Trichomonas vaginalis* (×3800), a protoctist parasite that infects humans; the rhesus macaque (*Macaca mulatta*); *Clostridium botulinum* (×16500), a bacterium that causes severe food poisoning; and the pinot noir vine (*Vitis vinifera*)

Examiner tip

You do not need to learn lots of examples of genomic comparisons. You need to be able to interpret data given about genetic information in order to draw conclusions. For example, if a gene is found in all organisms, then it is likely that the protein it codes for is essential for a fundamental life process.

Questions

1 Suggest why the term 'junk DNA' is no longer appropriate to describe the regions of DNA that do not code for proteins.

2 Why are mutations always seen in the phenotype in a haploid organism such as yeast, but not in diploid organisms such as humans?

3 Suggest an example of a gene that you would expect to find in:
   (a) all living organisms
   (b) mammals only
   (c) plants and other photosynthetic organisms.

By the end of this spread, you should be able to ...

❋ Outline how DNA fragments can be separated by size using electrophoresis.
❋ Describe how DNA probes can be used to identify fragments containing specific sequences.

Electrophoresis separates DNA fragments

(a) Samples are placed in wells cut into the gel at one end using a fine pipette

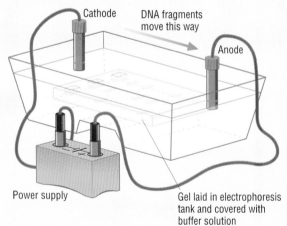

(b) Electrophoresis tank

(c) Electrophoresis gel showing separated DNA fragments, revealed by flooding with a DNA-binding dye.

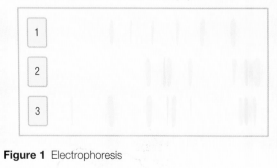

Figure 1 Electrophoresis

Electrophoresis is used to separate DNA fragments based on their size. The process is accurate enough to be able to separate fragments that are different by only one base in length. It is widely used in gene technology to separate DNA fragments for identification and analysis.

The technique uses a gel 'plate' or slab, containing agarose (a type of sugar), which is covered in buffer solution. Electrodes are attached to each end of the gel so that a current can be passed through it. The separation of strands of different lengths occurs because longer strands of DNA get caught up in the agarose gel and are slowed, whereas shorter strands can move more quickly through the gel.

The basic procedure is:
- DNA samples are treated with restriction enzymes to cut them into fragments (see spread 2.2.12).
- The DNA samples are placed into wells cut into one end (negative electrode end) of the gel.
- The gel is immersed in a tank of buffer solution and an electric current is passed through the solution for a fixed period of time, usually around 2 hours.
- DNA is negatively charged because of the many phosphoryl (phosphate) groups. It is attracted to the positive electrode, so the DNA fragments diffuse through the gel towards the positive electrode end.
- Shorter lengths of DNA move faster than longer lengths and so move further in the fixed time that current is passed through the gel.
- The position of the fragments can be shown by using a dye that stains DNA molecules.

The fragments may be lifted from the gel for further analysis. The technique used is called Southern blotting, named after the biologist Edwin Southern, who developed it.
- A nylon or nitrocellulose sheet is placed over the gel, covered in paper towels, pressed and left overnight (blotting).
- The DNA fragments are transferred to the sheet and can now be analysed.

The DNA fragments are not visible on the sheet. There are several methods available for showing up the separated strands. The simplest is to label the DNA with a radioactive marker before the samples are run. Placing photographic film over the nitrocellulose sheet shows the position of DNA samples in the finished gel.

If one particular fragment or sequence of DNA is being searched for, for example a particular gene, then a radioactive DNA probe can be used to check for the presence of that particular sequence. The use of DNA probes to identify specific sequences is described in Figure 2.

Module 2
Biotechnology and gene technologies
DNA manipulation – separating and probing

DNA probes

A DNA probe is a short single-stranded piece of DNA (around 50–80 nucleotides long) that is complementary to a section of the DNA being investigated. The probe is labelled in one of two ways:

- using a radioactive marker (usually by using ^{32}P in the phosphoryl groups forming the strand) so that the location can be revealed by exposure to photographic film
- using a fluorescent marker that emits a colour on exposure to UV light. Fluorescent markers are also used in automated DNA sequencing.

Copies of the probe can be added to any sample of DNA fragments and, because they are single-stranded, they will bind to any fragment where a complementary base sequence is present. This binding by complementary base pairing is known as **annealing**.

Probes are useful in locating specific sequences, for example:

- to locate a specific desired gene that is wanted for genetic engineering
- to identify the same gene on a variety of different genomes, from separate species, when conducting genome comparison studies
- to identify the presence or absence of an allele for a particular genetic disease.

DNA probes and disease diagnosis

Diagnoses of some genetic diseases and identification of symptomless carriers can be made by analysing the patient's DNA using DNA probes. Probes are made that are complementary to sequences found in faulty alleles of particular genes.

Scientists have been able to place a number of different probes on a fixed surface – known as a DNA microarray. Applying the DNA sample to the surface can reveal the presence of faulty or mutated alleles that match the fixed probes because the sample DNA will anneal to any complementary fixed probes.

In order to anneal, the sample DNA must be broken up into smaller fragments. It may also be amplified using PCR so that many copies of each fragment are present.

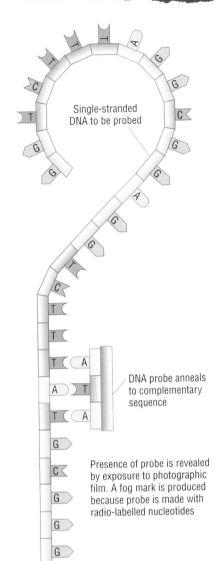

Figure 2 The action of DNA probes

Single-stranded DNA to be probed

DNA probe anneals to complementary sequence

Presence of probe is revealed by exposure to photographic film. A fog mark is produced because probe is made with radio-labelled nucleotides

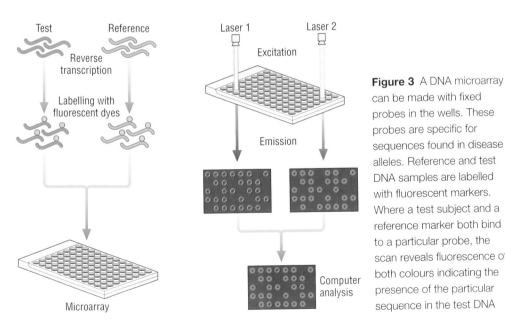

Figure 3 A DNA microarray can be made with fixed probes in the wells. These probes are specific for sequences found in disease alleles. Reference and test DNA samples are labelled with fluorescent markers. Where a test subject and a reference marker both bind to a particular probe, the scan reveals fluorescence of both colours indicating the presence of the particular sequence in the test DNA

Key definition

Electrophoresis is similar to chromatography. Separation of the different lengths of DNA fragments, in a mixture, is achieved because, as the negatively charged fragments move towards the positive electrode, shorter fragments pass through the gel more easily and so move further in a fixed time.

Examiner tip

Electrophoresis has a variety of applications. This means that examiners may ask you about how DNA strands can be separated in a variety of contexts. You should learn the basic procedure fully in preparation for your exam.

Questions

1. What type of bonding is responsible for annealing?
2. Why is electrophoresis described as 'similar to chromatography'?
3. Suggest why it is important that DNA probes are relatively short molecules.

By the end of this spread, you should be able to ...

✳ Outline how the polymerase chain reaction can be used to make multiple copies of DNA fragments.

The polymerase chain reaction – PCR

PCR is basically artificial DNA replication. It can be carried out on tiny samples of DNA in order to generate multiple copies of the sample. This is particularly useful in forensic investigations, where samples of DNA taken from crime scenes, for example in hair or blood, can be multiplied (referred to as **amplified**) in order to generate enough material for genetic profiling.

The sequencing reaction relies on the fact that DNA:
- is made up of antiparallel backbone strands
- is made of strands that have a 5′ (prime) end and a 3′ (prime) end
- grows only from the 3′ end
- base pairs pair up according to complementary base-pairing rules – A with T and C with G.

PCR is not identical to natural DNA replication
- It can only replicate relatively short sequences of DNA (a few hundred bases long), not entire chromosomes.
- The addition of **primer** molecules is required in order for the process to start.
- A cycle of heating and cooling is used in PCR to separate and bind strands; DNA helicase enzyme separates strands in the natural process.

PCR is a cyclic reaction
- The DNA sample is mixed with a supply of DNA nucleotides and the enzyme DNA polymerase.
- The mixture is heated to 95 °C. This breaks the hydrogen bonds holding the complementary strands together, so making the samples single-stranded.
- Short lengths of single-stranded DNA (around 10–20 bases) are added. These are called primers.
- The temperature is reduced to around 55 °C, allowing the primers to bind (hydrogen bonding) and form small sections of double-stranded DNA at either end of the sample.
- The DNA polymerase can bind to these double-stranded sections.

- The temperature is raised to 72 °C (the optimum temperature for the enzyme DNA polymerase). The enzyme extends the double-stranded section by adding free nucleotides to the unwound DNA (in the same way as in natural DNA replication).
- When the DNA polymerase reaches the other end of the DNA strand, a new double-stranded DNA molecule is generated.
- The whole process can be repeated many times so the amount of DNA increases exponentially (×2, ×4, ×8 and so on).

> **Key definition**
>
> **Primers** are short, single-stranded sequences of DNA, around 10–20 bases in length. They are needed, in sequencing reactions and polymerase chain reactions, to bind to a section of DNA because the DNA polymerase enzymes cannot bind directly to single-stranded DNA fragments.

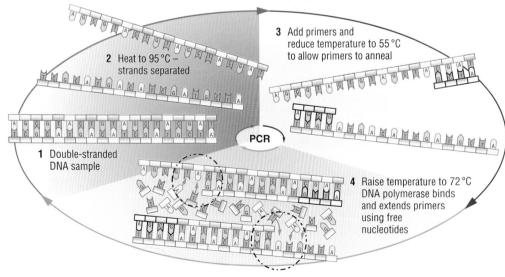

2 Heat to 95 °C – strands separated

3 Add primers and reduce temperature to 55 °C to allow primers to anneal

1 Double-stranded DNA sample

PCR

4 Raise temperature to 72 °C DNA polymerase binds and extends primers using free nucleotides

Figure 1 The polymerase chain reaction

Module 2
Biotechnology and gene technologies
Sequencing and copying DNA – 1

A little more about the enzyme …

The DNA polymerase enzyme used in PCR is described as 'thermophilic' because it is not denatured by the extreme temperatures used in the process. The enzyme is derived from a thermophilic bacterium, *Thermus aquaticus* (Taq), which grows in hot springs at a temperature of 90 °C.

STRETCH and CHALLENGE

Many non-protein coding sections of DNA are now known to code for the production of a variety of short mRNA strands. Some of these are 'antisense' strands, binding to the mRNA of other *coding* genes so that protein synthesis cannot take place. Others bind to part of a coding mRNA and trigger destruction of the mRNA. These processes are important in genome regulation. The 'silencing' of genes in this way is the subject of much research – for example genes responsible for laying down fat deposits have been artificially silenced in worms by adding RNA that leads to destruction of the coding mRNA. This artificial RNA interference is now known to have a natural role in genome control.

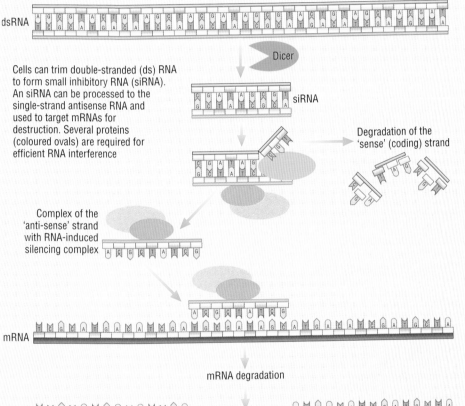

Cells can trim double-stranded (ds) RNA to form small inhibitory RNA (siRNA). An siRNA can be processed to the single-strand antisense RNA and used to target mRNAs for destruction. Several proteins (coloured ovals) are required for efficient RNA interference

Figure 2 RNA interference. Cells can trim double-stranded RNA to form small, inhibitory RNA (siRNA). An siRNA can be processed to the single-strand antisense RNA and used to target mRNAs for destruction. Several proteins (coloured ovals) are required for efficient RNA interference

Questions

A Explain why an antisense mRNA strand prevents a particular protein from being synthesised.

B Name two other non-coding RNA types that are produced using DNA as a template.

C How might investigations into natural RNA interference be of benefit in looking for treatments for cancer?

D Explain why the silencing of genes in an organism is an important part of cellular differentiation.

Question

1 Explain how the PCR enables forensic scientists to analyse minute samples of DNA found at the scenes of crime.

By the end of this spread, you should be able to ...

✳ Outline the steps involved in sequencing the genome of an organism.

Automated DNA sequencing is based on interrupted PCR and electrophoresis

Sequencing fragments of DNA was initially carried out in a slow and painstaking way using radioactively labelled nucleotides. The development of automated sequencing has led to a rapid increase in the number of organism genomes sequenced and published in recent years.

The reaction mixture (as with PCR) contains the enzyme DNA polymerase, many copies of the single-stranded template DNA fragment (the bit of DNA to be copied), free DNA nucleotides and primers. However, within the sequencing mixture, some of the free nucleotides carry a fluorescent marker. These nucleotides are modified and, if they are added to the growing chain, the DNA polymerase is 'thrown off' and the strand cannot have any further nucleotides added. Each nucleotide type has a different coloured fluorescent marker. The reaction proceeds as follows:

- The primer joins (anneals) at the 3′ end of the template strand, allowing DNA polymerase to attach.
- DNA polymerase adds free nucleotides according to base-pairing rules so the strand grows – this is essentially the same as natural DNA replication and PCR.
- If a modified nucleotide is added, the polymerase enzyme is thrown off and the reaction stops on that template strand.
- As the reaction proceeds, many molecules of DNA are made. The fragments generated vary in size. In some of them the template strand has only one additional nucleotide added before the polymerase is thrown off, in others the template strand is completed. In each case, the final added nucleotide is tagged with a specific colour.
- As these strands run through the machine (in the same way as DNA strands move in electrophoresis) a laser reads the colour sequence, from the strand with only a single nucleotide added, to the one with two nucleotides added, then three, then four and so on. The sequence of colours, and so the sequence of bases, can then be displayed.

The Human Genome Project findings

The Human Genome Project set its aim in 1990 to determine the whole human genome sequence. The DNA from a few individuals, taken from a pool of around 100 volunteers, was used. Laboratories around the world contributed to the project. In 2004, the completed human genome sequence was published. Sequencing, as described above, requires sections of DNA to be sequenced between 6 and 10 times in order to be confident that the base sequence information is accurate. Each contributing laboratory worked on different parts of the genome, sharing their sequence information with others to build the whole sequence. Probably the most fascinating fact is that at the outset, scientists expected to find around 100 000 coding genes on the genome. By 2004 the estimate was that humans only have around 25 000 – not many more than a worm! It seems now that human complexity has far more to do with regulation of gene expression than with the number of genes.

Module 2
Biotechnology and gene technologies
Sequencing and copying DNA – 2

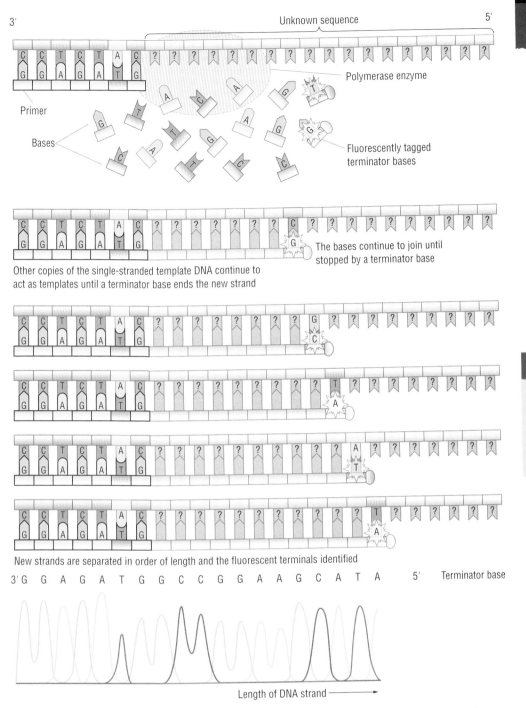

3' G G A G A T G G C C G G A A G C A T A 5' Terminator base

Figure 1 Automated sequencing: fragments of varying length are produced, with fluorescent marker as the last added base; then the sequence is shown by the order of colours as electrophoresis separates the fragments by length

Examiner tip

Pay close attention to the diagrams of electrophoresis and sequencing. You could be asked to label such diagrams or interpret the information in an electrophoresis or sequence run given as part of a question.

Questions

1 Suggest why DNA polymerase used in the PCR is Taq polymerase, obtained from thermophilic bacteria.

2 Explain why the reading sequence begins with the strand that has been terminated by the addition of a single terminator base to a fragment with primer attached.

173

(12) An introduction to genetic engineering

By the end of this spread, you should be able to . . .

* Define the term *recombinant DNA*.
* Explain that genetic engineering involves the extraction of genes from one organism and placing them into another organism.
* Describe how sections of DNA containing a desired gene can be extracted from a donor organism using restriction enzymes.
* Explain how isolated DNA fragments can be placed in plasmids, with reference to the role of ligase.
* State other vectors into which fragments of DNA may be incorporated.

What is genetic engineering?

Genetic engineering is a broad term that is used to describe a number of different processes for obtaining a specific gene and placing that gene in another organism (often of a different species). The organism receiving the gene (recipient) expresses the new gene product through the process of protein synthesis. Such organisms are described as **transgenic**.

Scientists often refer to the processes involved in genetic engineering as **recombinant DNA technology**, because these processes involve combining DNA, from different organisms or from different sources, in a single organism.

In genetic engineering the following steps are necessary:
1. The required gene is obtained.
2. A copy of the gene is placed (packaged and stabilised) in a **vector**.
3. The vector carries the gene to the recipient cell.
4. The recipient expresses the gene through protein synthesis.

A variety of approaches may be used at each stage. Table 1 shows an outline of the possible methods. These are described in more detail later.

Stage in engineering process	Methods possible
Obtaining the gene to be engineered	The mRNA produced from transcription of the gene can be obtained from cells where that gene is expressed. For example, the mRNA for insulin is obtained from β cells in islets of Langerhans in the pancreas. The mRNA can be used as a template to make a copy of the gene.
	The gene can be synthesised using an automated polynucleotide sequencer.
	A DNA probe can be used to locate the gene on DNA fragments and the gene can be cut from a DNA fragment using restriction enzymes.
Placing the gene in a vector	The gene can be sealed into a bacterial plasmid using the enzyme DNA ligase. This is by far the most common vector method used in genetic engineering.
	Genes may also be sealed into virus genomes or yeast cell chromosomes.
	Vectors often have to contain regulatory sequences of DNA. These ensure that the inserted gene is transcribed in the host cell.
Getting the gene into the recipient cell	The gene, once packaged in a vector, can form quite a large molecule that does not easily cross the membrane to enter the recipient cell.
	The methods used to get the vector into the cell depend on the type of cell and include: • Electroporation – a high-voltage pulse is applied to disrupt the membrane. • Microinjection – DNA is injected using a very fine micropipette into the host cell nucleus. • Viral transfer – the vector is a virus; this method uses the virus's mechanism for infecting cells by inserting DNA directly. • Ti plasmids used as vectors can be inserted into the soil bacterium *Agrobacterium tumefaciens*. Plants can be infected with the bacterium, which inserts the plasmid DNA into the plant's genome. • Liposomes – DNA is wrapped in lipid molecules. These are fat-soluble and can cross the lipid membrane by diffusion.

Table 1 Methods of genetic engineering

Module 2
Biotechnology and gene technologies
An introduction to genetic engineering

Restriction enzymes cut DNA backbones; ligase enzyme seals them

Recombinant DNA techniques often involve the cutting and sticking together of DNA strands. For example, a useful gene may need to be cut out of the chromosome on which it has been found, then sealed into a plasmid vector.

Enzymes known as **restriction enzymes** (restriction endonucleases) are used to cut through DNA at specific points. These enzymes were first extracted from bacterial cells, where they perform a natural defence function against infection by viruses. There are now more than 50 different commonly used restriction enzymes.

A particular restriction enzyme will cut DNA wherever a specific base sequence occurs and only where that sequence occurs. This sequence is called the **restriction site**, and is usually less than 10 base pairs long. In most of the restriction enzymes in use, the enzyme catalyses a hydrolysis reaction which breaks the phosphate–sugar backbones of the DNA double helix in different places. This gives a 'staggered cut' which leaves some exposed bases known as a **sticky end**.

When separate fragments of DNA need to be stuck together, an enzyme known as **DNA ligase** is used to catalyse a condensation reaction which joins the phosphate–sugar backbones of the DNA double helix together. This enzyme is the same as that used in natural DNA replication to seal DNA nucleotides together to form new DNA strands.

In order to join together DNA fragments from different sources both need to have originally been cut with the same restriction enzyme. This means that the sticky ends are complementary and allows the bases to pair up and hydrogen bond together. DNA ligase can then seal the backbone.

Where DNA fragments from different organisms are joined in this way, the resulting DNA is called **recombinant DNA**.

Key definition

A **sticky end** is formed when DNA is cut using a restriction enzyme. It is a short run of unpaired, exposed bases seen at the end of the cut section. Complementary sticky ends can anneal (bases pair together) as part of the process of recombining DNA fragments.

Examiner tip

Remember that cutting DNA into fragments is used in many DNA manipulation techniques, not just engineering. For example – DNA is cut in order to sequence sections, for genetic fingerprinting and to allow separation of fragments for analysis.

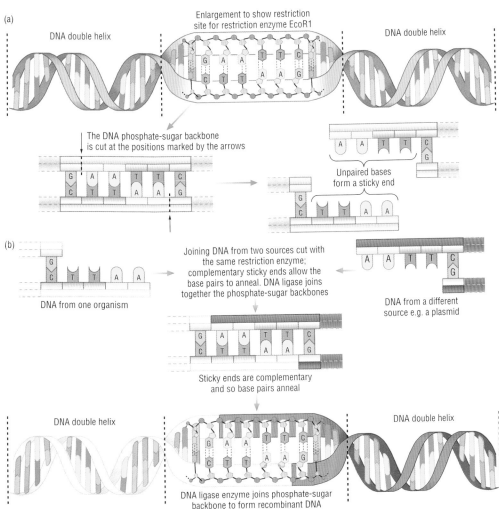

(a)
DNA double helix   Enlargement to show restriction site for restriction enzyme EcoR1   DNA double helix

The DNA phosphate-sugar backbone is cut at the positions marked by the arrows

Unpaired bases form a sticky end

(b)
DNA from one organism

Joining DNA from two sources cut with the same restriction enzyme; complementary sticky ends allow the base pairs to anneal. DNA ligase joins together the phosphate-sugar backbones

DNA from a different source e.g. a plasmid

Sticky ends are complementary and so base pairs anneal

DNA double helix   DNA double helix

DNA ligase enzyme joins phosphate-sugar backbone to form recombinant DNA

Figure 1 (a) Cutting DNA using a restriction enzyme produces a staggered cut and leaves sticky ends. **(b)** Joining DNA from two sources cut with the same restriction enzyme; complementary sticky ends allow the base pairs to anneal. DNA ligase joins together the phosphate–sugar backbones

Questions

1 Explain why different restriction enzymes have different restriction sites.

2 Explain why restriction enzymes are a useful defence mechanism for bacteria in order to resist infection by viruses.

3 If bacterial DNA contains base sequences that are the same as the restriction sites of their restriction enzymes, these sites are methylated (a -CH$_3$ group is added). Explain why this is necessary.

4 The restriction enzyme EcoR1 was the first restriction enzyme isolated from the bacterium *E. coli*. Suggest how restriction enzymes are named.

By the end of this spread, you should be able to . . .

✴ Explain how plasmids may be taken up by bacterial cells in order to produce a transgenic microorganism that can express a desired gene product.

✴ Describe the advantage to microorganisms of the capacity to take up plasmid DNA from the environment.

Why do we want to genetically engineer organisms?

There are two main reasons for carrying out genetic engineering.

1   Improving a feature of the recipient organism

- Inserting a gene into crop plants to give the plant resistance to herbicides (weed killers) allows farmers to use herbicides as the crops are growing and so increase crop yield.
- Inserting a growth-controlling gene, such as the myostatin gene, into livestock (farm animals) promotes muscle growth.

2   Engineering organisms that can synthesise useful products

- Inserting the gene for a human hormone, such as insulin or growth hormone, into bacteria and growing the bacteria produces large quantities of the hormone for human use.
- Inserting the gene for a pharmaceutical chemical into female sheep so that the chemical is produced in their milk means the chemical can then be easily collected.
- Inserting genes for beta-carotene production into rice so that the molecule is present in the edible part of the rice plant. Beta-carotene can be turned into vitamin A in people who eat it.

Bacterial cells and plasmids are often used in genetic engineering

Once a gene has been identified to be placed into another organism, it can be cut from DNA using a restriction enzyme and then placed in a vector. The vast majority of genetic engineering uses bacterial plasmids as the vector. A plasmid is a small circular piece of DNA. Plasmids are found in many types of bacteria and are separate from the main bacterial chromosome. Plasmids often carry genes that code for resistance to antibiotic chemicals.

If plasmids are cut with the same restriction enzyme as that used to isolate the gene, then complementary sticky ends will be formed. Mixing quantities of plasmid and gene in the presence of ligase enzyme means that some plasmids will combine with the gene, which then becomes sealed into the plasmid to form a **recombinant plasmid**. It is important to remember that many cut plasmids will, in the presence of ligase enzyme, simply reseal to reform the original plasmid.

Bacterial cells take up plasmid DNA – they become transformed and transgenic

Large quantities of the plasmid are mixed with bacterial cells, some of which will take up the recombinant plasmid. The addition of calcium salts and 'heat shock', where the temperature of the culture is lowered to around freezing, then quickly raised to 40 °C, increase the rate at which plasmids are taken up by bacterial cells. Even so, the process is very inefficient. Less than a quarter of 1% of bacterial cells take up a plasmid. Those that do are known as **transformed** bacteria. This transformation results in bacteria containing new DNA. By definition the bacteria are thus **transgenic**.

Bacterial conjugation and the advantages of taking up new DNA

Bacteria are capable of a process known as conjugation, where genetic material may be exchanged. In this process, copies of plasmid DNA are passed between bacteria, sometimes even of different species. Since plasmids often carry genes associated with resistance to antibiotics, this swapping of plasmids is of concern because it speeds the spread of antibiotic resistance between bacterial populations.

Key definition

Any organism is described as **transgenic** when it contains DNA that has been added to its cells as a result of genetic engineering.

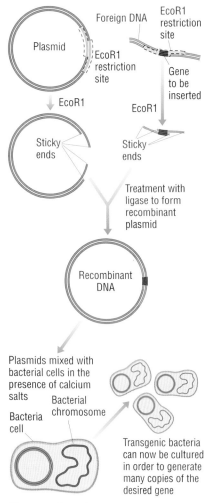

Figure 1 Plasmids containing a foreign gene can be placed into bacterial cells. The cells receiving the plasmids are transgenic

Module 2
Biotechnology and gene technologies
Genetic engineering and bacteria

Resistant strains of bacteria – such as MRSA (methicillin-resistant *Staphylococcus aureus*) – are causing healthcare problems because the bacterium is commonly found on human skin, where it is not a problem. The transfer of this bacterium to a wound, however, can lead to a very serious infection. Scientists are continually looking for new antibiotics to target these disease-causing organisms.

Examiner tip

Remember to link this work to your understanding of protein synthesis. The engineering of a gene is designed to ensure that the gene is expressed. The expression of a gene is the transcription of the gene to mRNA and translation to a protein product.

Clearly, the advantage to the bacteria of conjugation is that it may contribute to genetic variation and, in the case of antibiotic resistance genes, survival in the presence of these chemicals.

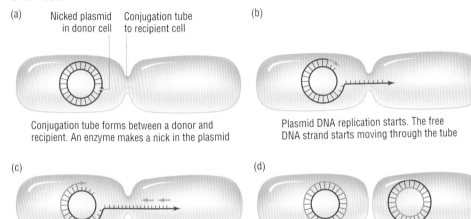

(a) Nicked plasmid in donor cell Conjugation tube to recipient cell

Conjugation tube forms between a donor and recipient. An enzyme makes a nick in the plasmid

(b) Plasmid DNA replication starts. The free DNA strand starts moving through the tube

(c) In the recipient cell, replication starts on the transferred DNA

(d) The cells move apart and the plasmid in each forms a circle

Figure 2 Bacterial conjugation. Note that the bacterium's chromosome is not shown in this diagram, only a plasmid. The chromosome does not pass from one cell to another in this way

Bacterial transformation and pneumonia in mice

In 1928, before the discovery of the role of DNA in living organisms, some pioneering work was done which demonstrated that bacteria can take up DNA from their surroundings and incorporate it into their genome. The experiments used two strains of the bacterium *Pneumococcus*:

- S-strain – which quickly kills mice on infection
- R-strain – which does not kill mice on infection.

It was found that only mice infected with the S-strain were killed by a protein that was toxic. The assumption was made that the S-strain must have the instructions to make the protein but the R-strain does not.

Further experiments revealed that injecting a mixture of living R-strain and dead S-strain bacteria killed the mice. On post-mortem examination, it was found that the mice contained living S-strain bacteria. These were described as *transformed* because somehow, the living R-strain bacteria had been transformed into the living S-strain type.

In 1944 the experimental work started by Griffith (see Figure 2) was followed up by Avery, McCarty and MacLeod. In a series of experiments they mixed separate parts of the S-strain bacteria (the cell wall fraction, the cytoplasmic fraction, the DNA fraction) with living R-strain bacteria. It was found that only the mixture containing S-strain DNA plus living R-strain bacteria resulted in death of mice and bacterial transformation. It was later confirmed that the R-strain bacteria were capable of taking up DNA from their surroundings, which in this case included DNA from S-strain bacteria with the gene for producing the toxin. This process of taking up DNA is called transformation and is another method by which bacteria can acquire DNA from each other.

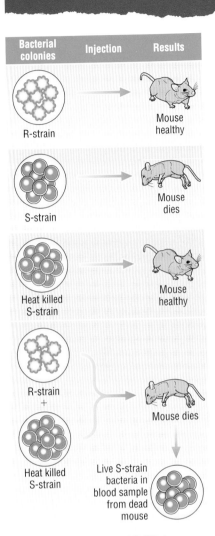

Bacterial colonies	Injection	Results

R-strain — Mouse healthy

S-strain — Mouse dies

Heat killed S-strain — Mouse healthy

R-strain + Heat killed S-strain — Mouse dies

Live S-strain bacteria in blood sample from dead mouse

Figure 3 An overview of Griffith's findings. The follow-up work of Avery, McCarty and MacLeod showed that only the DNA part of the S-strain needed to be injected with living R-strain in order to transform the bacteria

Questions

1 Suggest how the DNA content of the S-strain differs from that in the R-strain *Pneumococcus*.

2 Apart from DNA take-up as described above, what other process could give rise to genetic variation in bacterial cells?

Frederick Sanger

Fred Sanger, an English biochemist born in 1918, is the only scientist to have won two Nobel prizes in chemistry. In 1958 he won it for working out the sequence of amino acids in insulin in 1955. In 1980 he won another, together with Walter Gilbert and Paul Berg, for sequencing the genome of a phage virus (phage viruses are viruses that attack bacteria). This was pioneering work that helped to develop techniques that were then used for the Human Genome Project.

By the end of this spread, you should be able to . . .

✳ Outline the process involved in the genetic engineering of bacteria to produce human insulin.
✳ Outline how genetic markers in plasmids can be used to identify the bacteria that have taken up a recombinant plasmid.

Genetically engineered insulin

People who cannot produce the hormone insulin suffer from type I diabetes mellitus. Until the early 1980s, insulin for clinical use was extracted from the pancreatic tissues of slaughtered pigs. This is not identical to human insulin, is less effective and very expensive to produce, since only a very small amount of insulin is present in pancreatic tissue.

By the 1970s it was known that insulin is a polypeptide, consisting of 51 amino acids. The DNA code for such a polypeptide is very small (less than 200 bases long). This made it difficult to find in a genome of 300 million bases, particularly as, at that time, the human genome had not been mapped.

Scientists focused their attention on finding the mRNA for the gene. They used specialised centrifugation methods to separate mRNA of the right length from pancreatic tissue. Once they had found the mRNA, the enzyme *reverse transcriptase* was used to synthesise a complementary DNA strand. (This effectively gives us a copy of the template DNA – complementary to the coding strand. It is also a single strand of DNA, unlike the original gene.)

Adding *DNA polymerase* and a supply of DNA nucleotides to these single strands means the second strand is built on using the copied DNA as a template – just as in DNA replication. This produces a copy of the original gene called a cDNA gene. Unpaired nucleotides are added at each end to give sticky ends complimentary to those on the cut plasmid.

Plasmids are then cut open with a restriction enzyme and mixed with the cDNA genes. Some of the plasmids take up the gene. *DNA ligase* enzyme then seals up the plasmids which are now called **recombinant** plasmids because they contain a new piece of DNA.

The plasmids are then mixed with bacteria, some of which take up the recombinant plasmids.

The bacteria are then grown on an agar plate, where each bacterial cell grows to produce a mound of identical (cloned) cells, called a colony.

Not all bacteria take up a plasmid

There are three possible types of colony that may grow in this process:
- some from bacteria that did not take up a plasmid
- some from bacteria that have taken up a plasmid that has not sealed in a copy of the gene but has sealed up on itself to reform the original plasmid
- some – the ones we want – that have taken up the recombinant plasmid. We call these the **transformed** bacteria.

You can't tell the difference between them by looking at them!

Identification of transformed bacteria by replica plating

A complicated technique using radio-labelled antibodies that bind specifically to insulin was used in the original process of identifying the transformed bacteria. Modern methods of identification use plasmid vectors with genetic markers as follows:

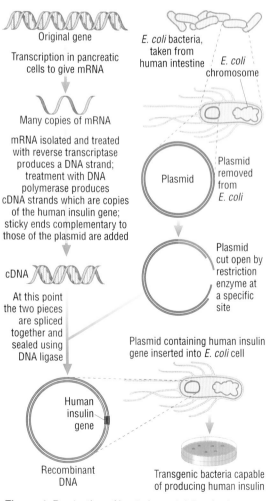

Figure 1 Production of bacteria containing the human insulin gene

Original gene

Transcription in pancreatic cells to give mRNA

Many copies of mRNA

mRNA isolated and treated with reverse transcriptase produces a DNA strand; treatment with DNA polymerase produces cDNA strands which are copies of the human insulin gene; sticky ends complementary to those of the plasmid are added

cDNA

At this point the two pieces are spliced together and sealed using DNA ligase

Human insulin gene

Recombinant DNA

E. coli bacteria, taken from human intestine

E. coli chromosome

Plasmid

Plasmid removed from E. coli

Plasmid cut open by restriction enzyme at a specific site

Plasmid containing human insulin gene inserted into E. coli cell

Transgenic bacteria capable of producing human insulin

Module 2
Biotechnology and gene technologies
Engineering case studies –
1: human insulin

- The original plasmids are chosen because they carry genes that make any bacteria receiving them resistant to two different antibiotic chemicals (usually ampicillin and tetracycline). These resistance genes are known as **genetic markers**. The bacteria (*E. coli* in this case) are susceptible to both of these antibiotic chemicals.
- The plasmids are cut by a restriction enzyme that has its target site (restriction site) in the middle of the tetracycline resistance gene, so that if the required gene is taken up, then the gene for tetracycline resistance is broken up and does not work. However, the gene for ampicillin resistance does still work.

A process of **replica plating** is then used:

- The bacteria are grown on standard nutrient agar, so all bacterial cells grow to form colonies.
- Some cells from the colonies are transferred onto agar that has been made with ampicillin, so only those that have taken up a plasmid will grow (see Figure 2).
- Some cells from these colonies are transferred onto agar that has been made with tetracycline so only those that have taken up a plasmid that does not have the insulin gene will grow.
- By keeping track of which colonies are which, we know that any bacteria that grow on the ampicillin agar, but not on the tetracycline agar, must have taken up the plasmid with the insulin gene.
- We can now identify the colonies we want and grow them on a large scale. These bacteria then produce insulin on a large scale which can be harvested for use.

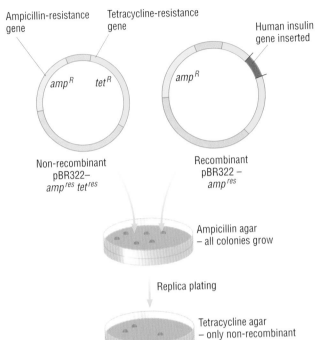

Bacteria with either the recombinant plasmid containing the insulin gene or the original plasmid without the insulin gene will grow on ampicillin agar

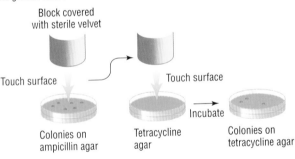

The replica plates are formed by transferring bacterial cells from colonies on one growth medium to another

Figure 2 Identification of transformed bacteria by replica plating – the colonies on the ampicillin plate but not the tetracycline plate are the transformed bacteria

---

Key definition

Replica plating refers to the process of growing bacteria on an agar plate, then transferring a replica of that growth to other plates, usually containing different growth promoters or inhibitors. Analysis of growth patterns on the replica plates gives information about the genetic properties of the growing bacteria.

Reverse transcriptase enzyme

Some viruses (known as retroviruses) carry their genetic material as RNA. When they infect a cell, they cause the enzyme reverse transcriptase to be synthesised on the host cell ribosomes. This enzyme copies the viral RNA back to form DNA – transcription in reverse. Scientists have made use of this naturally occurring viral enzyme in order to build copies of human genes from isolated mRNA as described above.

Examiner tip

The principles of genetically engineering the human insulin gene into bacterial cells, which can then be cultured to produce the polypeptide hormone for human use, can be applied to any similar product. You could be asked to suggest how a supply of any protein might be obtained using genetic engineering and the same broad answer would be appropriate.

Questions

1 Why was insulin mRNA extracted from pancreatic tissue?
2 Explain why a cDNA gene in a eukaryotic organism may not be exactly the same as the original gene on the chromosome.
3 Some people who have the condition type II diabetes mellitus are able to produce functioning insulin in their pancreas. Suggest what genetic malfunction is responsible for the condition in these people.

Engineering case studies – 2: Golden Rice™

By the end of this spread, you should be able to ...

＊ Outline the process involved in genetically engineering Golden Rice™.

Vitamin A deficiency

The World Health Organisation estimates that up to 500 000 people each year become irreversibly blind due to a lack of vitamin A in their diet. Over 120 million people, mainly, in Africa and South East Asia, are estimated to be affected in some way, with around 1–2 million deaths annually being in some part related to deficiency of vitamin A in the diet. Children under 5 years of age and pregnant women are particularly at risk.

As with any vitamin deficiency disease, the best way to overcome it is to take in a balanced diet. The countries where people are most affected by vitamin A deficiency are economically less developed, with substantial numbers of the population having little access to sufficient food. In many of these areas, programmes organised by the United Nations and various charities give vitamin and other food supplements to target groups. However, this has made little impact on the devastating effects of malnutrition on these populations.

The poorer populations of many of the countries where vitamin A deficiency is significant rely on rice as the main (staple) food.

Vitamin A is formed from beta-carotene in the human gut
Vitamin A (retinol) in the diet only comes from animal sources. In vegetarians and those without access to meat, vitamin A is derived from the intake of beta-carotene – known as a **precursor** – which is converted to active vitamin A in the gut. Beta-carotene is sometimes referred to as pro-vitamin A. Vitamin A is fat soluble, so the diet must also include some lipids if the vitamin is to be absorbed properly.

Eyesight	Forms part of the visual pigment rhodopsin
Cell growth and development	Involved in synthesis of many glycoproteins
Epithelial tissue	Needed for maintenance and differentiation of epithelial cells. This helps reduce risk of infection
Bones	Essential for growth of bones

Table 1 The functions of vitamin A

Rice has been engineered to be rich in beta-carotene
Rice plants (*Oryza sativa*) contain the genes that code for the production of beta-carotene. This molecule is a photosynthetic pigment molecule so is required in the green parts of the plant. Unfortunately, in the part of the plant that is eaten – the endosperm (grain) – the genes for beta-carotene production are switched off.

In 2000, scientists working in Switzerland published the results of an 8-year-long genetic engineering project. The project had worked to engineer rice plants so that beta-carotene accumulated in the endosperm. The accumulation of the molecule made the rice grains yellow–orange in colour, so the genetically engineered product was called **Golden Rice™**.

Engineering Golden Rice™
The metabolic pathway for synthesising beta-carotene is complex, but most of the enzymes of the pathway are already present in the endosperm. It was found that the insertion of two genes into the rice genome was needed in order for the metabolic pathway to be activated in the endosperm cells. The genes code for the following enzymes:
- *Phytoene synthetase*, the gene for which was extracted from daffodil plants
- *Crt 1 enzyme*, the gene for which was extracted from the soil bacterium *Erwinia uredovora*.

These genes were inserted into the rice genome near to a specific promoter sequence that switches on the genes associated with endosperm development. This meant they were expressed as the endosperm grew.

Figure 1 Golden Rice™ (right) is clearly different from the traditional variety (left)

Module 2
Biotechnology and gene technologies
**Engineering case studiest –
2: Golden Rice™**

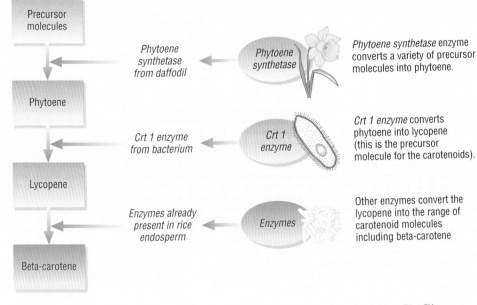

Figure 2 Simplified sequence showing the production of beta-carotene in Golden Rice™

Although this rice contains beta-carotene, its usefulness in dealing with vitamin A deficiency was questioned. It was estimated that someone would have to eat large amounts of rice in order to take in sufficient beta-carotene.

Further developments
Golden Rice has been cross-bred with natural rice varieties. These hybrids were grown in small-scale field trials in 2004 in Louisiana USA. The trials showed that they could produce 3–4 times more beta-carotene than the original Golden Rice™ variety.

In 2005, UK scientists of the biotechnology company Syngenta developed a new variety called Golden Rice 2. This variety accumulates around 20 times more beta-carotene in the endosperm than the original version. If successfully grown, this variety could deliver the required amounts of beta-carotene within a daily intake of 200–300 g of rice.

It is expected that full field trials (growing the rice in the natural environment) of Golden Rice™ will take place from the year 2011, after further food safety investigations have taken place.

A humanitarian triumph or a public relations exercise?
The researchers and biotechnology companies that have produced Golden Rice™ have offered *Humanitarian Use Licences* free of charge so that farmers can keep and replant crop seeds without having to pay a licence fee.

Critics of the use of genetically engineered crops, notably Greenpeace and Friends of the Earth, have accused these companies of using this as a public relations exercise to gain public acceptance of the use of genetically modified crops. Greenpeace argue that all use of genetically modified crops is unacceptable on the grounds that they believe:

- it will lead to a reduction in biodiversity
- the human food safety of engineered rice is unknown
- the genetically modified rice could breed with wild types and contaminate wild rice populations.

However, several thousand children each year become blind due to lack of vitamin A, and the countries of the developed world have not managed to solve this problem by other means.

Key definition
Golden Rice™ is said to be **biofortified** because it contains higher than normal concentrations of a particular nutrient, in this case beta-carotene.

Questions
1 Suggest why:
   (a) rice does not normally contain high levels of beta-carotene in the endosperm even though the genetic information to make it is already present in the plants
   (b) most of the enzymes associated with the beta-carotene pathways are present normally in the rice endosperm even though the complete pathway is not.
2 Why do people who are underweight suffer more quickly from vitamin A deficiency in the diet than people who are not underweight?
3 What do you think of the arguments put forward by Greenpeace against the use of genetically modified crops? Are they scientific? Can you think of some counterarguments?

By the end of this spread, you should be able to ...

* Explain the term *gene therapy*.
* Explain the differences between somatic cell gene therapy and germline cell gene therapy.

Gene therapy

The techniques of molecular genetic technology (gene technology) can be used to treat some genetic disorders. This is known as **gene therapy**. In its most basic sense, if we can get the working copy of a gene into cells that contain only dysfunctional copies of that gene, then transcription of the added working copy will mean that the individual may no longer have the symptoms associated with the genetic disorder. The developments brought about by the Human Genome Project have also led to further therapeutic possibilities including the use of RNA_i (interference RNA). This could silence genes by binding to mRNA. The only use for this at present is to treat cytomegalovirus infections in AIDS patients, by blocking replication of the cytomegalovirus.

Somatic cell gene therapy

As organisms grow, cells become specialised to function. Within specialised cells, certain genes are switched on and others are switched off. Although the cell still contains a full genome (set of genes), relatively few of them will be active in producing proteins.

* **Gene therapy by adding genes (augmentation):** Some conditions are caused by the inheritance of faulty alleles leading to the loss of a functional gene product (polypeptide). Engineering a functioning copy of the gene into the relevant specialised cells means that the polypeptide is synthesised and the cells can function normally.
* **Gene therapy by killing specific cells:** Cancers can be treated by eliminating certain populations of cells. Using genetic techniques to make cancerous cells express genes to produce proteins (such as cell surface antigens) that make the cells vulnerable to attack by the immune system could lead to targeted cancer treatments.

Germline cell gene therapy

All embryos begin when a sperm cell fertilises an egg cell, forming a zygote that undergoes cell division. Each cell of an early embryo is a stem cell. It can divide and specialise to become any cell type within the body. Each could also potentially become a new being, hence these cells are germline cells. Engineering a gene into sperm, egg, zygote or into all the cells of an early embryo means that as the organism grows, every cell contains a copy of the engineered gene. This gene can then function within any cell where that gene is required.

Some (transgenic) animals have been genetically engineered. The functioning allele they have received may also be passed on to the animals' offspring. This is not the case with somatic cell gene therapy.

Here, the genetic modification is restricted to somatic (body) cells, with no effect on the germline. An individual who has had gene therapy for a genetic disorder can still pass the allele for that disorder to his/her offspring. Although widely employed in experimental animals, germline gene therapy in humans is illegal and ethically unacceptable. This has been decided by ethics committees, such as the Clothier committee, who say that:

* an inadvertent modification of DNA introduced into the germline could create a new human disease or interfere with human evolution in an unexpected way
* permanent modifications to the human genome in this way raise difficult moral, ethical and social issues that need to be fully debated.

(a) A liposome is an artificial vesicle

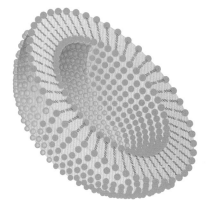

(b) Genes can be enclosed in the liposome and so are able to pass through the plasma membrane of the target cell

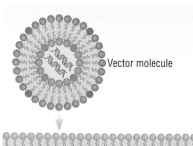

Vector molecule

Target cell

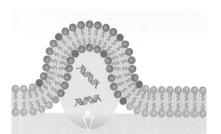

Functional protein   Functional protein

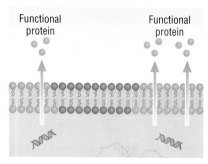

Figure 1 Gene therapy using a liposome vector

> **Examiner tip**
>
> Don't say genes are *replaced* during gene therapy. Only a copy of a functional allele can be *added*. To date, gene therapy is only used to treat recessive genetic disorders.

Somatic cell gene therapy	Germline cell gene therapy
The functioning allele of the gene is introduced into target cells – therefore techniques to get the gene to the target location are needed, or specific cells must be removed, treated and then replaced (this is called *ex vivo* therapy).	The functioning allele of the gene is introduced into germline cells – delivery techniques are more straightforward.
Introduction into somatic cells means that any treatment is short-lived and has to be repeated regularly. The specialised cells containing the gene will not divide to pass on the allele.	Introduction into germline cells means that all cells derived from these germline cells will contain a copy of the functioning allele. The offspring may also contain the allele.
There are difficulties in getting the allele into the genome in a functioning state. Genetically modified viruses have been tried but the host becomes immune to them so cells will not accept the virus vector on second and subsequent treatments. Liposomes are used but these may be inefficient.	Although more straightforward, it is considered unethical to engineer human embryos. It is not possible to know whether the allele has been successfully introduced without any unintentional changes to it, which may damage the embryo. This is further discussed in the next spread.
Genetic manipulations are restricted to the actual patient.	Genetic manipulations could be passed on to the patient's children.

Table 1 Some of the issues concerning gene therapy

Gene therapy for SCID (severe combined immunodeficiency)

There are ten different forms of this condition, which leads to complete dysfunction of the immune system. In 1972, one of the forms was found to be due to the presence of a defective gene for the enzyme **adenosine deaminase** (**ADA**). The lack of ADA leads to accumulation of metabolites toxic to T lymphocytes, so complete loss of T lymphocytes occurs. This form of the condition has a recessive inheritance pattern.

Gene therapy trials began in 1990:
- A **retrovirus**, which is capable of transferring its **DNA** into eukaryotic cells, is engineered to contain the normal human **ADA** gene.
- Bone marrow, containing **T cells**, is removed from the patient and exposed to the retrovirus in cell culture. Viral infection leads to uptake of the **ADA** gene.
- The transgenic cells formed are placed back into the patient's bone marrow where they establish a line of cells with functional **ADA**.

Trials are still ongoing, but it is hoped that gene therapy will eventually replace bone marrow transplant as the treatment for ADA-linked SCID. SCID patients can also be given daily injections of the enzyme adenosine deaminase.

Questions

1 In gene therapy, where is the desired working copy of the gene taken from?
2 Explain why gene therapy is potentially most useful in conditions associated with single gene mutations.

Key definitions

GMO – an organism that has undergone genetic engineering is a **genetically modified organism**.

A **transgenic** organism has received an allele of a gene from another organism, often of a different species.

Liposomes – small spheres of lipid bilayer containing a functioning allele. They can pass through the lipid bilayer of cells and therefore act as vectors to carry the allele into the cell.

One of the forms of SCID is X-linked

In April 2000, an article was published in *Science* magazine showing how two cases of X-linked SCID had been treated by gene therapy. Six months had passed since the therapy and both patients showed normal growth and development without other treatment. A further eight cases in the trial showed no improvement.

In October 2003 the results of another trial were published. Two of the patients in the trial had developed a T-cell cancer because insertion of the allele had occurred in their cells close to another gene (*LM02*) which is associated with leukaemia. With news of this devastating event, most X-linked SCID gene therapy trials were placed on hold worldwide. New trials have begun recently, using different vector systems in order to target the insertion of the allele more accurately.

(17) The rights and wrongs of genetic manipulation

By the end of this spread, you should be able to ...

✳ Outline how animals can be genetically engineered for xenotransplantation.
✳ Discuss the ethical concerns raised by the genetic manipulation of animals (including humans), plants and microorganisms.

A shortage of transplant organs

In some individuals, the failure of a particular organ results in the need for an organ transplant. However, there is a worldwide shortage of donor organs. It is estimated that around 60% of patients awaiting replacement organs die whilst on the waiting list. Even when transplantation is possible, transplanted organs are non-self ('foreign') tissue and can trigger an immune response. This results in rejection of the transplanted tissue. This is why compatibility of organs is checked and immuno suppressant drugs are usually needed following tranplant surgery.

Recent advances in understanding the mechanisms of transplant organ rejection mean we can now consider transplant to humans of organs from other species. This is known as **xenotransplantation**.

Engineered pigs as organ donors

The most significant first obstacle to overcome in using pig organs for xenotransplantation is that of immune rejection. In 2003 it was reported that pigs engineered to lack the enzyme α-1,3-transferase had been successfully developed. This enzyme is a key trigger for graft rejection in humans. In 2006, scientists in Middlesex, UK and Gdansk, Poland, reported that engineering of human nucleotidase enzyme (E5'N) into pig cells in culture reduced the activity of a number of immune cell activities involved in xenotransplant rejection. It is hoped that future developments will enable the use of animal organs and tissues for transplantation, so saving and improving the lives of many people.

Aside from the immune rejection issues, a variety of problems are associated with using pig organs for transplant to humans. These include physiological problems such as:
- differences in organ size
- the lifespan of most pigs is roughly 15 years, so a xenograft may age prematurely
- the body temperature of pigs is 39 °C (2 °C above the average human body temperature).

And ethical and wider medical problems:
- Some animal welfare groups strongly oppose killing animals in order to harvest their organs for human use.
- Religious beliefs – Orthodox Jewish and Muslim faiths prohibit eating pork.
- Medical concerns exist about possible disease transfer between animals and humans.

Ethical concerns raised by the genetic manipulation

Ethical concerns are those raised by the questions of what is right and wrong. In genetic manipulation, the capacity to move genes between organisms or to clone individual organisms or parts of organisms leads to the production of organisms which some people call 'unnatural'. In previous spreads, some of the numerous current and potential future benefits of genetic manipulation have been described. In any treatment that raises ethical concerns it is important to remember just how powerful the techniques for genetic manipulation really are.

Before considering the ethical issues of genetic manipulation, it is important to remember that humans have produced 'unnatural' organisms for centuries. This has been done mainly through selective breeding, where humans have selected organisms with particularly valuable traits and bred

---

Key definition

Xenotransplantation refers to transplantation of cell tissues or organs between animals of different species whereas **allotransplantation** refers to transplantation between animals of the same species.

Figure 1 Probably the most famous example of using animals to provide tissue for transplantation. In 1997 Dr Jay Vacanti grew a human ear from human cartilage cells grafted onto the back of a mouse, using a plastic support. The combination of genetic engineering techniques and stem cell technology will eventually lead to the ability to grow and transplant a range of compatible human tissues and organs

Examiner tip

Students often make unbalanced arguments in answering examination questions on ethics. Whatever your personal views, it is important that you give the case for and against a particular manipulation. Marks for the ethical concerns about genetic manipulation will not be given for vague statements about it being wrong to 'play god' or to use animals for our own purposes.

them over many generations. The domesticated varieties produced through selective breeding are far removed from their ancestral wild relatives. It is also very important that you place objections to genetic manipulation in their proper context. The media hype surrounding the use of the term 'Frankenfoods', for example, suggested that transfer of DNA into the human genome could occur from eating food containing DNA, with the suggestion that DNA as part of the diet is something unnatural. It is unfortunate that many of the loudest voices in opposing genetic manipulation are sometimes of those who appear to know little of the scientific background.

The rights and wrongs of genetic manipulation

The benefits and potential benefits of genetic manipulation have been discussed in previous spreads.

Genetic manipulation is a relatively young technology. Ethical concerns over many aspects of the use of such technologies are the subject of much debate. Positions range from those of animal welfare groups that argue the use of animals for all genetic manipulations is unethical, to those who identify the potential wider risks of genetic manipulation, which are substantially linked to the lack of long-term knowledge of the manipulations carried out. However, when any new technique is introduced, some people object. There were objections to use of cow pox to vaccinate against smallpox 200 years ago, objections to heart and organ transplants 40 years ago and objections to IVF (test tube babies) 30 years ago but now these routine medical procedures are widely accepted.

Some specific benefits and concerns listed in Table 1.

Questions

1 Explain why transplantation of tissues or organs between identical twins does not lead to rejection, whereas transplantation between siblings (brothers and sisters) usually does.

2 Suggest why 'genes for pest resistance could pass to other plant species, changing the stability of biological communities and possibly affecting many other organisms and food chains' as stated in the table.

3 Think of six natural events that are harmful, and six human activities that could be regarded as 'against nature' that are beneficial to us.

Organism	Example of benefit	Example of risk
Microorganisms	Genetically engineered microorganisms produce useful products such as human insulin and human growth hormone.	Engineered microorganisms may escape from containment and transfer genes (which may mutate with unknown effects) to other, pathogenic microorganisms. Genetic engineering often uses antibiotic resistance genes as markers. These genes could be passed to other microorganisms, leading to more widespread antibiotic resistance.
Plants	Accumulation of beta-carotene in endosperm of seeds (Golden Rice™) could combat vitamin A deficiency. Resistance to pesticides allows application of weedkillers and increase in yield. Resistance to pests increases yield.	Genes introduced to crop plants may pass to wild relatives. It is thought that this could result in less genetic variation and/or the production of less useful hybrid crops. Genes could pass to weed/unwanted species giving them herbicide or pesticide resistance, so forming 'super-weeds'. Genes for pest resistance could pass to other plant species, changing the stability of biological communities and possibly affecting many other organisms and food chains. Modified plants may be toxic to other organisms, or lead to allergic responses in humans. Plants resistant to pathogens could stimulate the more rapid evolution of attack mechanisms in these pathogens.
Animals	Pharmaceutical chemicals can be produced in milk (e.g. α anti-trypsin produced by female transgenic sheep and used to treat patients with hereditary emphysema). Increased milk or meat production. Production of compatible organs for transplantation to humans.	Animal welfare issues arise from genetic manipulations that might lead to animal suffering. Strong views about specific animals are held in some religions; cows are sacred to Hindus and pigs are considered unclean by orthodox Jews and Muslims.
Humans	Gene therapies treat some genetic disorders.	The main ethical objections to genetic manipulation in humans are to germline cell gene therapy because: • the effects of gene transfer are unpredictable. Even if the target disease is cured, further defects could be introduced into the embryo and then to their offspring • individuals resulting from germline gene therapy would have no say in whether their genetic material should have been modified • there are concerns that germline cell gene therapy could be used not only to eliminate disease, but also to enhance favourable characteristics. Such concerns include fears about 'designer children', with traits chosen by their parents. Concerns about possible eugenic uses (through the ability to manipulate the genetic properties of a population) have also been raised. Germline cell therapy is **not** practised in humans.

Table 1 Some benefits and risks of genetic engineering

Biotechnology and gene technologies summary

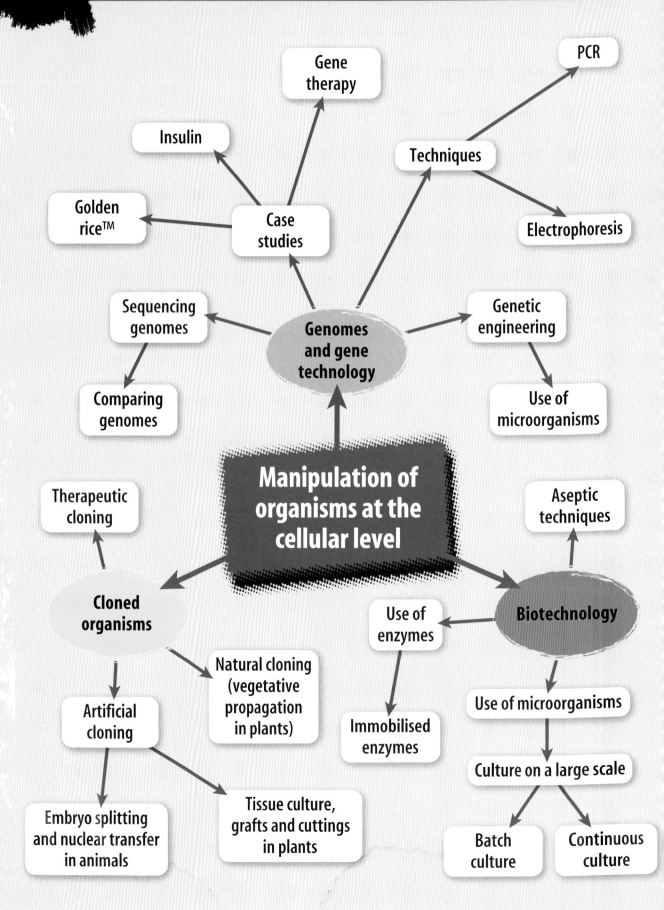

Practice questions

1 (a) State **one** place where meristems are found in flowering plants. [1]

(b) Explain why meristematic tissue is useful when a plant is to be genetically modified. [3]

2 (a) Explain what is meant by 'recombinant DNA'. [3]

(b) Outline the roles of restriction endonuclease enzymes in forming recombinant DNA. [5]

(c) Describe one use of genetic engineering in agriculture. [2]

(d) State **one** potential benefit and **one** potential hazard associated with the example you have described. [2]

3 Explain how lengths of RNA that are complementary to lengths of mRNA may suppress (silence) the expression of a gene. [5]

4 (a) State the meaning of 'immobilised enzyme'. [1]

(b) Describe **two** different methods of immobilising enzymes. [2]

(c) Suggest **three** practical advantages of using an immobilised urease reactor in a space station. [3]

5 (a) Explain why sterilised air has to be pumped into a batch fermenter being used to produce penicillin. [5]

(b) In such a batch fermenter, the mould *Penicillium*, from which the antibiotic penicillin is obtained, needs a source of carbon and a source of nitrogen.

(i) Explain why the mould fungus needs a source of carbon and a source of nitrogen. [4]

(ii) Explain how these sources of carbon and nitrogen are added. [3]

(c) Explain why a fermenter needs to have a water jacket around it. [4]

(d) Explain why the pH of the solution in a fermenter is monitored and controlled. [3]

(e) Other medically important products, such as insulin and growth hormone, are produced on a large scale using genetically modified microorganisms. Outline the advantages of using microorganisms for making these products. [5]

6 A 10-year study was carried out into the possibility of transgenic crops persisting in the wild in the event of their 'escaping' from cultivation. Four different transgenic crops and their non-transgenic counterparts were grown in 12 different natural habitats.

The four different transgenic crops were :
- oilseed rape tolerant of a herbicide
- maize tolerant of a herbicide
- sugar beet tolerant of a herbicide
- potato expressing an insecticide.

(a) Outline the advantages of growing **one** of the transgenic crops listed above. [3]

The mean percentage survival rate, at the end of the first growing season, of all the transgenic crops was less than that of their non-transgenic counterparts. Population sizes of all crops declined after the first year. In no case did transgenic crops survive significantly longer than their non-transgenic counterparts. All populations of oilseed rape, maize and sugar beet were extinct to all sites within four years. Non-transgenic potatoes survived in one habitat for longer than 10 years.

(b) Using the information above and your own knowledge, assess the potential danger to natural habitats of these transgenic crops **and** discuss the ethical implications of genetic engineering in agriculture. [6]

1 (a) Describe how plants may be produced commercially by:
   (i) taking cuttings; [3]
   (ii) grafting. [3]
Potatoes reproduce asexually by forming tubers on the end of underground stems or *stolons*. Stolons are formed from lateral buds. This process is shown in Figure 1.1.
An experiment was carried out to show the effect of light and dark on the formation of stolons and tubers in the potato. Twenty potato plants were grown. Ten of them were grown in such a way that the stolons were exposed to light; the other ten were grown with the stolons in the dark. The results are shown in Table 1.1.

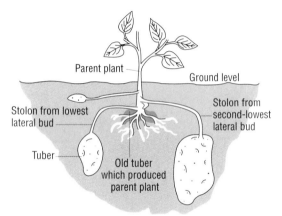

Figure 1.1

Table 1.1

Treatment	Number of stolons from lowest bud	Number of stolons from second-lowest bud	Total number of tubers from both stolons
Light	10	1	1
Dark	10	8	17

(b) With reference to Table 1.1, state the conclusions that can be drawn from this experiment. [4]
Young potato plants may also be grown by tissue culture.
(c) (i) Explain the advantages of producing potatoes by this method. [2]
   (ii) Suggest the possible consequences of the widespread cultivation of a small number of potato varieties. [2]
[Total: 14]
(OCR 2805 01 Jan02)

2 Figure 2.1 shows the growth in numbers of a population of microorganisms. The data were collected using the total (direct) count method.

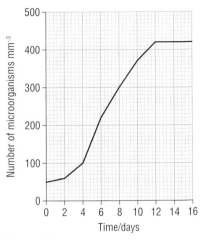

Figure 2.1

(a) (i) Describe how you would collect the data shown in Figure 2.1. [4]
   (ii) State **two** sources of error when collecting data on the growth of a population of microorganisms. [2]
   (iii) Give an explanation for the stationary phase between days 12 and 16. [4]
   (iv) Calculate the percentage increase rate of the population between day 6 and day 8. Show your working. [2]
(b) Describe how a prokaryote divides. [4]
[Total: 16]
(OCR 2805 01 Jan03)

3 (a) Duchenne muscular dystrophy (DMD) is a genetic disease caused by the absence of the protein dystrophin in muscle fibres. In the absence of dystrophin, muscle fibres gradually die.
A potential gene therapy for DMD involves injecting muscles with a viral vector carrying recombinant DNA (rDNA) for part of the normal allele for dystrophin. Outline the formation of recombinant DNA. [3]
(b) Mice with the symptoms of DMD were given this gene therapy shortly after birth. Each mouse was injected with the viral vector in a muscle of one hind limb. The corresponding muscle of the other hind limb was injected with a buffer solution to provide a control. The nuclei of muscle fibres that do not produce dystrophin move from the edge of the fibre to the centre. The fibres eventually die.
The percentage of muscle fibres with centrally placed nuclei was measured in fibres from treated and control muscles at different times after injection. The results are shown in Figure 3.1.

Answers to examination questions will be found on the Exam Café CD.

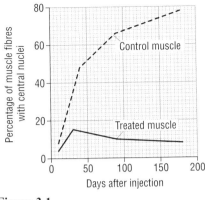

Figure 3.1

Using the information in Figure 3.1, describe the results of the experiment. [3]

[Total: 6]

(OCR 2805 02 Jun06)

4   (a) Give the term that corresponds to the definitions in **(i)** to **(iv)** below.
    (i) Large pressure cooker used to steam-sterilise equipment.
    (ii) Subunit of outer protein coat of a virus.
    (iii) Microorganism with an optimum growth temperature above 40 °C.
    (iv) Population phase in which the number of bacterial cells being produced equals the number of bacterial cells dying. [4]

(b) Figure 4.1 shows a laboratory fermenter (bioreactor) used by a student to **batch** culture microorganisms.

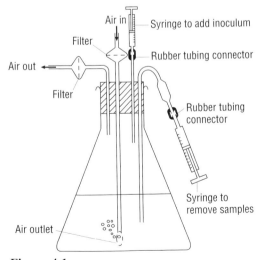

Figure 4.1

Explain how the student could modify the fermenter for **continuous fermentation**.
If you wish, you may add annotations to Figure 4.1 to help you in your answer. [4]

[Total: 8]

(OCR 2805 04 Jun06)

5   The enzyme amylase breaks down starch into a reducing sugar.

$$\text{starch} \xrightarrow{\text{amylase}} \text{maltose}$$

An experiment was carried out to investigate the activity of immobilised amylase. A solution of starch was passed through a column of immobilised enzyme as shown in Figure 5.1. Tests were carried out on the product collected at the end of the column.

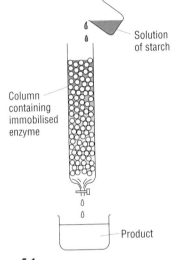

Figure 5.1

(a) Describe how amylase may have been immobilised. [1]
(b) Describe how the product could be tested to find out whether it had been contaminated by the enzyme. [3]
(c) To produce a greater yield, the product can be collected and passed through the column of immobilised enzyme. This process can be repeated.
Explain how it would be possible to determine experimentally how many times the process should be repeated to produce the optimal yield. [2]
(d) Explain **two** advantages of using immobilised enzymes in an industrial process. [4]

[Total: 10]

(OCR 2805 04 Jun03)

Module 3
Ecosystems and sustainability

Introduction

If you are a conservationist trying to conserve a particular species or sustainably manage a particular group of species in an ecosystem, your aim is to keep their population size at a high enough level for them to survive and reproduce. However, in any ecosystem, organisms do not work in isolation. They interact with other living organisms and with physical components of their environment. Therefore to conserve them effectively, you have to know the factors that affect the population size of the species you are interested in. In this module, you will learn more about how ecosystems work, and how to study them.

You may think that ecosystems pretty much stay the same all the time. However, many ecosystems are dynamic – they change all the time. Because all the populations of living organisms in an ecosystem interact with each other and with their physical environment, any small changes in one can affect the other. If the environment changes, population size may be affected, which itself may affect the environment. You will learn about the process of succession and how a more stable climax community is eventually reached.

You probably already know that energy is passed from one member of a food chain to another. The original source of energy is light energy from the Sun, which is converted by plants during photosynthesis. However, at each stage of the food chain, some energy is lost. We say that energy transfer between members of the food chain is inefficient. You will learn how this can limit the population size of organisms at different levels of the food chain. You will also learn about a number of other factors that may influence the population size of animals and plants.

Human activities can affect populations of animals and plants. You will learn how humans can manipulate the flow of energy through ecosystems, and how they can damage ecosystems. You will also learn how conservationists can help to manage ecosystems in a sustainable way.

Test yourself

1 What is an ecosystem?
2 What type of organisms are producers?
3 What do the arrows in a food chain represent?
4 Why do farmers use chemicals on their crops?
5 How do you estimate population size of a particular species of plant in a field?
6 What is conservation?
7 Why is conservation important?

Module contents

By the end of this spread, you should be able to . . .

* Define the term *ecosystem*.
* State that ecosystems are dynamic systems.
* Define the terms *biotic factor* and *abiotic factor* using named examples.
* Define the terms *producer*, *consumer*, *decomposer* and *trophic level*.

What is an ecosystem?

Any group of living organisms and non-living things occurring together, and the interrelationships between them, can be thought of as an **ecosystem**. Ecosystems can be on a large scale (like the African grassland in Figure 1) or on a smaller scale (like the pond in Figure 2).

Figure 1
African grassland

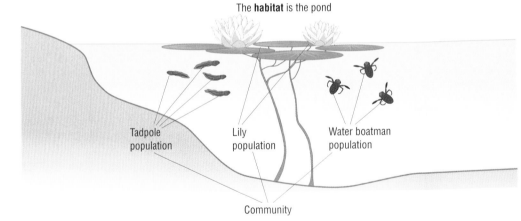

The **habitat** is the pond

Tadpole population

Lily population

Water boatman population

Community

Figure 2 The ecosystem of a pond

Key definitions

A **habitat** is the place where an organism lives.

A **population** is all of the organisms of one species that live in the same place at the same time, and that can breed together.

A **community** is all the populations of different species that live in the same place at the same time, and can interact with each other.

The components of an ecosystem include:
* **Habitat** – the place where an organism lives.
* **Population** – all of the organisms of one species, who live in the same place at the same time, and can breed together.
* **Community** – all the populations of different species who live in the same place at the same time, and can interact with each other.

The role that each species plays in an ecosystem is called its **niche**. Because each organism interacts with both living and non-living things, it is almost impossible to define its niche entirely. A description of its niche could include things like how and what it feeds on, what it excretes, how it reproduces, etc. It is impossible for two species to occupy *exactly* the same niche in the same ecosystem.

Depending on their niche, the living organisms in an ecosystem can affect each other. Such **biotic factors** include food supply, predation and disease. **Abiotic factors** describe the effects of the non-living components of an ecosystem – pH, temperature and soil type are all examples.

Ecosystems do not have clear edges. It is impossible to draw a line around a group of living things and to say that they interact only with each other, rather than other organisms outside that ecosystem. However, it is often useful to think of ecosystems as being 'closed', as it makes them easier to understand.

Ecosystems are dynamic

In most ecosystems, population sizes rise and fall, either very slightly or very noticeably. This is because the community of living things in an ecosystem interact with each other and with their physical environment. Any small changes in one can affect the other. For example:

- If a predator's population size goes up, the population size of the prey will go down (because more are being eaten more quickly).
- The nitrogen levels in soil can affect the population sizes of plants growing there. Nitrogen-fixing plants would grow successfully in nitrogen-deficient soil, but they would affect their environment by increasing the soil nitrogen levels. This change would then help other plants to grow there as well.

Energy and ecosystems

Matter is constantly recycled within an ecosystem – nutrient cycles, such as the nitrogen cycle and carbon cycle, are good examples. Energy is not recycled – it flows through the ecosystem (see Figure 3).

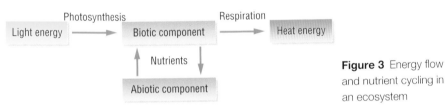

Figure 3 Energy flow and nutrient cycling in an ecosystem

All living organisms need energy. Via respiration, they release energy from organic molecules, such as glucose, in their food. This energy originally came from sunlight. At the start of nearly all food chains is a plant, which captures light energy through photosynthesis, and converts it to chemical energy stored in molecules like glucose.

- Because plants and other photosynthetic organisms, such as algae and some bacteria, supply chemical energy to all other organisms, they are called **producers**.
- Other organisms, like animals and fungi, are called **consumers**. **Primary consumers** are herbivores, who feed on plants, and who are eaten by carnivorous **secondary consumers**. These in turn are eaten by carnivorous **tertiary consumers**.
- Some living things called **decomposers** (bacteria, fungi and some animals) feed on waste material or dead organisms.

Questions

1 Look at the list below. Which of these are **(a)** biotic factors, **(b)** abiotic factors?

   Parasitism   Water pH   Light intensity   Competition

2 What is the difference between **(a)** a consumer and a producer, **(b)** a habitat and a niche?

3 Describe the niche of a rabbit that lives in a field.

4 Suggest why two species never occupy *exactly* the same niche in an ecosystem.

5 Thermal oceanic vents are examples of unusual ecosystems. There is no light on the ocean bed. What are the producers in this ecosystem and what form of energy do they use?

By the end of this spread, you should be able to . . .

✳ **Describe how energy is transferred through ecosystems.**
✳ **Outline how energy transfers between trophic levels can be measured.**
✳ **Discuss the efficiency of energy transfers between trophic levels.**

Transfer of energy in an ecosystem

A food chain shows how energy is transferred from one living organism to another (see Figure 1). The level at which an organism feeds is called its **trophic level**.

Key definition

The level at which an organism feeds in a food chain is called a **trophic level**.

Examiner tip

The arrows in a food chain show the direction of energy transfer, rather than just 'who eats what'.

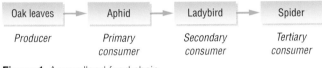

Oak leaves → Aphid → Ladybird → Spider

Producer — *Primary consumer* — *Secondary consumer* — *Tertiary consumer*

Figure 1 A woodland food chain

Within an ecosystem, living organisms are usually members of more than one food chain, and often feed at different trophic levels in different chains. Drawing these food chains together as a **food web** helps us to understand how energy flows through the whole ecosystem (see Figure 2).

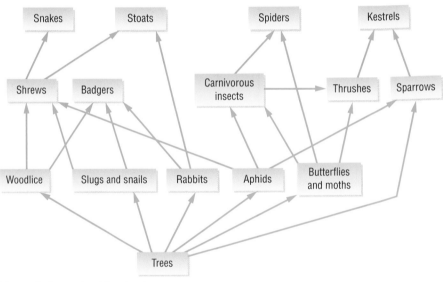

Figure 2 A woodland food web

Efficiency of energy transfer

At each trophic level, some energy is lost from a food chain, and is therefore unavailable to the organism at the next trophic level (see Figure 3):

- At each trophic level, living organisms need energy to carry out life processes. Respiration releases energy from organic molecules like glucose. Some of this energy is eventually converted to heat.
- Energy remains stored in dead organisms and waste material, which is then only available to decomposers, such as fungi and bacteria. This waste material includes parts of animals and plants that cannot be digested by consumers.

Because of this, there is less energy available to sustain living tissue at higher levels of the food chain, and so less living tissue can be kept alive. When the organisms in a food chain are about the same size, this means there will be fewer consumers at the higher

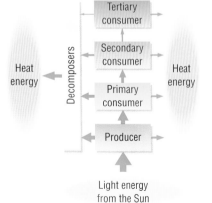

Figure 3 Energy loss through a food chain (the size of the arrows denotes the relative amounts of energy)

levels. Ecologists draw a **pyramid of numbers** to represent this idea (see Figure 4). The area of each bar in the pyramid is proportional to the number of individuals. Pyramids can be drawn for individual food chains, or for an ecosystem as a whole.

Measuring efficiency of energy transfer

Pyramids of biomass

Counting the number of organisms does not always provide an accurate picture about how much living tissue exists at each level. A better approach is to draw a **pyramid of biomass**, where the area of the bars is proportional to the **dry mass** of all the organisms at that trophic level. To do this properly, an ecologist would collect all the organisms and put them in an oven at 80 °C until all the water in them has been evaporated. Unfortunately, doing this is rather destructive to the ecosystem being studied, so ecologists often just measure the **wet mass** of the organisms and calculate the dry mass on the basis of previously published data.

Pyramids of energy

However, even pyramids of biomass still present problems, as different species may release different amounts of energy per unit mass. Because of this, ecologists sometimes prefer to construct a **pyramid of energy**. This involves burning the organisms in something called a calorimeter and working out how much heat energy is released per gram – this is calculated from the temperature rise of a known mass of water. Given that this too is destructive and also rather time-consuming, ecologists often revert to using pyramids of biomass instead.

Productivity

Even pyramids of energy have limitations.

- They only take a snapshot of an ecosystem at one moment in time.
- Because population sizes can fluctuate over time, this may provide a distorted idea of the efficiency of energy transfer.

As such, ecologists often look at the *rate* at which energy passes through each trophic level, drawing a pyramid of energy flow. This rate of energy flow is called **productivity**.

- Productivity gives an idea of how much energy is available to the organisms at a particular trophic level, per unit area (usually one square metre), in a given amount of time (usually one year) – it is measured in kilojoules or **megajoules** of energy per square metre per year.
- At the base of the food chain, the productivity of plants is called the primary productivity.
- The **gross primary productivity** is the rate at which plants convert light energy into chemical energy. However, because energy is lost when the plant respires, less energy is available to the primary consumer. This remaining energy is called the **net primary productivity** (NPP).

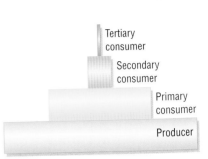

Figure 4 Pyramid of numbers

Key definitions

The rate at which energy passes through each trophic level in a food chain is a measure of the **productivity**.

A megajoule (MJ) is one million (10^6) joules or one thousand kilojoules.

Questions

1 Explain why ecologists may prefer to draw:
   (a) a food web instead of a food chain
   (b) a pyramid of energy instead of a pyramid of biomass.
2 Suggest why:
   (a) there are fewer individuals at higher trophic levels in a food chain
   (b) most food chains have no more than five stages.
3 Suggest why primary productivity is higher in tropical regions (closer to the equator) than in more temperate regions (closer to the poles).

By the end of this spread, you should be able to ...

✳ Explain how human activities can manipulate the flow of energy through ecosystems.

Primary productivity

Less than 1% of the sunlight energy reaching Earth is used for photosynthesis. The rest is reflected by clouds and the Earth's surface, used to heat the Earth's atmosphere or to evaporate water, or is of the wrong wavelength and is not captured by chlorophyll. The energy captured by leaves for photosynthesis is called the **primary productivity**. Some of this will be used by the plant and lost as **respiratory heat (R)**. The difference between primary productivity and R is the **net primary productivity (NPP)**. NPP is the rate of production of new biomass available for consumption by heterotrophs.

Improving primary productivity

When scientists measure NPP 'in the field', they find actual levels for crop plants are between 1% and 3%. By manipulating environmental factors, humans can increase NPP – making energy conversion more efficient, reducing energy loss and increasing crop yields.

- Light levels can limit the rate of photosynthesis, and hence NPP. Some crops are planted early to provide a longer growing season to harvest more light. Others are grown under light banks (Figure 1).

Key definitions

Primary productivity is the total amount of energy fixed by photosynthesis. It is the net flux of carbon from the atmosphere to plants, per unit time. It is a rate and may be measured in terms of energy per unit time, such as MJ m^{-2} yr^{-1}.

Net primary productivity is the rate at which carbohydrate accumulates in the tissue of plants of an ecosystem and is measured in dry organic mass, such as kg ha^{-1} yr^{-1}.

Net primary productivity = primary productivity – respiratory heat loss.

It is the amount of energy available to heterotrophs in the ecosystem. It is a fundamental ecological variable and is an important factor in determining the amount of biomass that a particular ecosystem can support.

Figure 1 High value crop plants are grown under light banks

- Lack of water is important in many countries. As well as irrigating crops, drought-resistant strains have been bred, e.g. drought-resistant barley in North Africa, wheat in Australia and sugar beet in the UK.
- Temperature can limit the speed of chemical reactions in a plant. Greenhouses can provide a warmer temperature for growing plants and therefore increase NPP (Figure 2). Planting field crops early to provide a longer growing season also helps to avoid the impact of temperature on final yield.
- Lack of available nutrients can slow the rate of photosynthesis and growth. Crop rotation (growing a different crop in each field on a rotational cycle) can help. Including a nitrogen-fixing crop, like peas or beans, in that cycle (see spread 2.3.6) replenishes levels of nitrates in the soil. Many field crops have been bred to be responsive to high levels of fertiliser.
- Pests, such as insects, caterpillars or nematodes, eat crop plants. They remove biomass and stored energy from the food chain, and lower the yield. Spraying with pesticides can help to reduce this loss. Some plants have also been bred to be pest-resistant, or have been genetically

Figure 2 Greenhouse cultivation of lettuce plants

modified with a bacterial gene (**Bt** gene) from *Bacillus thuringiensis*. In **Bt**-cotton (Figure 3) in the USA, the new gene confers resistance against bollworm, and in maize against corn-borers.

- Fungal diseases of crop plants can reduce NPP. Fungi cause root rot (reducing water absorption), damage xylem vessels (interfering with water transport), damage foliage through wilt, blight or spotting (interfering with photosynthesis directly), damage phloem tubes (interfering with translocation of sugars), or damage flowers and fruit (interfering with reproduction). Farmers spray crops with fungicides. Many crops have been bred to be resistant to fungal infections (e.g. Rhizomania resistance in sugar beet). Potatoes have been genetically modified to be resistant to potato blight.
- Competition from weeds for light, water and nutrients can reduce a crop's NPP. Farmers use herbicides to kill weeds.

Improving secondary productivity

Transfer of energy from producers to consumers is inefficient, as is transfer of energy from primary consumers to secondary consumers and beyond. Primary consumers don't make full use of plants' biomass – some plants die, consumers don't eat every part of the plant, and they don't digest everything they eat, egesting a lot of it in their faeces. Even when food is digested and absorbed, much of the stored energy is used to keep the animal alive, with only a small amount being stored when it grows. It is this small amount only that is available to the next consumer in the food chain – usually humans. However, it is still possible for humans to manipulate energy transfer from producer to consumer:

- A young animal invests a larger proportion of its energy into growth than an adult does. Harvesting animals just before adulthood minimises loss of energy from the food chain.
- In the past some farm animals have been treated with steroids to make them grow even more quickly, increasing the proportion of energy allocated to growth. However, this practice has been outlawed in the EU for many years.
- Selective breeding has been used to produce breeds with faster growth rates, increased egg production and increased milk production.
- Animals may be treated with antibiotics to avoid unnecessary loss of energy to pathogens and parasites.
- Mammals and birds waste a lot of energy walking around to find food, and keeping their body temperature stable. Approaches such as zero grazing for pig and cattle farming maximise energy allocated to muscle (meat) production by stopping the animals from moving about, by supplying food to them, and by keeping environmental temperature constant (Figure 4).

Although we say that transfer of energy from producers to consumers is inefficient and that grain could be used to feed humans directly as opposed to feeding cattle or pigs first, there are some areas where grain cannot be grown but animals can exist; for example, sheep can live on mountainsides. These areas are largely infertile and cannot be used to grow grain, but humans can eat the lamb produced.

Many people have serious concerns about modern farming practices and animal welfare. Deciding where the balance lies between welfare and efficient food production is a contentious topic that is constantly kept under review and should include informed public debate.

Figure 3 Genetically modified cotton is resistant to bollworm

Figure 4 Keeping cows indoors

Questions

1 Explain why modern agricultural practices may involve:
   (a) selective breeding   **(b)** steroid hormones   **(c)** genetic modification
   (d) greenhouses   **(e)** fertilisers   **(f)** pesticides.

2 Suggest which animals would be most efficient to farm: endotherms like birds and mammals (with a constant body temperature), or ectotherms like worms, fish and reptiles (whose body temperature varies). Explain your answer.

3 Rainforests are an important source of biodiversity. Some rainforests are being cleared to grow crops for animal feed. Because of this, some people have chosen to adopt a vegetarian diet. Explain their reasoning in terms of energy loss from the food chain.

By the end of this spread, you should be able to ...

* Describe one example of primary succession resulting in a climax community.

Changing ecosystems

Any change in a community of organisms can cause a change in their habitat. Any change in a habitat can also cause a change in the make-up of the community. These ideas can help explain why gradual directional changes happen in a community over time. Such a process of directional change is called **succession**.

How does succession happen?

The island of Surtsey in Iceland was created by a volcanic eruption in the 1960s, but is now home to a community of plants. Development of such a community from bare ground is known as primary succession, and comes about as follows:

- Algae and lichens begin to live on the bare rock. This is called a **pioneer community**.
- Erosion of the rock, and a build-up of dead and rotting organisms, produces enough soil for larger plants like mosses and ferns to grow (see Figure 1). These replace, or *succeed*, the algae and lichens.
- In a similar way, larger plants succeed these small plants, until a final, stable community is reached. This is called a **climax community**. In the UK, climax communities are often woodland communities.

Succession does not always start from bare ground. Secondary succession takes place on a previously colonised, but disturbed or damaged, habitat.

Succession on sand dunes

It can be difficult to understand how a habitat has changed, and how it's going to change. Sand dunes are interesting because they display all the stages of succession in the same place and at the same time. Look at the beach and sand dunes in Figure 2. Because the sea deposits sand on the beach, the sand nearest to the sea is deposited more recently than the sand further away.

Key definition

Succession is a directional change in a community of organisms over time.

Figure 1 Moss plants on the island of Surtsey

Figure 2 Sand dunes

This means that the sand just above the high water mark is at the start of the process of succession, whereas the sand much further away already hosts its climax community. By walking up the beach and through the dunes, it is possible to see each stage in the process of succession. The stages of succession are outlined below (see Figure 3). Eventually, a dune's community may develop into grassland, and then into woodland.

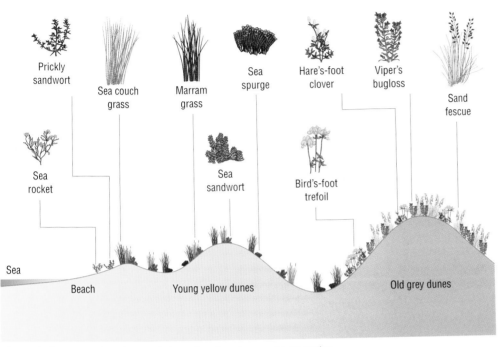

Figure 3 Cross-section of a sand dune showing stages of succession

1 Pioneer plants like sea rocket (*Cakile maritima*) and prickly sandwort (*Salsola kali*) colonise the sand just above the high water mark. These can tolerate salt water spray, lack of fresh water and unstable sand.
2 Wind-blown sand builds up around the base of these plants, forming a 'mini' sand dune. As plants die and decay, nutrients accumulate in this mini dune. As the dune gets bigger, plants like sea sandwort (*Honkenya peploides*) and sea couch grass (*Agropyron junceiforme*) colonise it. Because sea couch grass has underground stems, it helps to stabilise the sand.
3 With more stability and accumulation of more nutrients, plants like sea spurge (*Euphorbia paralias*) and marram grass (*Ammophila arenaria*) start to grow. Marram grass is special: its shoots trap wind-blown sand and, as the sand accumulates, the shoots grow taller to stay above the growing dune. This traps more sand.
4 As the sand dune and nutrients build up, other plants colonise the sand. Many, such as hare's-foot clover (*Trifolium arvense*) and bird's-foot trefoil (*Lotus corniculatus*), are members of the bean family (legumes). Bacteria in their root nodules convert nitrogen into nitrates (see spread 2.3.6). With nitrates available, more species, like sand fescue (*Festuca rubra*), and viper's bugloss (*Echium vulgare*), colonise the dunes. This stabilises them further.

Questions

1 Explain the meaning of the terms:
   (a) primary succession
   (b) pioneer community
   (c) climax community.
2 Compare and contrast the process of succession on bare rock and on a sand dune.
3 Explain why all stages of succession are visible in a sand dune.

⑤ Studying ecosystems

By the end of this spread, you should be able to . . .

✳ **Describe how the distribution and abundance of organisms can be measured, using line transects, belt transects, quadrats and point quadrats.**

Sampling

Ecologists usually study ecosystems to find out whether the abundance and distribution of a species is related to that of other species, or to environmental factors, such as light intensity or soil pH. Obviously it would be ideal to count every individual of every species, but in most habitats this is almost impossible. Ecologists get around this problem by taking samples from the habitat. This means selecting small portions of the habitat and studying them carefully.

Quadrats

Imagine that an ecologist wants to compare the abundance and distribution of plant species in two different fields. It is impossible to count all the individuals in the fields, so they decide to sample small parts of the habitat using a **quadrat**.

A quadrat is a square frame that defines the sample area (Figure 1). It is often 1 m square, and can have strings across every 10 cm, separating it into 100 smaller squares. You can collect two types of data using a quadrat. You can record simply *presence or absence of each species* (**distribution**), or you can *estimate or count the number of individuals* (**abundance**) of each species. For some plants, like grass and moss, it is difficult to count individuals, so ecologists tend to estimate **percentage cover**.

Estimating percentage cover is very difficult. You can estimate the cover in each of the smaller squares within the quadrat, but this is not very accurate. Using a point frame can help improve the estimate (Figure 1). Stand the point quadrat firmly on the ground, lower each needle downwards and record the species that the tip touches on its way down. The number of needles that touch each species is proportional to the percentage cover of that species. Don't forget to record bare ground!

Before starting to sample, it is important to decide where to place the quadrats, how many samples to take, and how big the quadrat should be.

- If you take samples only from one corner of the field, it may be that the soil in this corner is particularly rich in nitrates, and the species growing there are different from those in the rest of the field. To avoid biasing the sample, and to provide a sample which is representative of the whole habitat, either: (1) randomly position the quadrats across the habitat, using random numbers to plot coordinates for each one, or (2) take samples at regular distances across the habitat, so you sample every part of the habitat to the same extent.

- Just looking at one quadrat will not give an accurate representation of the whole habitat, but using 20 000 would take forever! To work out how many are needed, ecologists carry out a pilot study. They take random samples from across the habitat and make a cumulative frequency table, like the one shown in Table 1. They then plot cumulative frequency against quadrat number. The point where the curve levels off tells them the minimum number of quadrats to use. Ecologists often double this number.

- You can do something similar to work out how big your quadrats should be. Count the number of species you find in larger and larger quadrats. Plot quadrat area on the *x*-axis, against the number of species you find in each one on the *y*-axis. Read the optimal quadrat size at the point where the curve starts to level off.

Having collected the data, the equation below shows you how to estimate the size of each species' population in the whole habitat.

$$\text{population size of a species} = \frac{\text{mean number of individuals of the species in each quadrat}}{\text{fraction of the total habitat area covered by a quadrat}}$$

Examiner tip

A quadrat and a quadra*nt* are different. A quadrat is a square frame used for studying ecosystems, but a quadra*nt* is an instrument used to measure angles. So don't call a quadrat a quadrant!

Figure 1 Using a quadrat and point frame

Examiner tip

If you use a quadrat in your coursework, always state the size of quadrat you use.

Quadrat number	Number of new species found in this quadrat	Total number of different species found
1	5	5
2	4	9
3	4	13
4	3	16
5	3	19
6	2	21
7	2	23
8	2	25
9	1	26
10	1	27
11	1	28
12	0	28
13	0	28
14	0	28
15	0	28

Table 1 Cumulative frequency table of quadrat number against number of species found

Transects

It is also possible to look more systematically for changes in vegetation across a habitat. For example, you may want to look at the changes in abundance and distribution of species as you walk up a beach and through the sand dunes (spread 2.3.4), measuring the environmental conditions at the same time. This is done using a **transect**, which is a line taken across a habitat. It is usually easiest to stretch out a tape measure, and then take samples at regular intervals along the tape. The distance between samples will depend on the length of the line you want to look at, and the density of plants in the habitat. There are two approaches to using a transect.

- Line transect – at regular intervals, make a note of which species is touching the tape (Figure 2).
- Belt transect – at regular intervals, place a quadrat next to the line (interrupted belt transect), studying each, as described above. Alternatively, place a quadrat next to the line, moving it along the line after looking at each quadrat (continuous belt transect) (Figure 3).

Questions

1. Explain why you have to take samples from a habitat.
2. Suggest the advantages of taking those samples **(a)** at random, **(b)** at regular intervals through the habitat.
3. Explain **(a)** how you would estimate the abundance and distribution of species throughout a salt marsh, **(b)** how you would compare the abundance and distribution of species between a school playing field and a meadow.
4. Suggest the advantages and disadvantages of using a continuous belt transect compared to a line transect.

Examiner tip

Random sampling does not mean throwing the quadrat over your shoulder without looking. You need to lay out two tape measures on two edges of the study site, so that they look like axes on a graph. Use your calculator, or a pair of dice, or a random number table, to generate pairs of numbers. Use these pairs as coordinates to place your quadrats.

Figure 2 Using a line transect

Figure 3 Using a belt transect

By the end of this spread, you should be able to . . .

* **Describe the role of decomposers in the decomposition of organic material.**
* **Describe how microorganisms recycle nitrogen within ecosystems.**

Decomposing organic material

Spread 2.3.2 indicated how energy and materials are lost from a food chain when living things excrete waste or die. This dead and waste organic material can be broken down by decomposers – microorganisms such as bacteria and fungi.

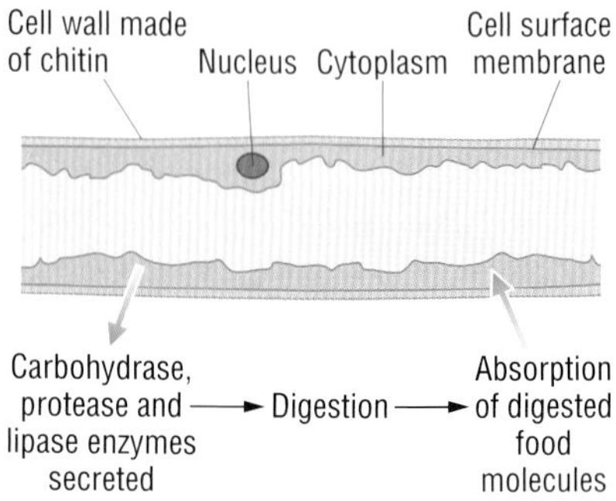

Bacteria and fungi involved in decomposition feed in a different way from animals. They feed **saprotrophically** so they are described as **saprotrophs**.

* Saprotrophs secrete enzymes onto dead and waste material.
* These enzymes digest the material into small molecules, which are then absorbed into the organism's body (Figure 1).
* Having been absorbed, the molecules are stored or respired to release energy.

If bacteria and fungi did not break down dead organisms, energy and valuable nutrients would remain trapped within the dead organisms. By digesting dead and waste material, microbes get a supply of energy to stay alive, and the trapped nutrients are recycled. Microorganisms have a particularly important role to play in the cycling of carbon and nitrogen within ecosystems.

Figure 1 Saprotrophic feeding by a fungus

Cell wall made of chitin · Nucleus · Cytoplasm · Cell surface membrane

Carbohydrase, protease and lipase enzymes secreted → Digestion → Absorption of digested food molecules

Recycling nitrogen within an ecosystem

Living things need nitrogen to make proteins and nucleic acids. Figure 2 shows how nitrogen atoms are cycled between the biotic and abiotic components of an ecosystem. Bacteria are involved in ammonification, nitrogen fixation, nitrification and denitrification.

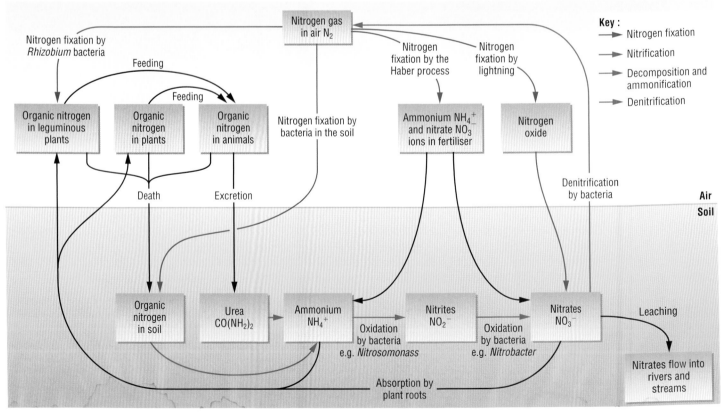

Figure 2 The nitrogen cycle

Nitrogen fixation

Although nitrogen gas makes up 79% of the Earth's atmosphere, it is very unreactive. This means it is impossible for plants to use it directly (even though it is so abundant). Instead, they need a supply of 'fixed' nitrogen such as ammonium ions (NH_4^+) or nitrate ions (NO_3^-). Nitrogen fixation can occur when lightning strikes or through the Haber process. However, these processes only account for about 10% of nitrogen fixation around the world.

Nitrogen-fixing bacteria account for the rest. Many of these live freely in the soil and fix nitrogen gas, which is in the air within soil, using it to manufacture amino acids.

Nitrogen-fixing bacteria, such as *Rhizobium*, also live inside the root nodules (Figure 3) of plants such as peas, beans and clover, which are all members of the bean family.
- They have a **mutualistic** relationship with the plant: the bacteria provide the plant with fixed nitrogen and receive carbon compounds, such as glucose, in return.
- Proteins, such as leghaemoglobin, in the nodules absorb oxygen and keep the conditions anaerobic. Under these conditions the bacteria use an enzyme, nitrogen reductase, to reduce nitrogen gas to ammonium ions that can be used by the host plants.

Nitrification

Nitrification happens when **chemoautotrophic** bacteria in the soil absorb ammonium ions.
- Ammonium ions are released by bacteria involved in putrefaction of proteins found in dead or waste organic matter.
- Rather than getting their energy from sunlight (like photoautotrophic bacteria, algae and plants), chemoautotrophic bacteria obtain it by oxidising ammonium ions to nitrites (*Nitrosomonas* bacteria), or by oxidising nitrites to nitrates (*Nitrobacter* bacteria).
- Because this oxidation requires oxygen, these reactions only happen in well-aerated soils.
- Nitrates can be absorbed from the soil by plants and used to make nucleotide bases (for nucleic acids) and amino acids (for proteins).

Figure 3 Root nodules of a plant from the legume (bean) family. Nitrogen-fixing bacteria live inside the nodules, which grow in response to chemical signals from the bacteria (×1.25)

Denitrification

Other bacteria convert nitrates back to nitrogen gas. When the bacteria involved are growing under anaerobic (without oxygen) conditions, such as in waterlogged soils, they use nitrates as a source of oxygen for their respiration and produce nitrogen gas (N_2) and nitrous oxide (N_2O).

Questions

1 Why are bacteria called saprotrophs?
2 Write down the names of three species of bacteria that are involved in the nitrogen cycle, together with their roles.
3 Explain the difference between *nitrogen fixation* and *nitrification*.
4 What type of respiration do denitrifying bacteria carry out?

By the end of this spread, you should be able to ...

∗ **Explain the significance of limiting factors in determining the final size of a population.**

∗ **Explain the meaning of the term *carrying capacity*.**

∗ **Describe predator–prey relationships and their possible effects on the population sizes of both the predator and the prey.**

Carrying capacity and limiting factors

In some circumstances, a species' population size may stay fairly stable over time. However, population size may also rise or fall quite suddenly, or oscillate up and down with a regular pattern. The size of a population depends upon the balance between the death rate (mortality) and the rate of reproduction. To understand what affects population size, it is important to understand how populations grow (Figure 1).

- At **a** (the lag phase), there may only be a few individuals, still acclimatising to their habitat. At this point, the rate of reproduction is low, and the growth in population size is slow.
- At **b** (the log phase), resources are plentiful and conditions are good. The rate of reproduction is fast and exceeds mortality. The population size increases rapidly.
- At **c** (the stationary phase), the population size has levelled out at the **carrying capacity** of the habitat – the habitat itself cannot support a larger population. In this phase, the rates of reproduction and mortality are equal. The population size therefore stays stable, or fluctuates very slightly up and down in response to small variations in environmental conditions each year.

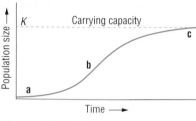

Figure 1 Population growth

> ### Key definition
>
> **Carrying capacity** is the maximum population size that can be maintained over a period of time in a particular habitat.

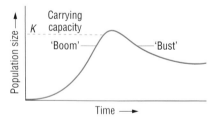

Figure 2 Population growth in *r*-strategist species

STRETCH and CHALLENGE

Species whose population size is determined by the carrying capacity are often called *k*-strategists. For these populations, limiting factors exert a more and more significant effect as the population size gets closer to the carrying capacity, causing the population size to gradually level out.

However, some species adopt a different type of population growth. In these cases, the population size increases so quickly that it can exceed the carrying capacity of the habitat before the limiting factors start to have an effect. Once the carrying capacity has been exceeded, there are no longer enough resources to allow individuals to reproduce or even to survive. Likewise, an excessive build-up of waste products may start to poison them, and they begin to die, entering a death phase (Figure 2).

This type of population growth is known as 'boom and bust', and the species that adopt it are known as *r*-strategists – the most important influence on population growth being the physical rate (*r*) at which individuals can reproduce. This type of growth is characteristic of species with short generation times (such as bacteria), and of pioneer species (see spread 2.3.4). Quick population growth means pioneer species colonise a disturbed habitat before *k*-strategists, dispersing to other habitats once limiting factors start to have an effect. In reality, *r*- and *k*-strategies represent two ends of a continuum of strategies adopted by living things.

Questions

A Look back at spread 2.3.4. What type of succession begins from a disturbed habitat?

B Why are *r*-strategists good at colonising disturbed habitats?

C Are *k*-strategists or *r*-strategists more likely to be members of a climax community? Explain your answer.

Module 3
Ecosystems and sustainability
What affects population size?

The habitat cannot support a larger population because of factors that limit the growth in population size. These are called **limiting factors**, and may include the availability of resources, such as food, water, light, oxygen, nesting sites or shelter. They may also include the effects of other species, including parasites and predators, or the intensity of competition for resources, both with individuals of the same species and individuals of other species. The carrying capacity is the upper limit that these factors place on the population size.

Predators and prey

A predator is an animal that hunts other animals (prey) for food. Predation can act as a limiting factor on a prey's population size, which in its turn can affect the predator's population size. Figure 3 shows how this happens:

1 When the predator population gets bigger, more prey are eaten.
2 The prey population then gets smaller, leaving less food for the predators.
3 With less food, fewer predators can survive and their population size reduces.
4 With fewer predators, fewer prey are eaten, and their population size increases.
5 With more prey, the predator population gets bigger, and the cycle starts again.

Figure 3 comes from an experiment conducted in a laboratory, where the predators only ate one type of prey, and predation was the main limiting factor on the prey's population. However, in the wild, predators often eat more than one type of prey, and there are a number of other limiting factors. Because of this, studies of predators and prey in the wild yield graphs of a similar, but not so well-defined, shape (Figure 4).

In areas with higher species diversity than the extreme environment inhabited by lynx and snowshoe hares, this pattern is even more difficult to see. The exact effects of other limiting factors are still a subject of intense debate. Most scientists agree that predation is responsible for a drop in the numbers of hares, but they think that lack of food is also responsible for the slowing of the rate of increase in hare population size.

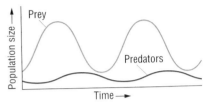

Key definition

Where the rate of a natural process is affected by a number of factors, the **limiting factor** is the one whose magnitude limits the rate of the process. It is often the factor in shortest supply.

Figure 3 Relationship between a population of a predator and its prey

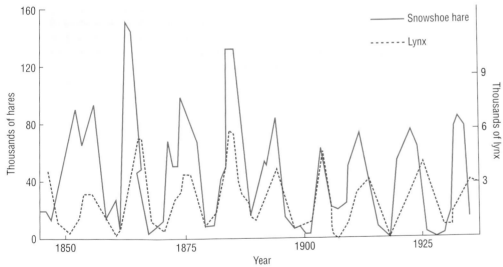

Figure 4 Relationship between the number of lynx and snowshoe hares in northern Canada, estimated from the number of pelts sold

Questions

1 List three biotic and two abiotic factors that could act as limiting factors.
2 For each of the following, compare the death rate and the reproduction rate:
   (a) lag phase
   (b) log phase
   (c) stationary phase.
3 Figure 3 derives from a laboratory experiment, and shows clearly how the populations of predators and prey can be dependent upon each other. Figure 4 comes from studies on wild populations. Explain why the relationship shown in Figure 4 is not as clear as that shown in Figure 3.
4 Explain the connection between limiting factors and carrying capacity.

By the end of this spread, you should be able to ...

❋ Explain, with examples, the terms *interspecific* and *intraspecific competition*.

Competition

Key definition

Competition happens when resources (like food or water) are not present in adequate amounts to satisfy the needs of all the individuals who depend on those resources.

If a resource in an ecosystem is in short supply, there will be **competition** between organisms for that resource. As the intensity of competition increases, the rate of reproduction decreases (because fewer organisms have enough resources to reproduce), whilst the death rate increases (because fewer organisms have enough resources to survive). There are two types of competition: **intraspecific** competition and **interspecific** competition.

Intraspecific competition

Intraspecific competition happens between individuals of the *same species*. As factors, such as food supplies, become limiting, individuals have to compete for them. Those individuals best adapted to obtaining food will survive and reproduce, while those not so well adapted will die or will fail to reproduce. As explained in spread 2.3.7, this slows down population growth and the population enters the stationary phase.

Although there are slight fluctuations in population size during the stationary phase, intraspecific competition keeps the population relatively stable.
- If the population size drops, competition reduces, and the population size then increases.
- If the population size increases, competition increases, and the population size then drops.

Interspecific competition

Interspecific competition happens between individuals of *different species*, and can affect both the population size of a species and the distribution of species in an ecosystem.

Some of the classic work on interspecific competition was carried out in 1934 by a Russian scientist called Georgyi Frantsevitch Gause. He grew two species of *Paramecium* (Figure 1), both separately and together. When together, there was competition for food,

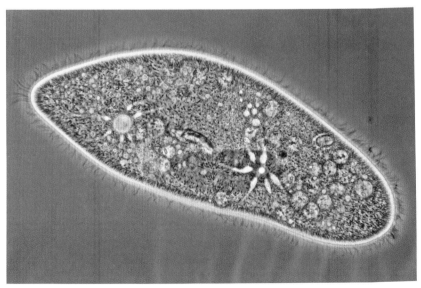

Figure 1 *Paramecium caudatum* (×540)

with *Paramecium aurelia* obtaining food more effectively than *Paramecium caudatum*. Over a period of 20 days, the population of *Paramecium caudatum* reduced and died out, whereas the population of *Paramecium aurelia* increased, eventually being the only species remaining (see Figure 2).

Gause concluded that more overlap between two species' niches would result in more intense competition. If two species have exactly the same niche, one would be out-competed by the other and would die out or become extinct in that habitat. This idea became known as the *competitive exclusion principle*, and can be used to explain why particular species only grow in particular places.

Often, though, it is not quite that simple. Other observations and experiments suggest that extinction is not necessarily inevitable.

- Sometimes, interspecific competition could simply result in one population being much smaller than the other, with both population sizes remaining relatively constant.
- It is also important to realise that in the laboratory it is easy to exclude the effects of other variables, so the habitat of the two species remains very stable. In the wild, however, a wide range of variables may act as limiting factors for the growth of different populations. These variables may change on a daily basis, or over the course of a year.

For example, experiments on competition between flour beetles *Tribolium confusum* and *Tribolium castaneum* initially confirmed the competitive exclusion principle – the *T. castaneum* population size increased, whilst the *T. confusum* population died out (Figure 3) – but even a small change in the temperature could change the outcome so that *T. confusum* survived instead.

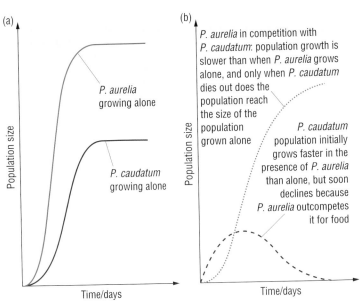

Figure 2 Population growth of two species of *Paramecium* grown **(a)** separately and **(b)** together

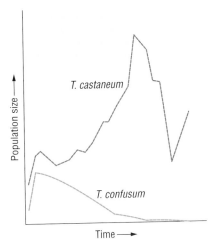

Figure 3 Changes in population size of *Tribolium castaneum* and *Tribolium confusum* over time

STRETCH and CHALLENGE

Competition between plants

We usually think of competition between species in relation to their niche – those better adapted survive and reproduce, whereas those less well adapted do not. However, some plants have a mechanism whereby they can interfere with neighbouring plants' physiology by releasing chemicals into their habitat. This phenomenon is called *allelopathy,* and is a form of competition because it stops their neighbours from using the resources in the habitat. The chemicals may inhibit growth, germination, or nutrient uptake.

Allelopathic chemicals can be found in almost any part of a plant. They can be released into the soil directly by the roots, or they may leach out of leaves and fruit when a plant sheds them. For example:

- Black walnut trees release a chemical called juglone, which inhibits respiration in neighbouring plants. This gives a young walnut tree better access to light, by preventing other plants from crowding it out.
- Sorghum (a cereal plant) also releases an allelopathic chemical (called sorgolene) through its roots. This disrupts photosynthesis and respiration in neighbouring plants, giving the sorghum greater access to resources such as light and water.

Question

A Explain why allelopathy is also known as chemical competition.

Questions

1 What is the difference between intraspecific and interspecific competition?
2 Explain how a species' population size can be affected by:
   (a) intraspecific competition
   (b) interspecific competition.
3 Explain why the competitive exclusion principle does not always apply in natural ecosystems.

By the end of this spread, you should be able to . . .

✳ Explain how the management of an ecosystem can provide resources in a sustainable way, with reference to timber production in a temperate country.

Sustainable management

In the past, our use of natural resources was mainly small scale. This caused minimal damage to our ecosystem, and the population sizes of species would stay fairly stable. In other words, our exploitation of natural resources was sustainable.

However, the human population size is getting larger increasingly quickly – it is expanding exponentially. Because of this, humans have had to use more intensive methods to exploit our environment for resources (see spread 2.3.3). Such approaches can disrupt or destroy ecosystems, reduce biodiversity, and even completely get rid of the resource we originally wanted to harvest.

One situation in which there is potential conflict between our need for resources and conservation is in wood and timber production. However, sustainable management and exploitation of forests and woodlands is possible. This can mean that biodiversity is maintained, and that wood and timber companies can have a financially secure and sustainable supply of wood. Such management can happen on both a small scale and a large scale.

Figure 1 A coppiced woodland

Key definition

Coppicing involves cutting a tree trunk close to the ground to encourage new growth.

Managing small-scale timber production

Coppicing is a traditional approach to obtaining a sustainable supply of wood. It is rather like using a tree as a wood factory; harvesting wood whilst keeping the tree alive. It involves cutting the trunk of a deciduous tree – one that loses its leaves in the winter – close to the ground. Once cut, several new shoots grow from the cut surface, and eventually mature into stems of quite narrow diameter (Figure 1). These can be cut and used for fencing, firewood or furniture. After cutting, new shoots start to grow again, and the coppice cycle continues.

Pollarding is like coppicing, but involves cutting the trunk higher up (see Figure 2). Pollarding is useful when the population size of deer is high, as they like to eat the emerging shoots from a coppiced stem. If cut higher up, the deer cannot reach the shoots.

To provide a continuous supply of wood, woodland managers divide a wood into sections and cut one section each year until they've all been cut. This is called *rotational coppicing*. By the time they want to coppice the first section again, the new stems have matured and are ready to be cut. The length of this rotation can vary. It depends on the time taken for stems to mature (which depends on the species), and on the dimensions of wood required. In each section, some trees are left to grow larger without being coppiced. These trees are called *standards*, and they are eventually harvested to supply larger pieces of timber.

Rotational coppicing is very good for biodiversity. Left unmanaged, woodland goes through a process of succession (see spread 2.3.4), blocking out light to the floor of the woodland, and reducing the number of species that can grow there. By using rotational coppicing, different areas of woodland provide different types of habitat, letting more light in and increasing the number and diversity of species.

Figure 2 A pollarded tree

STRETCH and CHALLENGE

Bradfield Woods

Bradfield Woods in Suffolk has been managed sustainably since 1252 and is one of the oldest managed woodlands in the UK. The woods were once owned by monks from a nearby abbey. They would still recognise the woods today, as the woodland is managed in the same way as it was then. It is home to over 370 species of plants and has thriving populations of badger and dormouse (Figure 3).

Figure 3 A hibernating dormouse. They live in the canopy of the woodland, eating insects, seeds and fruits

Bradfield Woods is managed by rotational coppicing on a 20–25 year cycle. Different areas are cut each year, providing a constant and sustainable source of wood, encouraging the growth of oxslips, anemones, violets and wood-spurge, and providing good cover for birds like nightingales, the Roe warbler and blackcaps. The majority of coppicing happens with hazel and ash trees, whilst oak trees are left as standards for timber (see Figure 4). Oak and ash trees on the edge of the wood are pollarded to prevent livestock eating the emerging shoots. The cut stems are used for firewood, fencing and roofing.

Question

A Explain what would happen to biodiversity in Bradfield Woods if the Suffolk Wildlife Trust stopped managing it.

Figure 4 Coppiced woodland with oak standards

Managing large-scale timber production

Large-scale production of wood for timber often involves clear-felling all the trees in one area. This can destroy habitats on a large scale and is now rarely practised in the UK. Clear-felling the trees can reduce soil mineral levels and leave soil susceptible to erosion. Soil may run off into waterways, polluting them. This is because trees usually:

- remove water from soil and stop soil being washed away by rain
- maintain soil nutrient levels through the trees' role in the carbon and nitrogen cycles.

Leaving each section of woodland to mature for 50–100 years before felling allows biodiversity to increase. However, such a timescale is not cost-effective. Modern sustainable forestry works on the following principles:

- Any tree which is harvested is replaced by another tree, either grown naturally or planted.
- Even with extraction of timber, the forest as a whole must maintain its ecological function regarding biodiversity, climate and mineral and water cycles.
- Local people should derive benefit from the forest.

Selective cutting involves removing only the largest, most valuable trees. This means that the habitat is broadly unaffected.

Sustainably managing woodland is a difficult job, and involves a balance between harvesting wood to make the woodland pay for itself, and conservation – both to enable a continued supply of wood and to maintain biodiversity. If each tree supplies more wood, fewer trees will need to be harvested. To achieve this, foresters:

- control pests and pathogens
- only plant particular tree species where they know they will grow well
- position trees an optimal distance apart. If trees are too close, this will cause too much competition for light, and they will grow tall and thin, producing poor quality timber.

Questions

1 What is meant by the phrase *sustainable management of wood production*?

2 Write down the differences and similarities between small-scale and large-scale production of wood.

By the end of this spread, you should be able to ...

* Explain that conservation is a dynamic process involving management and reclamation.
* Discuss the economic, social and ethical reasons for conservation of biological resources.
* Distinguish between the terms *conservation* and *preservation*.

Key definition

Conservation involves the maintenance of biodiversity, including diversity between species, genetic diversity within species, and maintenance of a variety of habitats and ecosystems.

Figure 1 Sugar beet researchers are studying wild sea beet (*Beta vulgaris* ssp. *maritima*) as a source of disease resistance for the commercial sugar beet crop

Figure 2 Aspirin was originally derived from the bark of willow trees (*Salix alba*)

Why is conservation important?

It is easy to think that **conservation** is about keeping things 'natural'. However, in the UK, very few habitats are truly natural. Most conservation programmes try to focus on maintaining **biodiversity**. This includes not just diversity between species, but also maintenance of genetic diversity within species, and maintenance of a range of habitats and ecosystems. Unfortunately, a steadily increasing human population can threaten biodiversity through:

* over-exploitation of wild populations for food (e.g. cod in the North Sea), for sport (e.g. sharks) and for commerce (e.g. pearls collected from saltwater oysters and freshwater clams): species are harvested faster than they can replenish themselves
* habitat disruption and fragmentation as a result of more intensive agricultural practices, increased pollution, or widespread building
* species introduced to an ecosystem by humans, deliberately or accidentally. These may out-compete native species, which may become extinct.

Ethics

Many conservationists would argue that these kinds of problems make conservation essential. They believe that every species has value in its own right, irrespective of whether it has financial value to humans. As such, every living thing has a right to survive, and humans have an ethical responsibility to look after them.

Although these arguments are laudable, they are subjective. The arguments in favour of human activities that work against conservation (e.g. burning fossil fuels, open-cast mining) are driven by economics. Expressing the value of conservation in economic terms is likely to be more effective in making governments give priority to conservation.

Economic and social reasons

Many species already have direct economic value when harvested. Others may also have direct value that is as yet unrecognised, and may provide benefit in the future.

* Many species provide a valuable food source, and were originally domesticated from wild species. Genetic diversity in wild strains may be needed in future to breed for disease resistance, drought tolerance or improved yield (Figure 1). Likewise, new species may be domesticated for food use.
* Natural environments are a valuable source of potentially beneficial resources. Many of the drugs we use today were discovered in wild plant species (Figure 2).
* Natural predators of pests can act as biological control agents. This can have advantages over the use of synthetic chemicals, although each situation is different. So far few such species are used.

Many species also have indirect economic value. For example, wild insect species are responsible for pollinating crop plants. Without them, harvests would fail and farmers go out of business. Likewise, other communities maintain water quality, protect soil and break down waste products. There is even evidence that a reduction in biodiversity may reduce climatic stability, resulting in drought or flooding and associated economic costs.

Ecotourism and recreation in the countryside also have significant social and financial value, which derives from the aesthetic value (they look nice!) of living things. Ecotourism depends on maintenance of biodiversity, and there is even a sizeable industry in natural history books, films and other media.

Although conservation involves management and reclamation, preservation can also be important to maintaining biodiversity. Preservation usually involves protecting areas of land, as yet unused by humans, in their 'untouched' form.

What does conservation involve?

Successful conservation requires consideration of the social and economic costs to the local community, as well as effective education and liaison with the community. Conservation can involve establishing protected areas like National Parks, green belt land or Sites of Special Scientific Interest (SSSIs). It can also involve giving legal protection to endangered species, or conserving them *ex-situ* in zoos or botanic gardens. However, maintaining biodiversity in dynamic ecosystems requires careful management to maintain a stable community, or even to reclaim an ecosystem by reversing the effects of human activity.

Some management strategies are outlined below; which strategies are adopted depends on the specific characteristics of the ecosystem and species involved.
- Raise carrying capacity by providing extra food.
- Move individuals to enlarge populations, or encourage natural dispersal of individuals between fragmented habitats by developing dispersal corridors of appropriate habitat.
- Restrict dispersal of individuals by fencing.
- Control predators and poachers.
- Vaccinate individuals against disease.
- Preserve habitats by preventing pollution or disruption, or intervene to restrict the progress of succession (see spread 2.3.4) by, for example, coppicing, mowing or grazing.

Sometimes, simple management is inappropriate, as disruption of a community may have gone too far. Understanding which species were part of the original community is not always clear, and succession is likely to take a long time before it allows such a community to survive again. Short-cutting that process requires detailed knowledge of all the species involved. Where environmental conditions have remained fairly stable, it is possible to clean up pollution, to remove unwanted species (see Figure 3), or to recolonise with the original species, for example from captive breeding programmes. However, it is often easier, and more successful, to 'replace' a disrupted community with a slightly different community, rather than to rehabilitate the original community.

Figure 3 South American coypus were inadvertently released from fur farms in southern England, causing significant damage to local plant communities. A capture programme was established to eradicate them

Questions

1 Conservation involves maintenance of biodiversity.
   (a) List the reasons why it is important to maintain biodiversity.
   (b) List the ways in which humans may reduce biodiversity.
2 Explain the differences between the direct and indirect financial value of a species.
3 Explain why:
   (a) it is impossible to conserve an ecosystem simply by putting a fence around it
   (b) reclamation of a habitat is so difficult.

(11) Humans and the Galapagos

By the end of this spread, you should be able to . . .

✳ **Outline, with examples, the effects of human activities on the animal and plant populations in the Galapagos Islands.**

Figure 1 The Galapagos Islands

Charles Darwin's visit to the Galapagos Islands (Figure 1) in 1835 provided the stimulus for his theory of natural selection. The islands' isolation and small population sizes provide optimal conditions for rapid evolutionary change. The Galapagos form part of one of the best-conserved tropical archipelagos and have high numbers of native species. These include the famous Darwin's finches, giant tortoises (Figure 2) and land iguanas (Figure 3). For these reasons, the United Nations allocated them World Heritage Site status in 1978. Unfortunately, 50% of vertebrate species and 25% of plant species on the islands are recognised as endangered, and conservation issues on the Galapagos reflect similar issues in the rest of the world. The population of the Galapagos has grown in response to a developing tourist trade, an expanding demand for marine products like sea cucumbers and lobster, and economic problems in mainland Ecuador in the 1990s. In 1980, the population was about 5000, and 4000 tourists visited the islands each year. In 2005, the population was 28 000 and 100 000 tourists visited the islands.

Figure 2 The giant tortoise

Figure 3 The land iguana

Habitat disturbance

This dramatic increase in population size has placed huge demands on water, energy and sanitation services, which the authorities are struggling to meet. More waste and pollution have been produced, and the demand for oil has increased. An oil spill in 2001 had an adverse effect on marine and coastal ecosystems. Increased pollution, building, and conversion of land for agriculture have caused destruction and fragmentation of habitats. Forests of *Scalesia* trees and shrubs (a genus unique to the islands) have been almost eradicated on Santa Cruz and San Cristobal to make way for agricultural land.

Over-exploitation of resources

In the nineteenth century, whaling boats and fur-traders took up residence, harvesting whales and seals to sell internationally. Species were harvested faster than they could replenish themselves. Giant tortoises were taken because they could survive on little food in the hold of a ship for a long time, before being killed and eaten. This had a catastrophic effect on tortoise populations, with 200 000 tortoises taken in less than half a century. The last remaining Pinta tortoise, nicknamed Lonesome George, now lives in the grounds of the Charles Darwin Research Station, which has begun a captive breeding programme to supplement tortoise numbers. Likewise, the more recent boom in fishing for exotic species in the 1990s had left populations seriously depleted. Depletion of sea cucumber populations has a drastic effect on under-water ecology, and the international market for shark fin has led to the deaths of 150 000 sharks each year around the islands. This includes 14 species listed as endangered.

Introduced species

With humans come non-native species that may have an impact on existing communities. Many species, such as goats, cats, and fruit and vegetable plants, were brought to the islands deliberately. Others, like insects, have been carried to the islands accidentally. As well as out-competing local species, alien species can eat native species, destroy native species' habitats, and bring diseases, such as avian malaria and bird flu, onto the islands.

The red quinine tree is an aggressively invasive species on Santa Cruz Island. It occupies the highlands, and spreads rapidly – it has wind-dispersed seeds. The ecosystem in the highlands has changed from low scrub and grassland to a closed forest canopy. Because of this, the native *Cacaotillo* shrub has been almost eradicated from Santa Cruz, and the Galapagos petrel has lost its nesting sites. The red quinine also successfully out-competes native *Scalesia* trees.

The goat has been one of the most damaging species to the Galapagos ecosystem. It eats Galapagos rock purslane, a species unique to the islands. It out-competes the giant tortoise for grazing, trampling and feeding on the tortoises' food supply, and changes the habitat to reduce the number of tortoise nesting sites. On Northern Isabella Island, the goat has also transformed forest into grassland, leading to soil erosion.

Cats hunt a number of species, including the lava lizard and young iguanas.

Introduced species are now the focus of conservationists' attention. In 1999, the Charles Darwin Research Station adopted strategies to prevent the introduction and dispersion of introduced species and to treat the problems caused by such species.

- They have instigated a quarantine system, where they search arriving boats and tourists for foreign species.
- Natural predators have been exploited to reduce the damage caused to ecosystems by pest populations – controlled release of a ladybird wiped out a scale insect, which was damaging plant communities.
- Culling has been successful against feral goats on Isabella Island and pigs on Santiago Island.

Any conservation plan for the Galapagos Islands has to take the concerns of local people seriously, and plan for their economic livelihood. Finding a balance between environmental, economic and social concerns is essential for conservation to be successful.

Because the majority of residents were not born on the islands, fostering a culture of conservation and educating new arrivals about the unique nature of the islands is a challenge. Strong leadership, combined with collaboration and compromise between organisations, is essential.

In fact, management of the Galapagos marine reserve (created in 1998, and extending 40 nautical miles around the islands) provides a model of how local stakeholders can work together to sustainably manage a resource. The reserve is managed by the National Park Service, the Charles Darwin Research Station, and representatives of local fishermen, the tourist industry and naturalist guides. At least 36% of coastal zones have been designated 'No-Take' areas, where no extraction of resources is allowed, and communities are left undisturbed.

Figure 4 *Scalesia* trees on Santa Cruz

Questions

1 List the main threats to native species on the Galapagos Islands.
2 Explain why goats pose such a problem to conservation on the Galapagos Islands.
3 Explain why management of the marine reserve around the Galapagos Islands provides a model for effective conservation.
4 Make a list of five 'dos' and 'don'ts' for conservation in the Galapagos Islands.

213

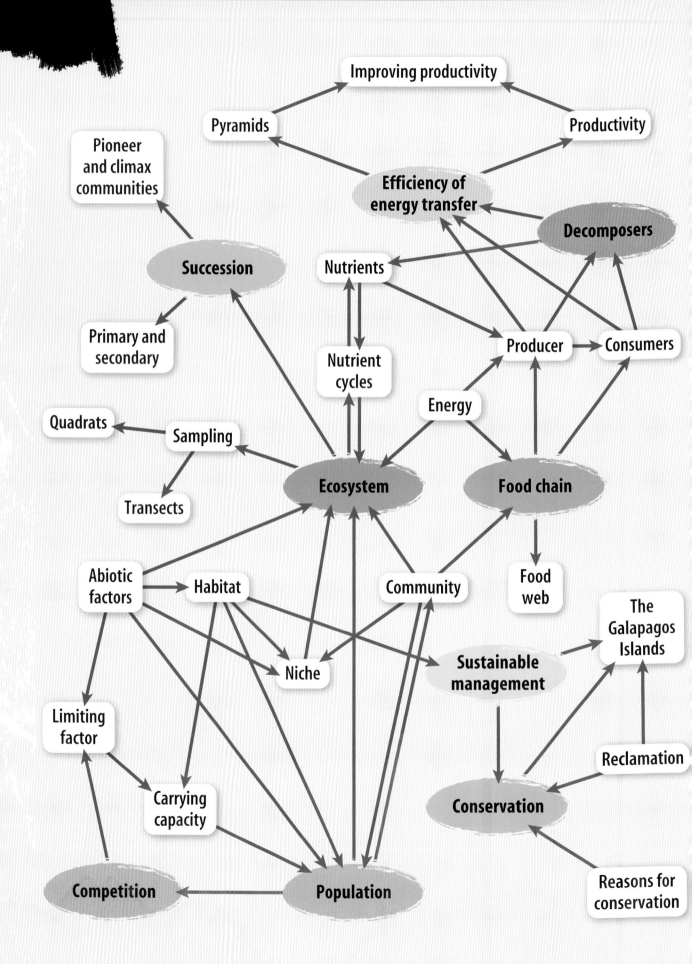

Practice questions

1 (a) What is meant by the following words?

    (i) Habitat
    (ii) Niche
    (iii) Community
    (iv) Population   [4]

(b) What are the roles of the following in an ecosystem?

    (i) Decomposer
    (ii) Producer   [2]

2 Explain

(a) why ecologists draw pyramids.   [1]

(b) the limitations of

    (i) pyramids of number
    (ii) pyramids of biomass
    (iii) pyramids of energy.   [3]

(c) the difference between gross primary productivity and net primary productivity.   [1]

3 Describe and explain three ways in which to increase net primary productivity.   [3]

4 Complete the following paragraph by filling in the spaces.

Ecosystems are dynamic. ………… flows through an ecosystem, while ………… are cycled within it. The ……… cycle is one good example. A ………… community colonises a piece of bare ground. This community changes the environment sufficiently for a different community of organisms to succeed them. The process repeats until a ………… community is reached. This community is influenced by ………… factors and abiotic factors. ……… competition (competition between species) influences the make-up of this community. Intraspecific competition can regulate a species' ………… size.   [8]

5 Explain how to use a

(a) quadrat

(b) point frame

(c) line transect

(d) belt transect.   [4]

6 Explain how bacteria are involved in

(a) ammonification
(b) nitrogen fixation
(c) nitrification
(d) denitrification.   [4]

7 (a) Explain the terms **(i)** 'limiting factor' and **(ii)** 'carrying capacity'.   [2]

(b) Sketch the relationship between the population size of predators and their prey, and explain the pattern.   [3]

8 (a) Define the term 'competition'.   [1]

(b) When two species' niches overlap, interspecific competition can happen. Explain the possible outcomes of such competition.   [3]

9 (a) Explain how rotational coppicing can provide a sustainable supply of timber.   [3]

(b) List five strategies used by foresters to harvest timber on a large scale whilst minimising the impact on biodiversity.   [5]

10 (a) What is conservation?   [1]

(b) Outline the economic, social and ethical reasons for conservation.   [3]

(c) Give two examples of conservation management strategies and two examples of reclamation strategies.   [4]

11 Give examples of how the following pose a threat to biodiversity in the Galapagos Islands:

(a) habitat disturbance

(b) introduced species

(c) over-exploitation of resources.   [3]

1 Lemmings are small mammals that live near the Arctic circle. Their populations show regular patterns of increase and decrease. In 2003, scientists published results based on a long-term project in East Greenland. They made the following observations.

- Population peaks occurred in regular four-year cycles.
- Four main predators feed on the lemmings: Arctic owls, Arctic foxes, long-tailed skuas and stoats.
- Stoats feed only on lemmings; the other predators feed on a range of prey species.
- Stoats reproduce more slowly than lemmings.

(a) Figure 1.1 shows the changes in the population of lemmings in the East Greenland project area from 1990 to 2002.

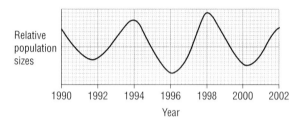

Figure 1.1

(i) Copy Figure 1.1, and sketch on the likely changes in the population size of stoats. [2]

(ii) Suggest three environmental conditions, other than climatic, that are required for a population explosion of lemmings. [3]

(b) With reference to the species studied in the East Greenland project, distinguish between *interspecific* and *intraspecific* competition. [3]

(c) The carrying capacities for lemmings and for the various predators in this area are all different. Explain the term *carrying capacity*. [2]

[Total: 10]

(OCR 2804 Jun06)

2 Figure 2.1 shows some of the stages that have occurred during succession at Glacier Bay in Alaska.

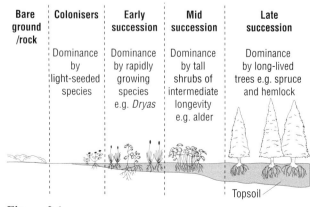

Figure 2.1

Using the information in Figure 2.1,

(a) explain what is meant by the term *succession*; [4]

(b) outline two changes which occur between mid and late succession; [2]

(c) suggest one biotic and one abiotic change which are not indicated in Figure 2.1. [2]

[Total: 8]

(OCR 2805/03 Jun06)

3 Preserving the diversity of life on Earth has come to be an accepted goal for many people. However, this goal can sometimes come into conflict with other goals, such as economic development. In 1980, the International Union for the Conservation of Natural Resources proposed a statement to form the basis for conserving biodiversity. One of the points included in the statement is:

'All species have an inherent right to exist. The ecological processes that support the integrity of the biosphere and its diverse species and habitats are to be maintained.'

Within the UK, many initiatives have been set up to maintain biodiversity. One is the Dartmoor biodiversity project, part of which is the habitat plan for moorland, which covers about 50% of the National Park. The action plan specifies many objectives, all of which aim to maintain the range of habitats on Dartmoor and ensure that all native plants and animals continue to breed successfully and maintain healthy populations.

(a) State what is meant by the term *biodiversity*. [2]

(b) Outline the ecological, economic and ethical reasons behind initiatives behind projects like the Dartmoor biodiversity project and other similar projects around the world. [8]

(c) Suggest four activities of conservation organisations which can contribute to the maintenance of biodiversity. [4]

[Total: 14]

(OCR 2805/03 Jun05)

4 **(a)** Define the term *interspecific competition*. [1]

The shag, *Phalacrocorax aristotelis*, and the cormorant. *Phalacrocorax canbo*, feed in the same waters and nest on the same cliffs. Table 4.1 shows the prey eaten by these two birds.

Drawn to same scale

Figure 4.1

Prey		% of prey taken by	
		Shag	**Cormorant**
surface swimming	sand eels	33	0
	herring	49	1
bottom feeding	flat fish	1	26
	shrimps, prawns	2	33

Table 4.1

Answers to examination questions will be found on the Exam Café CD.

(b) State why the results for each species of bird do not add up to 100%. [1]

(c) With reference to Figure 4.1 and Table 4.1, describe how the behaviour of shags and cormorants avoids direct competition. [4]

(d) Suggest a resource for which these two species show interspecific competition. [1]

[Total: 7]

(OCR 2804/01 Jan05))

5 The climax vegetation in tropical areas with abundant rainfall is rainforest. Although rainforests now cover less than 4% of the land surface of the Earth, they account for more than 20% of the planet's net carbon fixation. By comparison, temperate forests are about half as productive (per unit area), while boreal forests (forests of northern latitudes) and grasslands are only a quarter as productive. A 13 km² rainforest preserve in Costa Rica has 450 species of trees, more than 1000 other plant species, 400 species of birds, 58 species of bats and 130 species of amphibians and reptiles.

Figure 5.1 shows a diagram of a typical area of tropical rainforest.

Figure 5.1

(a) List **three** reasons why tropical rainforests have been destroyed, so that they now cover only 4% of the land surface of the Earth. [3]

(b) In this question, one mark is available for the quality of use and organisation of scientific terms.
Making use of the information in the passage and Figure 5.1, describe the important features of tropical rainforests **and** explain why their disappearance is a cause of considerable concern. [8]
Quality of written communication [1]

[Total: 12]

(OCR 2805/03 Jan05 4a and b)

6 The cyclamen mite is a pest of strawberry crops in California. Populations of these mites are usually kept under control by a species of predatory mite of the genus *Typhlodromus*.
An experiment was carried out to investigate the effectiveness of predation in controlling cyclamen mites.

Both predator and prey mites were released on a group of strawberry plants in a greenhouse and the numbers of both types of mite were monitored over a period of 12 months. The results are summarised in Figure 6.1. A second investigation was carried out on a crop of strawberry plants growing in a field. The plants were sprayed periodically with parathion, an insecticide that reduces the number of predators, but does not affect the cyclamen mite. The effects of this on the numbers of cyclamen mites is summarised in Figure 6.2.

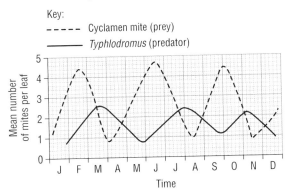

Figure 6.1

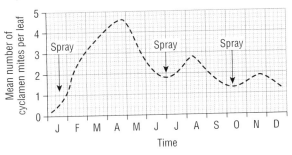

Figure 6.2

(a) The results shown in Figure 6.1 illustrate many of the features of a typical predator–prey relationship. Describe and explain these typical features. [4]

(b) (i) Sketch a curve on Figure 6.2 to show the likely effect of spraying on the population of the predatory mite. [2]

(ii) Suggest **two** reasons for the gradual decrease in the numbers of cyclamen mites over the year, as shown in Figure 6.2. [2]

(c) Many Californian strawberry growers keep the cyclamen mite under control by ensuring that there are healthy populations of the *Typhlodromus* mite.

(i) State the name given to the type of pest control. [1]

(ii) Explain why many would regard the use of predatory mites as preferable to the application of insecticides. [5]

(d) Suggest **two** methods of pest control other than the use of predatory mites or insecticides. [2]

[Total: 16]

(OCR 2805/03 Jan05)

217

Module 4
Responding to the environment

Introduction

All living organisms respond to changes in their environment which act as stimuli. The responses are significant in survival of the species and have been selected for through evolutionary processes affecting the development of all organisms.

In this module you will learn that a coordinated response, whatever the organism, requires the ability to detect the stimulus and to respond appropriately, in a coordinated way. Coordinating mechanisms range from the hormone-based responses of plants to light and gravity to the complex patterns seen in mammals which are orchestrated by the nervous system and endocrine system working together. In all cases, coordination is a complex and continuous process.

You will learn that there is a wide range of responses in organisms. Plant responses are often growth responses in relation to the stimulus, whereas animal responses are often much more complicated, involving the action of muscles and glands in order to coordinate movement of the whole organism and the homeostatic mechanism involved in movements and control within the organism.

In primates, the development of the cerebrum has been accompanied by development of extended care of the newborn and the learned behaviours associated with social life and problem solving.

This module also explores our developing understanding of how human behaviours are linked to the genetic principles learned in the last module. This is exemplified by consideration of the effect of dopamine and its receptors on human behaviour patterns.

Test yourself

1 How does natural selection operate in relation to organisms' responses to their environment?
2 Why does coordination need to be continuous?
3 How does homeostasis link to behaviours in organisms?
4 What role do simple reflex arcs play in behavioural responses?
5 Why do we refer to coordinated behaviours in all organisms as being 'essential to life'?
6 List all of the effectors that you have learned about in previous studies in biology.
7 Primates are not the only organisms that have social behaviour – can you name some others?

Module contents

By the end of this spread, you should be able to . . .

* Explain why plants need to respond to their environment in terms of the need to avoid predation and abiotic stress.
* Define the term *tropism*.

Plants respond to stimuli

It seems obvious that animals respond to the **biotic** (living) and **abiotic** (non-living) components of their environment. Plants also respond to external stimuli. Responding to the environment in this way may help the plant to avoid stress or to avoid being eaten, and to survive long enough to reproduce.

Tropisms are directional growth responses of plants. They include:
* Phototropism – shoots grow towards light (they are positively phototropic), which enables them to photosynthesise.
* Geotropism – roots grow towards the pull of gravity. This anchors them in the soil and helps them to take up water, which is needed for support (to keep cells turgid), as a raw material for photosynthesis and to help cool the plant. There will also be minerals, such as nitrates, in the water, needed for the synthesis of amino acids.
* Chemotropism – on a flower, pollen tubes grow down the style, attracted by chemicals, towards the ovary where fertilisation can take place (see Figure 1).
* Thigmotropism – shoots of climbing plants, such as ivy, wind around other plants or solid structures and gain support.

What controls plant responses?

Hormones coordinate plant responses to environmental stimuli. Like animal hormones, plant hormones are chemical messengers that can be transported away from their site of manufacture to act in other parts (**target cells** or **tissues**) of the plant (although some hormones stay in the cells that make them and exert their effect there). They are often referred to as plant growth regulators rather than hormones because, unlike animal hormones, they are not produced in endocrine glands, but by cells in a variety of tissues in the plant.

When hormones reach their target cells, they bind to receptors on the plasma membrane. Specific hormones have specific shapes (you can see the molecular structure of one hormone in Figure 2), which can only bind to specific receptors with complementary shapes on the membranes of particular cells. This specific binding makes sure that the hormones only act upon the correct tissues. Some plant hormones are commercially important and you will learn more about how they are used in spread 2.4.4.

Hormones can move around the plant in any of the following ways:
* active transport
* diffusion
* mass flow in the phloem sap or in xylem vessels.

You can see the main types of hormone and their effects in Table 1. However, some hormones can have different effects on different tissues; some can amplify each others' effects (this is called synergy), and some can even cancel out each other's effects (antagonism). Hormones can influence cell division, cell elongation, or cell differentiation.

Key definition

A **tropism** is a directional growth response in which the direction of the response is determined by the direction of the external stimulus.

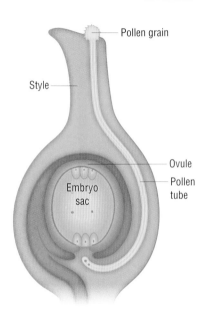

Figure 1 Longitudinal section showing a pollen tube growing from a pollen grain towards the ovary of a flower

Labels: Pollen grain, Style, Ovule, Pollen tube, Embryo sac

Examiner tip

Don't confuse '-*tropic*' with '-*trophic*'. The first describes a growth response towards or away from a stimulus (e.g. photo*tropic* response). Tro*phic* is connected with how living things feed (e.g. plants are auto*trophic*).

Examiner tip

If a plant responds towards a stimulus, it is a *positive tropic* response. If a plant responds away from a stimulus, it is still a *tropic* (directional) response but *negative* rather than positive.

Module 4
Responding to the environment
Why plants respond to the environment

Hormone	Effects
Auxins e.g. IAA (indole-3-acetic acid)	Promote cell elongation; inhibit growth of side-shoots; inhibit leaf abscission (leaf fall)
Cytokinins	Promote cell division
Gibberellins	Promote seed germination and growth of stems
Abscisic acid	Inhibits seed germination and growth; causes stomatal closure when the plant is stressed by low water availability
Ethene	Promotes fruit ripening

Table 1 Plant hormones and their effects

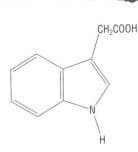

Figure 2 Molecular structure of IAA (indole-3-acetic acid)

STRETCH and CHALLENGE

Plants show other responses, besides tropisms, to environmental stimuli. Non-directional responses are called nasties (nastic responses). The sensitive plant, *Mimosa pudica*, responds to touch with a sudden drooping of the leaves (see Figure 3). This response is an example of thigmonasty.

Questions
A How might thigmonastic responses in plants help them to survive?
B Explain how such a response might have evolved (been selected for).

Many plants in temperate zones respond to day length (the photoperiod); it influences their flowering. Some are called short-day plants because they flower in spring or autumn. Some are called long-day plants because they flower in summer. Research shows that it is the length of darkness that is really crucial – short-day plants need a long period without light, each night.

Questions
C Suggest how flower growers can make plants flower out of their normal season.
D Suggest why research into photoperiodism could be beneficial in agriculture, to help increase the yield from food crops.

Figure 3 Thigmonastic response of the sensitive plant

Questions
1 Suggest why plant growth regulators are called hormones, although they are not produced in endocrine glands.
2 Explain why only certain tissues in a plant will respond to a particular plant hormone/growth regulator substance.
3 Ethene is a gas. It is produced by flowers, stems and leaves and causes ripening of fruit and ageing of flowers. Ripe fruit also produces it. Bananas produce lots of it. Why is it not advisable to place a vase of cut flowers next to a fruit bowl containing bananas?

By the end of this spread, you should be able to ...

* Explain how plant responses to environmental changes are coordinated by hormones, with reference to responding to changes in light direction.
* Outline the role of hormones in leaf loss in deciduous plants.

Plant growth

In contrast to animal cells, the cell wall around a plant cell limits the cell's ability to divide and expand. Because of this, growth only happens in particular places in the plant, where there are groups of immature cells that are still capable of dividing. These places are called **meristems** (see Figure 1).

* **Apical meristems** are located at the tips or apices (singular: apex) of roots and shoots, and are responsible for the roots and shoots getting longer.
* **Lateral bud meristems** are found in the buds. These could give rise to side shoots.
* **Lateral meristems** are found in a cylinder near the outside of roots and shoots and are responsible for the roots and shoots getting wider.
* In some plants, **intercalary meristems** are located between the nodes (where the leaves and buds branch off the stem). Growth between the nodes is responsible for the shoot getting longer.

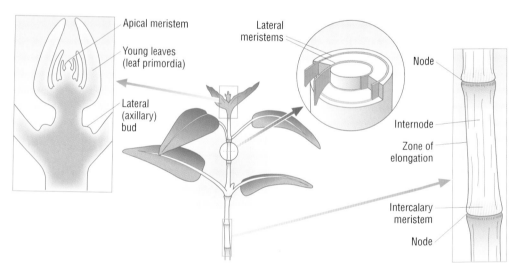

Figure 1 Meristems in a plant

You can see a longitudinal section of a shoot in Figure 2. Cell division happens closest to the apex and then cell elongation happens just behind the apex. Auxins (such as IAA) are produced at the apex. The auxin travels, either by diffusion or active transport, to the cells in the zone of elongation, causing them to elongate, making the shoot grow.

Auxins stimulate shoot growth by causing cell elongation. The extent to which cells elongate is proportional to the concentration of auxins. Auxin increases the stretchiness of the cell wall by promoting the active transport of hydrogen ions, by an ATPase enzyme on the plasma membrane, into the cell wall. The resulting low pH provides optimum conditions for wall-loosening enzymes (expansins) to work. These enzymes break bonds within the cellulose (at the same time the increased hydrogen ions also disrupt hydrogen bonds within cellulose) so the walls become less rigid and can expand as the cell takes in water.

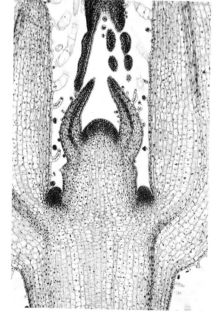

Figure 2 Longitudinal section of a shoot

Module 4
Responding to the environment
How plants respond to the environment

What causes phototropisms?

In a phototropic response, a shoot bends towards a light source. This happens because the shaded side elongates faster than the illuminated side, which pushes the end of the shoot towards the light. Evidence from experiments involving cereal seedlings in particular suggests that the light shining on one side of the shoot causes the auxins to be transported to the shaded side, where they promote an increase in the rate of elongation (see Figure 3), making the shoot bend towards the light.

How the light causes redistribution of auxin is still uncertain. Two enzymes have been identified (phototropin 1 and phototropin 2). Their activity is promoted by blue light (wavelength 400–450 nm). Hence, there is a lot of phototropin 1 activity on the light side, but progressively less activity towards the dark side. This gradient is thought to cause the redistribution of auxins.

Examiner tip

Remember that auxin causes cell elongation, not cell division.

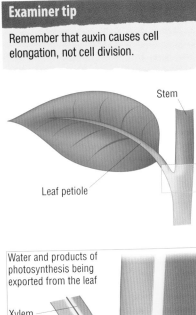

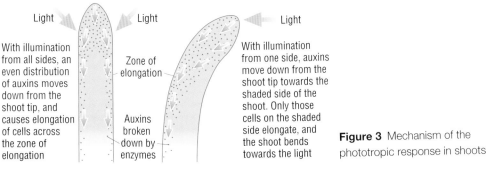

With illumination from all sides, an even distribution of auxins moves down from the shoot tip, and causes elongation of cells across the zone of elongation

Zone of elongation

Auxins broken down by enzymes

With illumination from one side, auxins move down from the shoot tip towards the shaded side of the shoot. Only those cells on the shaded side elongate, and the shoot bends towards the light

Figure 3 Mechanism of the phototropic response in shoots

Shedding leaves

Cytokinins stop the leaves of deciduous trees senescing (ageing – turning brown and dying) by making sure the leaf acts as a *sink* for phloem transport – this means the leaf is guaranteed a good supply of nutrients. However, if cytokinin production drops, that supply of nutrients dwindles and senescence begins. This is usually followed by leaves being shed (abscission).

Usually auxin inhibits abscission by acting on cells in the abscission zone (see Figure 4). However:

- Leaf senescence causes auxin production at the tip of the leaf to drop.
- This makes cells in the abscission zone more sensitive to another growth substance called ethene.
- A drop in auxin concentration also causes an increase in ethene production.
- This in turn increases production of the enzyme cellulase, which digests the walls of the cells in the abscission zone, eventually separating the petiole from the stem.

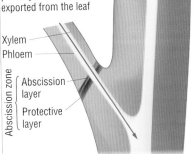

Water and products of photosynthesis being exported from the leaf

Xylem
Phloem

Abscission zone

Abscission layer
Protective layer

Figure 4 The leaf petiole showing the abscission zone

Examiner tip

Researchers used to think that abscisic acid was involved in abscission – hence the name. Since then, they've found that in most plants this is not true.

STRETCH and CHALLENGE

Questions

A Tropic responses are described as growth responses. Discuss whether they really do involve growth.

B Roots are negatively phototropic. However, auxins still control their response, and are still thought to travel to the more shaded side of the root. Look at Figure 5, and suggest how auxin in the roots may be responsible for negative phototropisms.

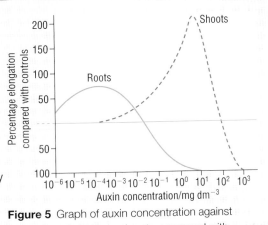

Figure 5 Graph of auxin concentration against elongation of shoots and roots compared with controls

Questions

1 What effect do auxins have on cells?

2 Explain how auxin is involved in:
   (a) shoot growth
   (b) leaf abscission.

3 Explain how positive phototropism in plant shoots and negative phototropism in plant roots would be selected for.

2.4 ③ Controlling plant growth

By the end of this spread, you should be able to . . .

✳ **Evaluate the experimental evidence for the role of auxins in the control of apical dominance and the role of gibberellin in the control of stem elongation.**

Apical dominance

Key definition

The growing apical bud at the tip of the shoot inhibits growth of lateral buds further down the shoot. This is called **apical dominance**.

If you break the shoot tip (apex) off a plant, the plant starts to grow side branches from lateral buds that were previously dormant (Figure 1). Researchers suggested the explanation that auxins usually prevent the lateral buds from growing. This is called **apical dominance**. When the tip (the source of auxin) is removed, the auxin concentration in the shoot drops and the buds grow. To test their hypothesis, they applied a paste containing auxins to the cut end of the shoot, and the lateral buds did not grow.

When looking at experimental results, it is important to think about the strength of evidence they provide to support or reject a hypothesis. For example, instead of auxin being involved, some other factors may produce the effect:

- Upon exposure to oxygen, cells on the cut end of the stem could have produced a hormone that promoted lateral bud growth.

Because of this, Ken Thimann and Folke Skoog applied a ring of auxin transport inhibitor below the apex of the shoot. The lateral buds grew. Based on this result, they suggested that normal auxin concentrations in lateral buds inhibit growth, whereas low auxin concentrations promote growth.

- This seems like a sensible explanation, but be wary of assuming that a simple association between auxin concentrations and growth inhibition confirms that the hormone directly *causes* the pattern of growth.
- The two variables (auxin concentration and growth inhibition) may have no effect on each other, but could coincidentally both be affected by a third variable!

In fact, several years later, Gocal disproved a *direct* causative link. He found that auxin concentrations in lateral buds of the kidney bean plant actually increased when the shoot tip was cut off. Scientists now think that two other hormones may be involved:

- Abscisic acid inhibits bud growth. High concentrations of auxin in the shoot may keep abscisic acid levels high in the bud. When the tip is removed (the source of auxin), the abscisic acid concentrations drop and the bud starts to grow.

Figure 1 Apical dominance: the apical bud has been removed from the shoot on the right

- Cytokinins promote bud growth – directly applying cytokinin to buds can override the apical dominance effect. High concentrations of auxin make the shoot apex a sink for cytokinins produced in the roots – this means that most of the cytokinin goes to the shoot apex. When that is removed, cytokinin spreads more evenly around the plant, promoting growth in the buds.

You have already seen an example of a similar type of response in coppicing (spread 2.3.9).

Module 4
Responding to the environment
Controlling plant growth

Gibberellins and stem elongation

In Japan, there is a plant disease called *Bakanae* (which translates as 'foolish seedling'!). It is caused by a fungus and makes rice grow very tall. Attempts to isolate the fungal compounds involved identified a family of compounds called gibberellins. One of these was gibberellic acid (GA_3). Having identified it, scientists began testing it on lots of different plants. When they applied it to dwarf varieties of plants (such as maize and peas – like the ones originally studied by Gregor Mendel), or rosette plants (such as cabbages), these plants grew taller.

These results seem to suggest that gibberellic acid is responsible for plant stem growth. However, such a conclusion is too hasty. It is always important to think about whether an experiment has actually investigated a natural phenomenon – just because GA_3 *can* cause stem elongation, it does not necessarily mean that it *does so* in nature. An experiment like this needs to work within concentrations of gibberellins naturally found in plants, and in parts of the plant that gibberellin molecules would normally reach.

It sounds impossible to set up an experiment that meets these criteria, but researchers found a way to do it. They compared GA_1 concentrations (another member of the gibberellin family) of tall pea plants (homozygous for the dominant **Le** allele), and dwarf pea plants (homozygous for the recessive **le** allele), which were otherwise genetically identical (Figure 3). They found that plants with higher GA_1 concentrations were taller.

Figure 2 Effect of gibberellins on cabbage growth

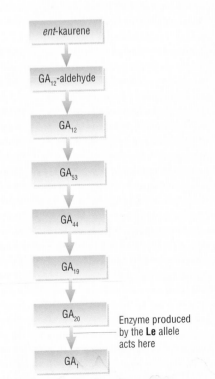

Figure 3 Dwarf and tall pea plants

le le   Le Le or Le le

However, to show that GA_1 *directly* causes stem growth, the researchers needed to know how GA_1 is formed (see Figure 4). They worked out that the **Le** allele was responsible for producing the enzyme that converted GA_{20} to GA_1. They then did something very clever. They chose a pea plant with a mutation that blocks gibberellin production between *ent*-Kaurene and GA_{12}-aldehyde in the pathway shown in Figure 4. Those plants produce no gibberellin and only grow to about 1 cm in height. However, if you graft (spread 2.2.2) a shoot onto a homozygous **le** plant (which cannot convert GA_{20} to GA_1), it grows tall. The shoot has no GA_{20} of its own, but it has the enzyme to convert GA_{20} to GA_1, and it can use the unused GA_{20} from the normal plant. Because the shoot grew tall, this confirmed that GA_1 caused stem elongation.

Further studies have shown that gibberellins cause growth in the internodes by stimulating cell elongation (by loosening cell walls) and cell division (by stimulating production of a protein that controls the cell cycle). Internodes of dwarf peas have fewer cells and shorter cells than those of tall plants, and mitosis in the intercalary meristems (see spread 2.4.2) of deep-water rice plants increases with gibberellin treatment.

Questions

1 Explain why an association or correlation between two variables does not necessarily mean that one directly *causes* a change in the other.
2 Compare and contrast the action of gibberellins and auxins.
3 In deep-water rice plants, lower oxygen levels stimulate production of ethene, which reduces abscisic acid levels. Abscisic acid is an antagonist of gibberellin. Using this information, explain how deep-water rice plants keep their upper foliage above water.

ent-kaurene

GA_{12}-aldehyde

GA_{12}

GA_{53}

GA_{44}

GA_{19}

GA_{20} — Enzyme produced by the **Le** allele acts here

GA_1

Figure 4 Synthesis pathway for gibberellins

By the end of this spread, you should be able to . . .

* Describe how plant hormones are used commercially.

Auxins

Artificial auxins can be used to prevent leaf and fruit 'drop', and to promote flowering for commercial flower production. Surprisingly, in high concentrations auxins can also *promote* fruit drop. This is useful if there are too many small fruits that will be difficult to sell – the plant then produces fewer, larger fruits. Other uses of auxins are in shown in Table 1.

Figure 1 Dipping a cutting into rooting powder

Taking cuttings	Dipping the end of a cutting in rooting powder before planting it encourages root growth (Figure 1). Rooting powder contains auxins, a fungicide and talcum powder!
Seedless fruit	Treating unpollinated flowers with auxin can promote growth of seedless fruit (parthenocarpy). Applying auxin promotes ovule growth, which triggers automatic production of auxin by tissues in the developing fruit, helping to complete the developmental process.
Herbicides	Artificial auxins are used as herbicides to kill weeds. They are transported in the phloem to all parts of the plant and they can act within the plant for longer because they are not a close fit to the enzymes that break them down. They promote shoot growth so much that the stem cannot support itself, buckles and dies.

Table 1 Commercial uses of auxins

Gibberellins

Fruit production

- Gibberellins delay senescence in citrus fruit, extending the time fruits can be left unpicked, and making them available for longer in the shops.
- Gibberellins acting with cytokinins can make apples elongate to improve their shape.
- Without gibberellins, bunches of grapes are very compact; this restricts the growth of individual grapes. With gibberellins, the grape stalks elongate, they are less compacted and the grapes get bigger.

Brewing

To make beer you need malt, which is usually produced in a malthouse at a brewery (Figure 2). When barley seeds germinate, the aleurone layer of the seed produces amylase enzymes that break down stored starch into maltose. Usually, the genes for amylase production are switched on by naturally occurring gibberellins. Adding gibberellin can speed up the process. Malt is then produced by drying and grinding up the seeds.

Figure 2 Barley grains in a malthouse at a brewery

Sugar production

Spraying sugar cane with gibberellins stimulates growth between the nodes, making the stems elongate. This is useful because sugar cane stores sugar in the cells of the internodes (the sub-sections of the stems in Figure 3), making more sugar available from each plant. Spraying with gibberellins can increase sugar yield by up to 4.5 tonnes per hectare.

Figure 3 Sugar cane

Plant breeding

- A plant breeder's job is to produce plants with desired characteristics by breeding together other plants, usually over many generations. However, in conifer plants this can take a particularly long time because conifers spend a long time as juveniles before becoming reproductively active. Gibberellins can speed up the process by inducing seed formation in young trees.
- Seed companies who want to harvest seeds from biennial plants (which flower only in their second year of life), such as sugar beet, can add gibberellins to induce seed production in the first year.

Stopping plants making gibberellins is also useful. Spraying with gibberellin synthesis inhibitors can keep flowers short and stocky (desirable in plants like poinsettias), and ensures that the internodes of crop plants stay short, helping to prevent lodging. Lodging happens in wet summers – stems bend over because of the weight of water collected on the ripened seed heads, making the crop difficult to harvest (see spread 2.1.21).

Cytokinins

Because cytokinins can delay leaf senescence, they are sometimes used to prevent yellowing of lettuce leaves after they have been picked.

Cytokinins are used in tissue culture to help mass-produce plants. They promote bud and shoot growth from small pieces of tissue taken from a parent plant. This produces a short shoot with a lot of side branches, which can be split into lots of small plants. Each of these is then grown separately.

Ethene

Because ethene is a gas, and cannot be sprayed directly, scientists have developed 2-chloroethylphosphonic acid, which can be sprayed in solution, is easily absorbed and slowly releases ethene inside the plant. Commercial uses of ethene include:

- speeding up fruit ripening in apples, tomatoes and citrus fruits
- promoting fruit drop in cotton, cherry and walnut
- promoting female sex expression in cucumbers, reducing the chance of self-pollination (pollination makes cucumbers taste bitter) and increasing the yield
- promoting lateral growth in some plants, yielding compact flowering stems.

Restricting ethene's effects can also be useful. Storing fruit in a low temperature with little oxygen and high carbon dioxide concentration prevents ethene synthesis and thus prevents fruit ripening. This means fruits can be stored for longer – essential when shipping unripe bananas from the Caribbean. Other inhibitors of ethene synthesis, such as silver salts, can increase the shelf life of cut flowers.

Figure 4 Preparing bananas for shipping

Questions

1 Which hormones are involved in:
   (a) slowing senescence
   (b) fruit ripening
   (c) encouraging root growth?
2 Which hormones can help clone daughter plants from a parent plant?
3 Instead of using hormones to produce seedless fruit, plant breeders sometimes take a different approach and breed genetically parthenocarpic varieties. A good example is bananas, which have triploid cells, stopping them from forming male and female gametes. Suggest one advantage and one disadvantage of each approach.
4 Putting a silver coin into the water of a vase of flowers is said to keep the flowers fresh for longer. Suggest how this might work.

By the end of this spread, you should be able to . . .

* Describe, with the aid of diagrams, the gross structure of the human brain and outline the functions of the cerebrum, cerebellum, medulla oblongata and hypothalamus.
* Describe the role of the brain and nervous system in coordinated muscular movement.

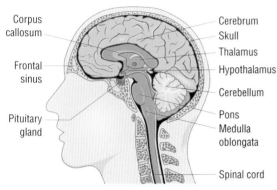

Figure 1 The human brain shown in cross-section within the skull

Key definition

The **cerebrum** is the largest and most recognisable part of the brain. It is responsible for the elements of the nervous system that are associated with being 'human', including thought, imagination and reasoning.

The structure of the human brain

What makes us human – the structure and function of the cerebrum

The **cerebrum**, the largest part of the human brain, is divided into two hemispheres. These are connected via the corpus callosum. The outermost layer, which has a surface area of around 2.5 m², is folded and consists of a thin layer of nerve cell bodies known as the cerebral cortex. This region of the brain is more highly developed in humans than in any other organism. It is in control of what are often described as the 'higher brain functions' including:

* conscious thought and emotional responses
* the ability to override some reflexes
* features associated with intelligence, such as reasoning and judgement.

The cerebral cortex is subdivided, for reference, into areas responsible for specific activities and body regions.

* **Sensory areas** receive impulses indirectly from the receptors.
* **Association areas** compare input with previous experiences in order to interpret what the input means and judge an appropriate response.
* **Motor areas** send impulses to effectors (muscles and glands).

The motor areas on the left side of the cerebral cortex control the muscular movements on the right side of the body and vice versa.

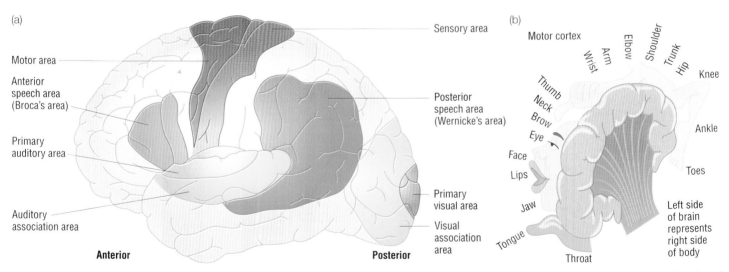

Figure 2 (a) The left cerebral cortex labelled to show locations of some of the sensory and motor areas; **(b)** there is a relationship between the size of motor area in the brain and the complexity of movements

Coordinated motor responses involve the cerebellum

Whilst the conscious decision to move voluntary muscles is initiated in the cerebral hemisphere, this alone cannot operate the human body in a coordinated way. The fine control of muscular movements such as those involved in walking, riding a bicycle, driving a car and playing musical instruments requires a significant level of nonconscious operation. For example:

- muscular activities associated with responding to changes in body position to remain balanced and upright
- sensory activities such as judging the position of objects and limbs
- the tensioning of muscles in order to manipulate tools/instruments effectively
- feedback information on muscle position, tension and fine movements
- operation of antagonistic muscles to coordinate contraction and relaxation.

Neurones from the **cerebellum** carry impulses to the motor areas so that motor output to the effectors can be adjusted appropriately in relation to these requirements. This explains why we often go onto 'auto-pilot' when carrying out such activities, as they are said to become programmed into the cerebellum. It also explains how we can catch a ball by judging its speed and trajectory and close our fingers around it.

The cerebellum contains over half of all of the nerve cells in the brain. It plays a key role in coordinating balance and fine movement. To do this it processes sensory information from the following locations:

- the retina
- the balance organs in the inner ear
- specialised fibres in muscles called 'spindle' fibres which give information about muscle tension
- the joints.

Other brain regions

Region of brain	Outline of function
Medulla oblongata	Controls non-skeletal muscles (i.e. cardiac and involuntary muscles). This means that it is effectively in control of the autonomic nervous system (see spread 2.4.6). Regulatory centres for a number of vital processes are found in the medulla oblongata including: • the cardiac centre, which regulates heart rate • the respiratory centre, which controls breathing and regulates the rate and depth of breathing.
Hypothalamus	Controls most of the body's homeostatic mechanisms. Sensory input from temperature receptors and osmoreceptors is received by the hypothalamus and leads to the initiation of automatic responses that regulate body temperature and blood water potential. The hypothalamus also controls much of the endocrine function of the body because it regulates the pituitary gland.

Table 1 Functions of other brain regions

STRETCH and CHALLENGE

Understanding language and the capacity to speak

Two areas of the brain, which in most people are found in the left cerebral hemisphere, are associated with understanding language and speaking. Broca's area is named after Paul Broca who, in the 1860s, examined the brains of patients with an inability to speak (aphasia). Wernicke's area is named after Karl Wernicke who, also in the 1860s, identified another region, damage to which caused problems with understanding language.

The two areas are connected by a bundle of neurones. Damage to these neurones leads to an individual being able to understand language but unable to repeat words.

Questions

A Suggest and explain whether (i) Broca's area, (ii) Wernicke's area is a sensory, motor or association area.
B List the areas of the brain that would be involved in (i) reading out loud, (ii) singing along to a song being played on the radio.

Key definitions

The **cerebellum** controls the coordination of movement and posture.

The **hypothalamus** controls the autonomic nervous system and the endocrine glands.

The **medulla oblongata** controls the action of smooth muscle in the gut wall, and controls breathing movements and heart rate.

Examiner tip

The specification requires an understanding in outline of the regions of the brain. You could be asked to label diagrams, for example, giving the location of the various regions. You should ensure that you can link the brain region activities to other information, for example to muscular contraction and the structure and function of neurones and synapses.

Questions

1 Suggest what effect damage to the cerebellum might have on a person.
2 Explain why a cerebrovascular accident (stroke) leading to damage to the left side of the brain can result in paralysis of the right arm and leg.
3 Figure 2 shows the left cerebral hemisphere – how do you know this?

By the end of this spread, you should be able to . . .

* Discuss why animals need to respond to their environment.
* Outline the organisation of the nervous system in terms of central and peripheral systems in humans.
* Outline the organisation and roles of the autonomic nervous system.

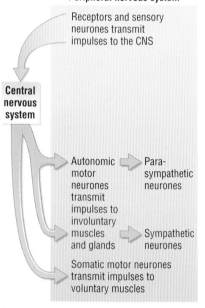

Figure 1 The organisation of the nervous system

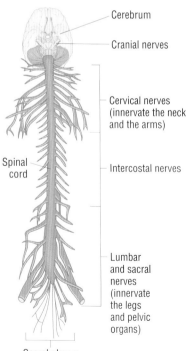

Figure 2 The cranial and spinal nerves of the peripheral nervous system

Responding to the environment

One of the characteristics of living things is that they are able to respond to changes in both the internal and external environment. All organisms need a method of communication between their sensors and their responsive systems (effectors). Animals need to be able to coordinate a vast set of responses if they are to stay alive, from the coordinated voluntary muscle actions required in order to run away from a predator, to the fine control of balance, posture and temperature regulation. The coordinated response, using nerves and hormones, to stress situations is discussed in spread 2.4.10. We shall first look at how the nervous system is arranged.

Subdividing the nervous system

The nervous system coordinates the actions of the body through electrical impulses. As we shall see later, it works in conjunction with the endocrine system. Both are essential in maintaining life in humans. It is structurally and functionally divided into subsystems. This helps us describe nervous actions and to understand coordination processes.

The central nervous system and peripheral nervous system

The central nervous system (CNS) consists of the brain and spinal cord. It is made up of grey matter (billions of non-myelinated nerve cells) and white matter (longer, myelinated axons and dendrons that carry impulses). The presence of myelin makes the long fibres appear white.

The peripheral nervous system is made up of the neurones that carry impulses into and out of the CNS.

Sensory and motor systems of the peripheral nervous system

Sensory neurones carry impulses from the many receptors, in and around the body, to the CNS. Motor neurones carry impulses from the CNS to the effector organs. Many neurones are bundled together and covered in connective tissue to form nerves. The motor system is further subdivided:

* **Somatic** motor neurones carry impulses from the CNS to skeletal muscles, which are under voluntary (conscious) control.
* **Autonomic** motor neurones carry impulses from the CNS to cardiac muscle, to smooth muscle in the gut wall and to glands, none of which are under voluntary control.

The autonomic nervous system

Autonomic translates as 'self-governing', referring to the fact that the system operates to a large extent independently of conscious control. The system is responsible for controlling the majority of homeostatic mechanisms and so plays a vital role in regulating the internal environment of the body within set parameters. In addition, the system is capable of controlling the heightened responses associated with the stress response (as we shall see later).

Module 4
Responding to the environment
Organising the nervous system

The organisation of the autonomic nervous system

The autonomic nervous system differs from the somatic nervous system in a number of ways.

- Most autonomic neurones are non-myelinated whilst most somatic neurones are myelinated.
- Autonomic connections to effectors always consist of at least two neurones (whereas somatic connections to effectors consist of only one). The two neurones connect at a swelling known as a ganglion.
- Autonomic motor neurones occur in two types: **sympathetic** and **parasympathetic**.

Sympathetic and parasympathetic subsystems

Sympathetic and parasympathetic systems differ in both structure and action. They are often referred to as **antagonistic** systems because in many cases the action of one system opposes the action of the other. It is important to remember that under normal, resting conditions, impulses are passing along the neurones of both systems at a relatively low rate. Changes to internal conditions, or stimulation of the stress response, lead to an altered balance of stimulation between the two systems, which leads to an appropriate response. The balance of stimulation is controlled by subconscious parts of the brain, as we shall see later.

Parasympathetic	Sympathetic
Most active in sleep and relaxation	Most active in times of stress
The neurones of a pathway are linked at a ganglion within the target tissue. So pre-ganglionic neurones vary considerably in length.	The neurones of a pathway are linked at a ganglion just outside the spinal cord. So pre-ganglionic neurones are very short.
Post-ganglionic neurones secrete acetylcholine as the neurotransmitter at the synapse between neurone and effector.	Post-ganglionic neurones secrete noradrenaline at the synapse between neurone and effector.
Effects of action include: decreased heart rate, pupil constriction, decreased ventilation rate, and sexual arousal.	Effects of action include: increased heart rate, pupil dilation, increased ventilation rate, and orgasm.

Table 1 Antagonistic actions of the sympathetic and parasympathetic nervous systems

Key definition

The **central nervous system** consists of the brain and spinal cord. The **peripheral nervous system** consists of all of the sensory and motor neurones that are outside the central nervous system – connecting the receptors and effectors to the CNS.

Examiner tip

Remember that sympathetic and parasympathetic stimulation occurs all the time. Information from external and internal receptors is fed into the brain continuously and used to alter the balance of stimulation between the two systems in order to ensure an appropriate response to all situations.

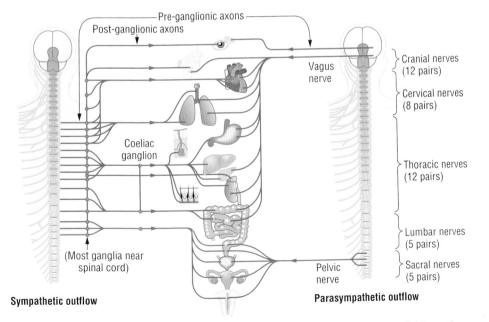

Figure 3 The autonomic nervous system showing sympathetic and parasympathetic divisions; the post-ganglionic axons of the parasympathetic neurones are very short and are in the walls of the organs

Questions

1 Explain why the control of pupil diameter in the eye is described as an autonomic reflex.
2 If a person exercises by running for 2 minutes, a number of autonomic responses will take place during and after the exercise. Using information from this and other spreads, list the responses and state whether these are sympathetically or parasympathetically controlled.

231

By the end of this spread, you should be able to ...

✱ **Describe how coordinated movement requires the action of skeletal muscles about joints, with reference to the elbow joint.**

✱ **Compare and contrast the action of synapses and neuromuscular junctions.**

As described in the previous spreads, coordinated movements require the action of the brain in sending impulses along motor neurones to voluntary muscles. Voluntary muscles are attached to the bones of the skeleton by tendons, such that contraction of the muscles moves the bones at the joints. Tendons are made of tough, inelastic collagen which is continuous with the muscle and the periosteum (the connective tissues covering the bone).

Action of muscles

Muscles are only capable of producing a force when they contract. So the movement of any bone at a joint requires the coordinated action of at least two muscles. As one muscle is stimulated to contract, the other muscle of the pair must relax to allow for smooth movement. Muscles working in pairs opposite each other are described as **antagonistic**. However, the movement of bones at many joints requires a wider range of actions and is under the control of groups of muscles called synergists.

Movement at the elbow joint

The elbow joint is an example of a **synovial** joint. These joints occur where a large degree of movement is required. The synovial fluid is a lubricant. It eases the movement of the bones at the joint. The biceps and triceps muscles act antagonistically in order to move the forearm at the elbow, as shown in Figure 1.

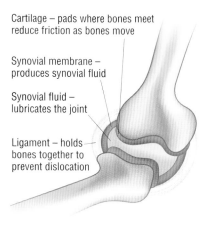

Shoulder blade

Tendon

Humerus

Biceps muscle

Triceps muscle

Radius   Ulna

Figure 1 The biceps and triceps muscles act antagonistically to move the forearm at the elbow

Cartilage – pads where bones meet reduce friction as bones move

Synovial membrane – produces synovial fluid

Synovial fluid – lubricates the joint

Ligament – holds bones together to prevent dislocation

Figure 2 The elbow is an example of a synovial joint

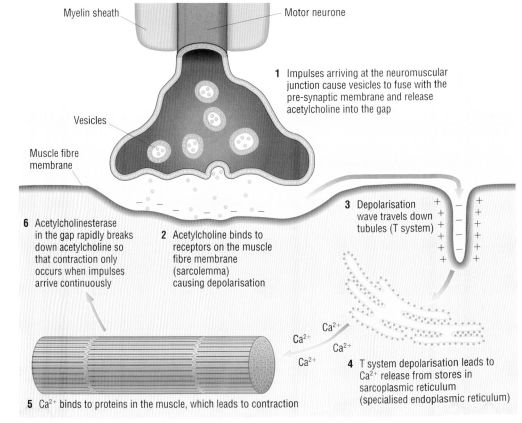

Myelin sheath

Motor neurone

1 Impulses arriving at the neuromuscular junction cause vesicles to fuse with the pre-synaptic membrane and release acetylcholine into the gap

Vesicles

Muscle fibre membrane

3 Depolarisation wave travels down tubules (T system)

6 Acetylcholinesterase in the gap rapidly breaks down acetylcholine so that contraction only occurs when impulses arrive continuously

2 Acetylcholine binds to receptors on the muscle fibre membrane (sarcolemma) causing depolarisation

Ca^{2+} Ca^{2+}

Ca^{2+} Ca^{2+}

Ca^{2+}

4 T system depolarisation leads to Ca^{2+} release from stores in sarcoplasmic reticulum (specialised endoplasmic reticulum)

5 Ca^{2+} binds to proteins in the muscle, which leads to contraction

Figure 3 Operation of the neuromuscular junction

Control of contraction

The nervous system controls muscle action because motor neurones are connected to muscle cells at a neuromuscular junction. Impulses arriving at the neuromuscular junction stimulate contraction. A neuromuscular junction is very similar in structure and operation to a synapse between neurones.

The motor unit

Some muscular movements require a stronger contraction than others. Consider the differences in the strength of contraction of leg muscle for walking compared with the strength of contraction for running, or the difference between holding an egg and crushing a tin can. The brain controls the strength of contraction because many motor neurones stimulate a single muscle. Each one branches to neuromuscular junctions, causing the contraction of a cluster of muscle cells – known as a motor unit. The more motor units stimulated, the greater the force of contraction. This is known as **gradation of response**.

Key definition

A **neuromuscular junction** is a specialised synapse which occurs at the end of a motor neurone where it meets the muscle fibre. Release of acetylcholine (neurotransmitter), following depolarisation at the neuromuscular junction, stimulates contraction of the muscle fibre.

The end of the motor neurone is often referred to as an end plate or motor end plate.

Investigating contraction in isolated muscles – twitch and tetanus

Investigations into the action of whole muscles is often carried out using the calf (gastrocnemius) muscle removed, with the nerve still attached, from a frog. The muscle is attached to spring-loaded pins and to a datalogger, which records contractions following electrical stimulation of the attached nerve. The muscle is kept alive by immersion in a solution of salts (electrolytes) known as Ringer's solution.

Examiner tip

Ensure that you can compare *and* contrast synapses and neuromuscular junctions. They have many similarities but there are significant differences too.

A single electrical stimulus of sufficient strength causes a muscle twitch (a quick contraction of the muscle followed by immediate relaxation). Increasing the strength of the stimulus increases the force of contraction of the twitch, up to a maximal response. Further increase in stimulus strength leads to the maximal response twitch.

Two large, separate stimuli, if far apart, give two separate twitches, each to the maximal response. However, if stimuli are applied close together (i.e. at an increased frequency), the response becomes overlapped and is more powerful than a single maximal response. This is called summation.

Repeated large stimuli give a sustained and powerful contraction known as a tetanus.

The brain controls all voluntary muscle action by setting the frequency of impulses at a level to achieve controlled contraction and relaxation in antagonistic muscles.

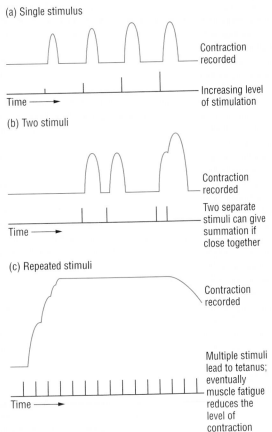

(a) Single stimulus

Contraction recorded

Increasing level of stimulation

Time →

(b) Two stimuli

Contraction recorded

Two separate stimuli can give summation if close together

Time →

(c) Repeated stimuli

Contraction recorded

Multiple stimuli lead to tetanus; eventually muscle fatigue reduces the level of contraction

Time →

Figure 4 The response of isolated whole muscles to electrical stimulation

Questions

1 A motor unit can consist of between three and 200 muscle cells. Suggest what types of muscles have many motor units, each stimulating very few muscle fibres.

2 Explain why the elbow joint is described as a hinge joint.

3 Draw a table to compare and contrast the features of synapses and neuromuscular junctions.

By the end of this spread, you should be able to . . .

✳ Outline the structural and functional differences between voluntary, involuntary and cardiac muscle.

Muscle structure

Muscles are composed of cells that are elongated to form fibres. Muscle cells are able to contract and relax. All muscles produce a force on contraction because they contain filaments made of the proteins **actin** and **myosin**. The mechanism by which these protein filaments operate to cause contraction of voluntary muscle is described in detail on the next spread.

There are three types of muscle:

- **involuntary** muscle – also known as smooth muscle
- **cardiac** muscle
- **voluntary** muscle – also known as striated or skeletal muscle.

The three muscle types have distinctly different structures and functions.

Involuntary (smooth) muscle

Smooth muscle is innervated by neurones of the autonomic nervous system. This means that contractions of this type of muscle are not under voluntary (conscious) control (see Table 1).

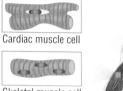

Cardiac muscle cell

Skeletal muscle cell

Smooth muscle cell

Figure 1 Three types of muscle

Location	Arrangement of muscle cells	Action
Walls of the intestine	Circular and longitudinal bundles	Peristalsis – moves food along the intestine.
Iris of the eye	Circular and radial bundles	Controls the intensity of light entering the eye: • Contraction of radial muscles dilates the pupil. • Contraction of circular muscle constricts the pupil.
Wall of arteries and around arterioles; wall and cervix of uterus	Circular bundles	Important in temperature regulation, regulation of local blood pressure and the redirecting of blood to voluntary muscles during exercise: • Contraction of muscle narrows vessel diameter so reducing blood flow. • Relaxation causes dilation, increasing blood flow.

Table 1 Examples of involuntary muscle location, arrangement and action

Under microscopic examination, involuntary muscle does not appear striated like voluntary and cardiac muscle. Muscle cells are referred to as being 'spindle-shaped'. They contain small bundles of actin and myosin, and a single nucleus. Cells in the relaxed state are around 500 μm long and 5 μm wide. Contraction is relatively slow, but this muscle also tires very slowly.

(a)

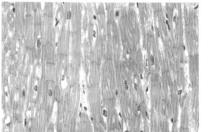

(b)

Circular layer of smooth muscle

Longitudinal layer of smooth muscle

The circular layer runs around the intestine and its contraction causes segmentation

The longitudinal layer runs along the intestine; it causes wave-like contractions

Figure 2 (a) Light micrograph of involuntary muscle (×400); **(b)** Smooth muscle arrangement in the gut

Cardiac muscle

Cardiac muscle forms the muscular part of the heart. There are three types:

- atrial muscle
- ventricular muscle
- specialised excitatory and conductive muscle fibres.

Module 4
Responding to the environment
Three types of muscle

Atrial and ventricular muscle contract in a way similar to that of skeletal muscle but with a longer duration of contraction. The excitatory and conductive fibres contract feebly but conduct electrical impulses and control the rhythmic heartbeat.

Some cardiac muscle fibres are capable of stimulating contraction without a nerve impulse. This type of contraction is described as **myogenic**. However, neurones of the autonomic system carry impulses to the heart to regulate the rate of contraction. Sympathetic stimulation increases its rate, whereas parasympathetic stimulation decreases its rate.

The sinoatrial node, in the wall of the right atrium, is made of specialised excitatory and conductive fibres. It has the greatest ability for self-excitation and the electrical activity generated there immediately spreads into the atrial wall. A layer of non-conducting fibres separates atria and ventricles, so electrical activity can only spread to the ventricles at the atrioventricular (AV) node. The AV node conducts the activity to the ventricle tips, via the Purkyne fibres.

Cardiac muscle fibres are made of many individual cells connected in rows. The dark areas are *intercalated discs,* which are cell membranes. These membranes fuse in such a way that there are gap junctions with free diffusion of ions and so action potentials pass very easily and quickly between cardiac muscle fibres, through the latticework of interconnections. Cardiac muscle, when viewed under the microscope, is striated.

Contraction and relaxation of cardiac muscle is continuous throughout life. This type of muscle contracts powerfully and without fatigue.

Voluntary (skeletal or striated) muscle

The action of voluntary muscles leads to movement of the skeleton at the joints. This moves the limbs, as described in previous spreads.

Muscle cells form fibres, of about 100 µm in diameter, containing several nuclei. Each fibre is surrounded by a cell surface membrane called the *sarcolemma*. Muscle cell cytoplasm is known as *sarcoplasm*, and contains organelles, including:

- many mitochondria
- an extensive sarcoplasmic reticulum (specialised endoplasmic reticulum)
- a number of myofibrils. These are the contractile elements and each consists of a chain of smaller contractile units called sarcomeres. Within the myofibrils are two types of protein myofilaments – thin actin and thick myosin – which run the length of the cell.

Microscopic examination of voluntary muscle shows a striped or banded pattern. The bands are given names as shown in Figure 4; these are further discussed in the next spread. This type of muscle contracts quickly and powerfully, but it fatigues quickly, unlike involuntary and cardiac muscle.

> **Key definition**
>
> **Sarcomere** – the smallest contractile unit of a muscle.

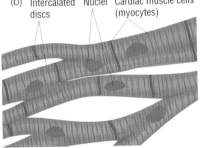

Figure 3 (a) Light micrograph of cardiac muscle (×400); **(b)** diagrammatic representation of cardiac muscle

> **Examiner tip**
>
> Note that you are required to know the structural *and* functional differences between the muscle types. You should learn specific examples of each as part of your revision.

Questions

1. Blood vessels such as arterioles contain circular smooth muscle. Contraction of this muscle constricts the vessel. Why do blood vessels not need longitudinal muscle to act against the circular muscle in order to cause dilation?

2. Draw a simple table to show the structural and functional differences between the three types of muscle – cardiac, involuntary and voluntary.

3. Suggest the advantage of the electrical activity of the heart being able to pass from atria walls to ventricle walls only at the AV node.

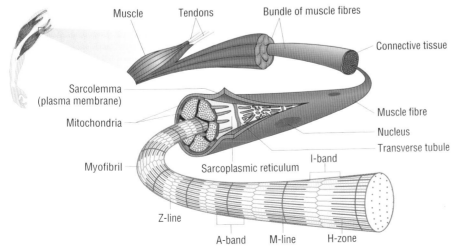

Figure 4 The structure of skeletal muscle

By the end of this spread, you should be able to . . .

* Explain, with diagrams and photographs, the sliding filament model of muscular contraction.
* Outline the role of ATP in muscular contraction, and how the supply of ATP is maintained in muscles.

The sarcomere

The striped appearance of voluntary muscle under the microscope is different when muscles are relaxed and when they are contracted. The span from one Z-line to the next is known as the sarcomere, and in a relaxed state is around 2.5 μm in length. Z-lines are closer together during contraction because the lengths of the I-band and H-zone are reduced. The A-band does not change in length during contraction.

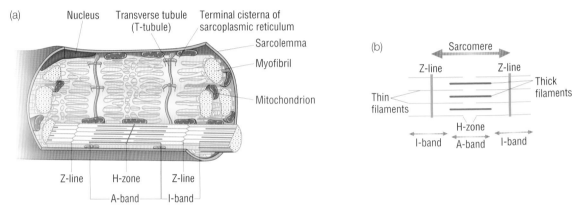

Figure 1 **(a)** Diagram showing a section of a myofibril to reveal sarcomere layout; **(b)** diagram showing the bands and zones present

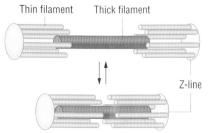

Figure 2 **(a)** Thin filament structure and **(b)** thick filament structure

Figure 3 One thick filament is surrounded by six thin filaments

Two types of protein filament are involved in contraction

The protein filaments that are found in muscle cells are of two types (Figures 2 and 3):

* Thin filaments are two strands, made chiefly of the protein actin (F actin), coiled around each other like a twisted double string of beads. Each strand is composed of G actin (globular protein) subunits. Tropomyosin (a rod-shaped protein) molecules coil around the F actin, reinforcing it. A troponin complex is attached to each tropomyosin molecule. Each troponin complex consists of three polypeptides. One binds to actin, one to tropomyosin (so this keeps tropomyosin in place around the actin filaments) and one to calcium ions.
* Thick filaments are bundles of the protein myosin. Each myosin molecule consists of a tail and two protruding heads. Each thick filament consists of many myosin molecules whose heads stick out from opposite ends of the filament.

The power stroke

In muscle contraction the following stages occur.
1 Myosin head groups attach to the surrounding actin filaments forming a **cross-bridge**.
2 The head group then bends, causing the thin filament to be pulled along and so overlap more with the thick filament. This is the **power stroke**. ADP and P_i (inorganic phosphate) are released.
3 The cross-bridge is then broken as new ATP attaches to the myosin head.
4 The head group moves backwards as the ATP is hydrolysed to ADP and P_i. It can then form a cross-bridge with the thin filament further along (stage 1) and bend (stage 2) again.

Module 4
Responding to the environment
The sliding filament model

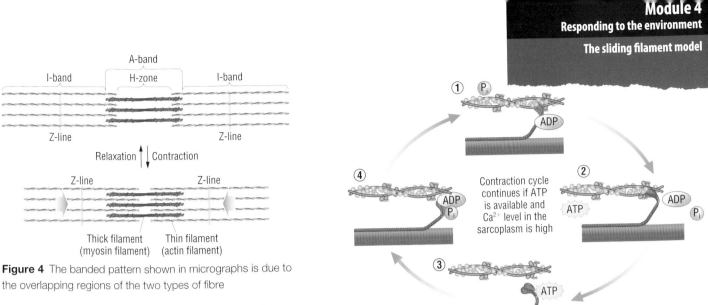

Figure 4 The banded pattern shown in micrographs is due to the overlapping regions of the two types of fibre

In a contracting muscle, several million cross-bridges are continuously being made and broken, causing the thin filaments to slide past the thick filaments and so shorten the sarcomere. This shortens the whole length of the muscle.

Calcium ions allow muscles to contract

The binding sites for the myosin head group on the actin fibre are covered by the tropomyosin subunits. This means that a myosin head group cannot attach to any such binding site, cross-bridges cannot form and muscle contraction cannot occur. When an action potential arrives, via a neurone, at the neuromuscular junction, calcium ions are released from the sarcoplasmic reticulum in the sarcomeres. These calcium ions diffuse through the sarcoplasm and bind to the troponin molecules. This binding changes the shape of the troponin, which moves the tropomyosin away from the binding sites on the actin. The actin–myosin binding sites are uncovered and so cross-bridges can form. This allows the power stroke and muscle contraction to occur.

When nervous stimulation stops, calcium ions are actively transported back into the sarcoplasmic reticulum by carrier proteins on the membrane. This leads to muscle relaxation.

The role of ATP in the power stroke

When the myosin head group attaches to the actin binding site and 'bends', the molecules are in their most stable form. Energy from ATP is required in order to break the cross-bridge connection and re-set the myosin head forwards. The myosin head group can then attach to the next binding site along the actin molecule and bend again.

Maintenance of ATP supply

There is only sufficient ATP available in a muscle fibre to support perhaps 1–2 seconds worth of contraction. In order to allow for continued contraction, ATP must be regenerated as quickly as it is used up. There are three mechanisms by which the ATP supply is maintained.

- Aerobic respiration in muscle cell mitochondria: the level at which this process can regenerate ATP is dependent on the supply of oxygen to the muscles and the availability of respiratory substrate.
- Anaerobic respiration in the muscle cell sarcoplasm: this process is quite quick, but leads to the production of lactic acid, which is toxic. The lactic acid produced enters the blood, where it leads to the stimulation of increased blood supply to the muscles.
- Transfer from creatine phosphate in the muscle cell sarcoplasm: the phosphate group from creatine phosphate can be transferred to ADP to form ATP very quickly by the action of the enzyme creatine phosphotransferase. The supply of creatine phosphate is sufficient to support muscular contraction for a further 2–4 seconds.

Figure 5 Cycle of contraction: attachment, bending and detachment of head groups, their reattachment, and movement of actin filaments leads to muscle contraction

Key definition

A **cross-bridge** is the name given to the attachment formed by a myosin head binding to a binding site on an actin filament.

Examiner tip

The different arrangements of myofibrils, in contracted and relaxed muscle fibres, often come up in exams. You need to be able to show myofibril (actin and myosin) arrangements in cross-section and in transverse section and interpret whether these show contracted or relaxed states.

Remember a muscle fibre is a cell. Myofibrils are organelles.

Questions

1 List three uses of ATP in muscle cells.
2 Suggest why anaerobic exercise, such as weight lifting, does not help in weight loss.
3 Suggest which process for ATP regeneration is most important for:
  (a) a marathon runner
  (b) a 100 m sprinter.

Pulling together – muscles, nerves and hormones

By the end of this spread, you should be able to . . .

* State that responses to environmental stimuli in mammals are coordinated by the nervous and endocrine systems.
* Explain how, in mammals, the 'fight or flight' response to environmental stimuli is coordinated by the nervous and endocrine systems.

Responding to environmental stimuli

Mammals possess complex nervous and endocrine systems. The external and internal environments are continuously monitored by the many sensors of both systems. Stimuli feed information into the systems. This ranges from visual stimulation of the retina in the eyes to blood glucose levels being detected by the islets of Langerhans in the pancreas.

The responses of mammals to environmental stimuli are coordinated and balanced to ensure survival. This may be in the short term, e.g. in homeostatic mechanisms of temperature control; and in the longer term, e.g. in the behaviours associated with reproduction.

The coordination of responses to stimuli from the external environment is mainly the result of brain activity to assess the most appropriate responses. The brain also regulates a number of endocrine responses through the action of the hypothalamus and its control on the pituitary gland.

The 'fight or flight' response

The perception of a threat to the safety of a mammal leads to a number of physiological changes that prepare the organism to deal with the threat, either by escaping or by challenging the threat directly.

Physiological changes

- Pupils dilate.
- Heart rate and blood pressure increase.
- Arterioles to the digestive system and skin are constricted whilst those to the muscles and liver are dilated.
- Blood glucose levels increase.
- Metabolic rate increases.
- Erector pili muscles in the skin contract, making hairs stand up.
- Ventilation rate and depth increase.
- Endorphins (natural painkillers) are released in the brain.
- Sweat production increases.

Coordination of the physiological changes

Perception of the threat might come in the form of visual or auditory stimuli. The sight of a charging lion would elicit the 'fight or flight' response in many animals. The cerebral understanding of a threat activates the hypothalamus, which in turn stimulates increased activity in the sympathetic nervous system and triggers the release of the hormone adrenaline from the adrenal medulla (the central part of the adrenal glands) into the blood.

The hypothalamus also releases corticotropin-releasing factor (CRF) into the pituitary gland, stimulating the release of the hormone adreno-corticotropic hormone (ACTH) from the (anterior) pituitary gland. This hormone stimulates the release of a number of different (corticosteroid) hormones from the adrenal cortex (the outer part of the adrenal glands). Some of these hormones help the body to resist **stressors**.

The combined effects of increased sympathetic nervous system activity and the release of adrenaline and other hormones into the blood are responsible for the physiological changes of the 'fight or flight' response.

Key definition

The **'fight or flight' response** refers to the full range of coordinated responses of animals to situations of perceived danger. The combined nervous and hormonal response has dramatic effects on the whole organism, making it ready for actions that lead to confrontation of the danger or escape from it.

Examiner tip

Remember that sympathetic and parasympathetic stimulation in the nervous system goes on continuously. The fight or flight response results from the balance of stimulation being shifted to increase activity of the sympathetic nervous system and decrease activity of the parasympathetic nervous system.

Key definition

A **stressor** is a stimulus that causes the stress response. It causes wear and tear on the body's physical or mental resources.

Module 4
Responding to the environment
Pulling together – muscles, nerves and hormones

(a)

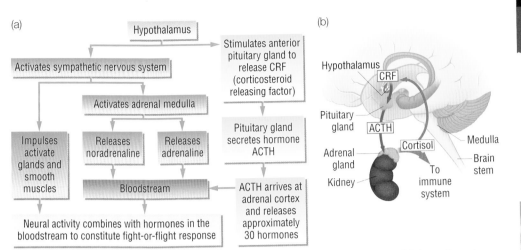

(b)

Figure 1 (a) Coordination of nerves and hormones in the 'fight or flight' response; **(b)** the action of hormones in the 'fight or flight (stress) system

Timing of the fight or flight response

On perception of threat, there is almost always a period of heightened awareness. If this threat is perceived to be from another animal, the threatened animal assesses, for example, the size and aggression level of the other. This is an autonomic assessment seen, for example, in dogs and is the basis of threat displays like the baring of teeth, flattening of the ears, growling and raised fur on the back of the neck.

Figure 2 Threat display by a dog

STRETCH and CHALLENGE

One of the advantages to humans of a large brain is the ability to imagine and to consider possibilities that are not real. Whilst this is essential for creative processes it does have a downside. Humans are capable of imagining threatening situations as part of the capacity to consider solutions to problems. However, the hypothalamus always interprets the cerebral judgement of threat in the same way – by setting off the automatic fight or flight response. In situations of imagined threat, the aspects of the fight or flight response that gear the body up to physical action are inappropriate.

In modern human societies, the result of repeated inappropriate fight or flight stimulation has a variety of effects. These vary from positive effects associated with the drive to succeed, to inappropriate agression and levels of continued stress which can lead to a variety of health issues including digestive system problems and immune system problems, high blood pressure and a variety of mental health issues.

Questions

A How is it possible for an imagined threat to stimulate the fight or flight response?

B Explain why imagination is said to be vital for creative processes.

C Post-traumatic stress disorder (PTSD) has been identified in servicemen and women long after they return from active duties. Suggest how the fight or flight response is implicated in this disorder.

D Why are digestive disorders a feature of prolonged stress in humans?

E Suggest other human functions and capabilities that may be affected by long-term stress.

The fight or flight response

This was first described by Walter Cannon in 1915: 'Animals automatically react to threats with a general discharge of the sympathetic nervous system, preparing the animal for fighting or fleeing.' Cannon was the first to use the word 'stress' in a biological rather than an engineering context.

The theory was updated in various ways, firstly because animals may fight in some situations but flee in others and secondly, not all animals fight or flee. There is a range of responses to threat in different animals, from animals which remain absolutely still in certain threat situations (e.g. possums 'playing dead') to others that have evolved a mechanism for rapid colour change (some fish and reptile species) designed to give camouflage.

Question

1 Construct a table showing how each of the physiological changes associated with the fight or flight response listed above prepares the body for action.

By the end of this spread, you should be able to . . .

* Explain the advantages to organisms of innate behaviour.
* Describe escape reflexes, taxes and kineses as examples of genetically determined behaviours.

What is meant by behaviour?

Behaviour is described as the responses of an organism to its environment which increase its chances of survival. An organism must be able to detect changes in the environment, which form stimuli, then carry out an appropriate response through the operation of effectors. A stimulus can lead to a reflex response. This is a simple form of genetically determined behaviour. More complicated behaviours in animals are the result of a combination of genetically determined (therefore fixed) and learned (therefore adapted) responses to stimuli.

Innate behaviours	Learned behaviours
Genetically determined and so environment has no impact on behavioural response. Passed on to offspring via reproduction.	Determined by the relationship between the genetic make-up of the individual and environmental influences. Not passed on to offspring via reproduction but may be by teaching.
Rigid and inflexible.	Can be altered by experience.
Patterns of behaviour are the same (stereotypical) in all members of a species.	Considerable variety is shown between members of a species.
Unintelligent in the sense that the organism probably has no sense of the purpose of the behaviour.	Learned behaviours form the basis of all intelligent and intellectual activity.

Table 1 A comparison of learned and innate behaviours

Key definition

Innate behaviour is any animal response that occurs without the need for learning. It is an inherited response, similar in all members of the same species and is always performed in the same way in response to the same stimulus.

Innate and learned behaviours

Innate behaviours are those that are inherited in the genome of an organism. Learned behaviours result from responses that are adapted with experience of the environment. Table 1 summarises some of the main differences.

Examples of innate behaviour

Invertebrates rely for their survival on three types of innate behaviour. These behaviours allow them to escape predators, locate and stay in a suitable habitat, and locate food. Invertebrates tend to have very short life spans, live solitary lives and do not take care of their offspring. Each of these factors means that innate behaviours are more suitable as survival mechanisms than learned behaviours.

Reflexes

Many invertebrates have an escape reflex, the function of which is to avoid predators. Earthworms withdraw underground in response to vibrations on the ground. Escape reflexes, like the spinal reflexes seen in humans, are involuntary responses which follow a specific pattern in response to a given stimulus.

Kineses

A kinesis (plural kineses) is an orientation behaviour where the rate of movement increases when the organism is in unfavourable conditions. The behaviour is 'non-directional', meaning that the response is to change the rate of movement overall in relation to the intensity of the stimulation, not in any particular direction. Woodlice avoid predation and drying out by living in damp, dark areas. If placed in dry/bright conditions, woodlice will move around rapidly and randomly until they are in more suitable conditions (damp and dark), when they will move more slowly or even stop moving. This is purely a physiological response; they do not actively seek a dark, damp area.

Examiner tip

Although innate behaviours are important to invertebrates, remember that all animals, including humans show innate behaviour patterns.

Also remember that an innate behavioural response will only survive if it confers an advantage to the species, so examiners could link questions about behaviour to those about natural selection.

Taxes

A taxis (plural taxes) is a 'directional' orientation response. The direction of movement is described in relation to the stimulus which triggers the behavioural response:

- Positive phototaxis is towards, and negative phototaxis is away from, light stimulus.
- Positive chemotaxis is towards, and negative chemotaxis is away from, a specific chemical.

An example of a taxis is seen in the nematode worm *Caenorhabditis elegans*. Chemoreceptors in its lips sense chemical signals in the air, and the animal can be observed to move its head from side to side in order to compare signal strengths and detect the direction of a chemical gradient before moving its whole body up (positive chemotaxis) or down (negative chemotaxis) the gradient.

More complex innate behaviours

The linking together of a series of innate behaviours gives some complex behaviour patterns in invertebrates. The waggle dance is a display used by worker honey bees to communicate the direction and distance of a food source to other worker bees (Figure 1).

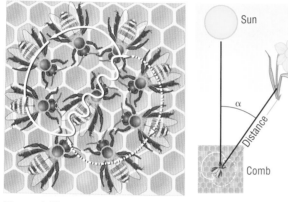

The angle between the waggle part of the dance and the vertical represents the angle between the sun and hence the direction of the flower field horizontally.

The duration of the waggle part of the dance signifies the distance. Approximately one second of dance = 1 km distance.

Figure 1 The waggle dance is an example of complex behaviour in invertebrates

Fixed action patterns (FAP)

A stimulus is required to initiate an instinctive behavioural response. Stimuli lead to releaser mechanisms within the brain (essentially brain activity that leads to a fixed pattern of neuronal output) which, in turn, produce the response (FAP).

The work of Niko Tinbergen with three-spined sticklebacks (fish) demonstrated how a series of stimuli and responses is responsible for the coordination of courtship and mating behaviours between male and female sticklebacks.

For this work, Tinbergen shared the 1973 Nobel Prize with Konrad Lorenz and Karl von Frisch.

Tinbergen noted that the female will follow almost any small red object to the nest, and once within the nest, neither the male nor any other red object need be present. Any object touching her near the base of her tail will cause her to release her eggs.

Case study: sand wasps and fixed action patterns

Sand wasps (*Sphex languedocian*) are insect predators that nest in the soil. Fertilised females dig a shaft down into the soil, then outwards from the main shaft to form small compartments into which their eggs will be laid. When the nest is complete the female goes hunting for prey (perhaps a cricket or a small caterpillar) which she stings to paralyse it. She brings the prey back to the entrance of the nest, moves down into the shaft and turns around, then comes back out to pull the prey down. She lays an egg near the prey's body then goes out to hunt and repeat the process. Her behaviour is highly rigid and stereotypic, consisting of a series of fixed action patterns. The French naturalist J. Henri Fabré conducted a classic experiment on this insect and recorded his findings as follows:

'At the moment when the *Sphex* is making her domiciliary visit, I take the grasshopper, left at the entrance to the dwelling, and place it a few inches further away. The *Sphex* comes up ... looks here and there in astonishment, and, seeing the game too far off, comes out of the hole to seize it and bring it back to its right place. Having done this she goes down again, but alone. I play the same trick again, and the wasp has the same disappointment on her arrival at the entrance. The victim is once more dragged to the hole, but the wasp always goes down alone. This goes on as long as my patience is not exhausted.'

Figure 2 A sand wasp with prey

Questions

1 Adaptive behaviour is behaviour that increases the chances of an organism's survival to adulthood. Explain how such behaviour may be selected for.

2 Why does the short life span of invertebrates mean that innate responses are more important to their survival than learned responses?

3 Suggest a list of stimuli and responses that would explain the behaviour of the sticklebacks.

By the end of this spread, you should be able to ...

* ✳ **Explain the meaning of the term *learned behaviour*.**
* ✳ **Describe habituation, imprinting, classical and operant conditioning, latent and insight learning as examples of learned behaviours.**

What is learned behaviour?

Learned behaviour is described as that which shows adaptation in response to experience. This type of behaviour is of greatest survival benefit to animals:

* with a longer lifespan and so time to learn
* with an element of parental care of the young, which involves learning from parents
* living for a part of the time at least with other members of the species in order to learn from them.

The main advantage of learned behaviour over innate behaviour is that it is adapted in response to changing circumstances or environments.

Classifying learned behaviours

Habituation

Animals learn to ignore certain stimuli because repeated exposure to the stimulus results in neither a reward nor a punishment.

Examples of habituation are found in most animals, including invertebrates. It is why birds learn to ignore scarecrows. It is also important in screening out the many non-dangerous stimuli in the environment such as the sound of wind and waves. Habituation allows humans living near rail or road links to sleep without constant awakening in response to noise stimuli. It avoids wasting energy in making escape responses to non-harmful stimuli.

Imprinting

This involves young animals becoming associated with (imprinting on) another organism – usually the parent. In pioneering work by Konrad Lorenz, goslings were shown to follow the first moving thing they see on hatching. After that, they will only follow (and learn from) objects that look like the first object.

Imprinting only occurs in a **sensitive period** (also known as receptive period). This is around 36 hours after hatching in goslings, less in young chickens. It is significant in helping the young to learn skills from the parents. Skills include flight in birds and knowing to seek out the appropriate type of organism for mating.

Classical conditioning

Classical conditioning was described by the Russian scientist Pavlov in probably the most famous study into animal behaviour. He was, in fact, studying the process of salivation, the amount of saliva produced and its composition. He had observed that when dogs were shown food, or when they smelt food, they salivated. This is a normal reflex action. It is a response to an *unconditioned stimulus* – the sight or smell of food. As he wished to collect their saliva, he fed the dogs in order to stimulate them to salivate. He rang a bell when he was about to give the dogs food. He noticed that the dogs soon began to salivate on hearing the bell, even if they could not see or smell food.

The ringing is known as a **conditioned stimulus** which leads to a new reflex action called a **conditioned response**. This is **classical conditioning**, where animals can learn to relate a pair of events and respond to the first in anticipation of the second. This type of learning is passive and involuntary. Pavlov used other conditioned stimuli including the ticking of a metronome and even gunshots in order to elicit the conditioned response.

Key definition

Learned behaviour refers to animal responses that change or adapt with experience. There is a range of learned behaviours identified, from simply learning not to respond to a repeated stimulation, to the ability to consider a problem and formulate a solution.

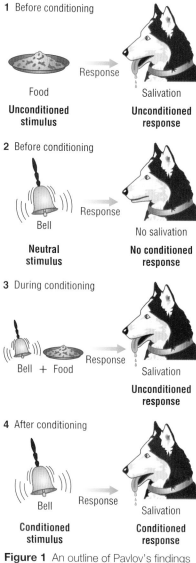

1 Before conditioning

Food — Response → Salivation
Unconditioned stimulus — **Unconditioned response**

2 Before conditioning

Bell — Response → No salivation
Neutral stimulus — **No conditioned response**

3 During conditioning

Bell + Food — Response → Salivation
Unconditioned response

4 After conditioning

Bell — Response → Salivation
Conditioned stimulus — **Conditioned response**

Figure 1 An outline of Pavlov's findings with dogs

Module 4
Responding to the environment
Behaviour can be learned

Operant conditioning

The scientist B.F. Skinner became interested in creating a specific behavioural reaction to a stimulus by adding an element of reward or punishment. In several experiments with rats and pigeons, using a **Skinner box**, he showed that animals in the box would at first accidentally press a lever which resulted in the reward of a food pellet. This reward led to increasing frequency of pressing the lever because the animals had learned to associate the operation (hence **operant conditioning**) of pressing the lever with the reward of food. A variety of rewards and punishments (sometimes referred to as **reinforcers**) can be used in conditioning animal behaviours. The training of dogs is substantially based on the reward of attention from owners. Monkeys can be conditioned with 'social rewards' such as seeing other monkeys. This type of learning is active and to an extent voluntary. In natural circumstances, we often refer to operant conditioning as **trial and error learning**.

Figure 2 A rat in a Skinner box

Latent (exploratory) learning

Animals will explore new surroundings and retain information about their surroundings that is not of immediate use but may be essential to staying alive at some future time. Young rabbits explore the surroundings of their burrows, learning the features of the environment. This knowledge can be life-saving if it helps the rabbit escape a predator in later life.

Insight learning

Insight learning is regarded as the highest form of learning. It is based on the ability to think and reason in order to solve problems or deal with situations in ways that do not resemble simple fixed, reflex responses or the need for repeated trial and error. Once solved, the solution to the problem is remembered. In Wolfgang Köhler's work, chimpanzees (*Pan satyrus*) were presented with bananas hung out of reach and a set of boxes. The chimpanzees were able to stack the boxes on top of each other to reach the bananas. Since then, behaviour among other apes, such as gorillas, orang-utans and gibbons, has been studied.

Figure 3 A chimpanzee building up boxes in order to reach bananas

Examiner tip

You should not try to learn lots of examples of behaviours from the types listed. Exam questions will feature organisms that you have probably never heard of. It is more important that you can identify what type of learned behaviour has taken place from the information that you are given in the question.

Question 1 at the end of this spread is an example of this.

Questions

1 Birds quickly learn to avoid eating cinnabar moth larvae. The moth larvae are bright orange and black and have a very bad taste.
   (a) What type of learning is this?
   (b) Some other insects that are possible prey for birds and don't taste bad have orange and black colouring. Suggest how natural selection has led to them having this colouring.
2 Suggest how latent learning may have played a part in the chimpanzees being able to reach the bananas offered by them by Köhler.

243

By the end of this spread, you should be able to . . .

* Describe, using one example, the advantages of social behaviour in primates.

Primates are mammals. They include the apes and monkeys as well as the more primitive lemurs. Most primates live in family groups where the young remain until they reach sexual maturity – in chimpanzees this may be as long as 12 years. The organisation of the groups usually shows a **hierarchy** where different individuals have different status and roles within the group. Such hierarchies lead to social control within the group which protects all group members.

Primates have large brains (compared to body size) with a highly developed cerebral cortex. This is linked to social development and interaction. It is thought that all **social behaviours** in primates are derived from the extended dependency period of the offspring.

Social organisation in gorillas
Mountain gorillas (*Gorilla gorilla beringei*) live in stable groups (called a troop) of around 10 individuals. This usually consists of one mature dominant male (a silverback), a number of adult females and their offspring. The dominant male protects the other members of the group, leads them in search of food and is the only male that mates with the mature females. As younger males reach sexual maturity they leave the group to live alone until they are mature enough to attract females. As younger females mature they either stay with the same group or join another.

Social behaviours in gorillas
As with all primates, grooming for gorillas is an important social activity. One individual picks the parasites from the fur of another. This occurs between all members of the group, reinforcing relationships between individuals.

Care of young offspring is the role of the mother. During the first 5 months the infant remains in constant contact with the mother, suckling at hourly intervals. By the age of 12 months, infants will venture as far as 5 m from the mother.

During this period the female protects the young gorilla as it learns the social and other skills necessary to live independently. Further learning takes place after the age of two, as juvenile gorillas play together and imitate the adult behaviour of foraging for food. The silverback is important in the development of young gorillas from the age of 3 to 6 years, both in terms of protection from older males within the group, and in play as a source of learning new skills.

Communication systems exist. A variety of calls, displays and grunts are used to signal danger to other members of the group, to issue threats to predators or other groups, and in play fighting displays as juveniles learn how to behave as adults. Facial expressions are also important in gorillas and other large primates, especially in terms of recognition of other members of the group.

The advantages of social behaviour
Many organisms demonstrate social behaviour, from insects, such as bees, to humans. The advantages of social behaviour in primates include:
* Females give birth to only one (or very few) infants at a time. The maternal care and group protection enhances the survival rate of the young.
* The young learn through observation of and play with the other members of the group. Learned behaviour is vital to the survival of primates.

Key definition
A **hierarchy** within a group exists where individuals have a place in the order of importance within the group. This is often shown by individuals higher up in the hierarchy receiving more food, or having rights of access to mate with other individuals.

Social behaviour refers to that of organisms of a particular species living together in groups with relatively defined roles for each member of the group.

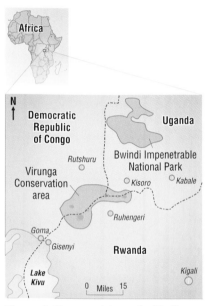

Figure 1 The mountain gorilla (*Gorilla gorilla beringei*) is the largest of all the great apes. They exist in two small areas in the western Rift Valley in central Africa – the Virunga Volcanoes that straddle Rwanda, Uganda and the Democratic Republic of Congo. Only around 650 individuals remain in the wild

Figure 2 Mountain gorilla behaviour

- The final relatively large brain size slows the maturation of primates. The security of a group enhances the survival and learning of immature young.
- Knowledge and protection of food sources is shared with the group.
- Greater ability to detect and deter predators is achieved by groups of individuals working together.

STRETCH and CHALLENGE

The infants of most primates spend the early part of their lives in constant contact with the mother. This enhances the speed of learning from the mother and aids the social development of the infant through the ability to learn social structures and activities in safety. This is often referred to as 'carry behaviour'. Less developed animals leave their young in nests, returning to them to feed them at intervals. The young are born more fully developed, able to cope with being left and are quickly able to follow the mother.

In many wealthier human societies, parenting of young follows patterns not seen in primates:

- bottle feeding at regular set and increasing intervals rather than breastfeeding
- sleeping separately from mother from birth
- fixed periods of waking interaction with parents
- sometimes long and regular periods of separation from parents at an earlier stage than would be seen in similar primate species.

Some experts now suggest more primate-like care of human infants is physically and psychologically better for the infants. They call this approach 'kangaroo mother care'. They suggest two actions:

1 Skin to skin contact between mother and infant

The separation of newborns from their mother results in the protest despair response (recognised in orphans after WW2). It is characterised by lowered body temperature and heart rate and high levels of stress hormones in the blood. Continued skin contact has been shown to reduce stress hormone levels, raise and stabilise body temperature and increase blood oxygenation levels.

2 Breastfeeding on demand for the first months of life

Human milk provides the best balance of nutrients for the growth pattern of the infant. It also provides immunity to a number of infections as antibodies pass from mother to infant. There is also strong evidence that a number of illnesses associated with gut formation and allergy are linked to the replacement of human milk with formula milk, and a suggestion that formula milk is associated with the risk of later developing type 2 diabetes.

Questions

A Suggest how human milk differs from that of a non-primate mammal such as a cow.

B Primate gestation periods are relatively short compared to the lifespan of the animal. Suggest:
  (i) why this is necessary in terms of birth
  (ii) how this is linked to social behaviour.

C The body temperature of a mother has been shown to vary, by as much as 3°C, when in contact with her baby. The temperature variations are antagonistic to the baby's temperature fluctuations. Suggest how fixed action patterns could explain this feature.

D Many human societies behave in social groups in ways very similar to other primates.
  (i) Explain why the study of such groups is useful in researching the advantages of kangaroo mother care.
  (ii) Suggest why the findings from these societies may not be directly transferable to other societies.

Examiner tip
You are required to be able to use one example of a primate to demonstrate the advantages of social behaviour. Remember that social behaviour is not exclusive to primates; however, the main advantages to primates of social behaviour are concerned with extended care of the young, which is not common to other types of social behaviour.

Questions

1 Suggest why in gorillas it is necessary for all sexually mature males and most sexually mature females to leave the social group they were born into.

2 All social behaviours in insects are innate, whereas many social behaviours in primates are learned. Suggest an advantage of the type of social behaviour shown in each type of organism.

By the end of this spread, you should be able to ...

* Discuss how the links between a range of human behaviours and the dopamine receptor DRD4 may contribute to the understanding of human behaviour.

Structure and activity

Dopamine acts as a neurotransmitter and a hormone. It is a precursor molecule in the production of adrenaline and noradrenaline. Abnormally low levels of dopamine are associated with Parkinson's disease, the treatment of which involves clinical administration of the dopamine precursor L-dopa. Unfortunately, the raised dopamine levels resulting from treatment and sometimes occurring naturally have been linked to the development of mental health conditions such as schizophrenia. It has also been noted that patients with Parkinson's disease who are treated with L-dopa are prone to behavioural changes such as compulsive gambling.

Dopamine increases general arousal and decreases inhibition, leading to an increase in creativity in conjunction with cerebral activity.

The range of activities affected by dopamine is partly due to the number of and variation within dopamine receptors. There are five different dopamine receptors referred to as DRD1 to DRD5. Each of these is coded for by a separate gene. Binding of dopamine to its receptor is involved in a number of processes, including the control of motivation and learning, and is linked to regulatory effects on other neurotransmitter release. A number of antipsychotic drugs work by blocking dopamine receptors.

The DRD4 receptor gene

There are currently over 50 known variants of the *DRD4* gene. The variants differ in a specific sequence known as a variable number tandem repeat. A short section of nucleotides shows a different number of repeats in each variant.

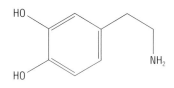

Figure 1 The structure of dopamine

Key definition

DRD4 is one of five genes that code for dopamine receptor molecules. Dopamine can bind to each of these receptor molecules but they cause differing effects because they lead to different cellular responses.

A number of studies have shown that some of these variants are implicated in certain human behavioural conditions. It is thought that the inheritance of particular variants of the *DRD4* gene affects the levels and action of dopamine in the brain.

Attention-deficit hyperactivity disorder (ADHD)

Drugs such as methylphenidate (Ritalin) used to treat ADHD affect dopamine levels in the brain. In a number of studies, a particular dopamine receptor variant of DRD4, has been shown to be more frequent in individuals suffering from ADHD.

Addictive and risk behaviours

A number of studies have suggested that particular variants of the DRD4 receptor gene are implicated in increased likelihood of addictive behaviours, including smoking and gambling. A study into the effects of administering L-dopa to one group of individuals and haloperidol (a drug that blocks dopamine receptors) to another group showed not only a difference in general arousal, but also a significant difference in the risk-taking levels of the individuals.

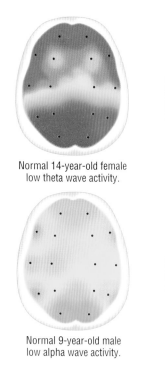

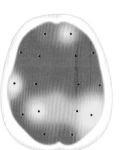

Normal 14-year-old female low theta wave activity.

ADHD 14-year-old female high theta wave activity.

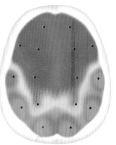

Normal 9-year-old male low alpha wave activity.

ADHD 9-year-old male high alpha wave activity.

Figure 2 Level of brain excitation in children with ADHD

Module 4
Responding to the environment
Human behaviour, dopamine and DNA

Other neurotransmitter-influenced behavioural conditions

The interest in *DRD4* and dopamine has stimulated research into other possible links between behaviour and genetic constitution. Obsessive–compulsive disorder (OCD) is thought to result from a deficiency in the levels of the neurotransmitter serotonin.

In 2007 a genome-wide scan for DNA sequences related to OCD was carried out when the DNA from 1008 people from 219 families was analysed. Eight genetic markers that appear to be linked to OCD were found.

Genes for specific behaviours?

The development in understanding of how genes and their alleles may affect the human brain and the behaviour of humans has been transformed by the Human Genome Project. The implications for society, medicine and health are potentially far-reaching. However, the association of specific genes or groups of genes with particular behaviours must be approached with caution.

If the human genome has taught us nothing else, it has opened our eyes to the fact that we have relatively few genes, with a whole range of, as yet undiscovered, control mechanisms. Single gene variants may influence specific behaviours – but the key word here is influence. The complexity of human genetics and behaviour is such that a simple gene-to-behaviour link is unlikely to be the case for even the most simple behavioural mechanisms.

The environment also plays a part. Whereas scientists used to talk about the 'nature–nurture debate', they now realise that genes and environment work together. It is often the case that nature loads the gun and the environment pulls the trigger. Behaviour cannot be attributed to either genes or environment, but is a product of both.

Identical twins show different levels of gene expression

In the UK the Twins Early Development Study (TEDS), a **longitudinal study** of several hundred sets of identical and non-identical twins, entered its 13th year in 2008. In a pilot test, the study identified differing levels of expression of some genes in the saliva of identical twins. The development of understanding of the ways behaviour might affect the DNA gives us a further complication in understanding human behaviours.

Key definition

Psychosis refers to a mental health condition, characterised by an impaired grasp on reality, diminished impulse control and disorder of perception (such as hallucinations).

A **longitudinal study** is an investigation in which the same individuals are studied repeatedly over a long period of time in order to gather relevant data about progression of the factors under investigation.

Examiner tip

The research on gene–behaviour links is progressing rapidly and examiners often use up to date research to set questions. You should ensure that you can give examples that link behaviour to genetic influence, for example the use of dopamine receptor targeted drugs to treat **psychosis**.

Questions

1 The term variants has been used here to note the different forms of the dopamine receptor gene *DRD4*. Suggest why these are not referred to as alleles.
2 Explain why it is unlikely that a single gene for a condition such as ADHD will be found, whereas a gene responsible for a condition such as cystic fibrosis has been found.
3 Why is evidence of different levels of gene expression in identical twins described as a 'further complication' in our understanding of human behaviour?

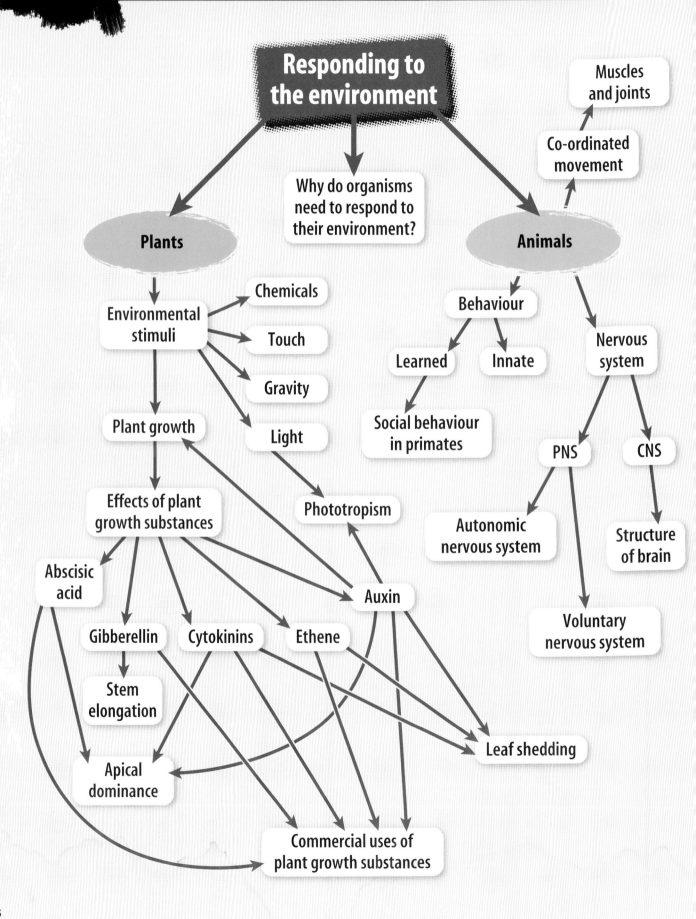

Responding to the environment

Why do organisms need to respond to their environment?

Plants

Environmental stimuli
- Chemicals
- Touch
- Gravity
- Light

Plant growth

Effects of plant growth substances

Abscisic acid

Gibberellin

Cytokinins

Ethene

Auxin

Phototropism

Stem elongation

Apical dominance

Leaf shedding

Commercial uses of plant growth substances

Animals

Muscles and joints

Co-ordinated movement

Behaviour
- Learned
  - Social behaviour in primates
- Innate

Nervous system
- PNS
  - Autonomic nervous system
  - Voluntary nervous system
- CNS
  - Structure of brain

Practice questions

1. Reflexes can be automatic, stereotyped responses to stimuli that can also be conditioned.

   Explain the meaning of the terms:
   (a) automatic
   (b) stereotyped
   (c) conditioned   [3]

2. (a) Explain the role of acetylcholine in synapses.   [3]

   (b) Explain the role of acetylcholinesterase in synapses.   [3]

3. The mammalian nervous system is made up of the central nervous system and the peripheral nervous system. State the components of these two systems.   [8]

4. Troponin and tropomyosin are two proteins that control the process of muscle contraction. When muscle is stimulated to contract, calcium interacts with these proteins to allow myosin molecules to make contact with actin molecules.

   (a) Describe the events that occur when muscle contracts. In your answer explain the roles of calcium ions, tropomyosin, troponin, actin, myosin and ATP.   [8]

   (b) Explain why muscle cells contain glycogen granules. [4]

5. (a) State the meaning of the terms:
      (i)   tropism
      (ii) positive phototropism
      (iii) negative geotropism   [3]

   (b) Explain why the following may be useful to a plant:
      (i) geotropism
      (ii) phototropism   [2]

   (c) Compare plant and animal hormones.   [5]

6. (a) Explain how auxin produces shoot growth.   [2]

   (b) Explain the mechanism of phototropism.   [2]

   (c) Explain how a plant sheds its leaves.   [3]

7. Complete the table to show the areas of the mammalian brain and their functions

Area of brain	Function
	Co-ordination of posture
	Control of heart rate
	Control of temperature regulation
	Control of speech and voluntary movements
	Interpreting a photograph

                                                                        [5]

8. Ten young lambs were placed in an enclosure. An animal behaviour scientist entered the enclosure carrying an umbrella, which he opened and closed repeatedly, in front of the lambs. At first the lambs backed away from the umbrella but eventually they ignored it.

   (a) (i) State the type of behaviour shown by the lambs at the end of the experiment.   [1]

       (ii) Suggest **two** advantages to the lambs of adopting this behaviour.   [2]

   When a larger group of lambs was placed in a field, near to a busy road, they at first reacted nervously whenever a noisy vehicle drove past. Eventually they ignored this unpleasant stimulus.

   Once a day the farmer drove to the field to fill up food containers for the sheep. After two weeks the lambs would run to the food containers as soon as the farmer's vehicle approached the field, even though they could not see it.

   (b) Explain the learning processes shown in the lambs' responses to

   (i) the general road traffic   [2]

   (ii) the farmer's vehicle   [4]

1 Figure 1.1 shows a neuromuscular junction.

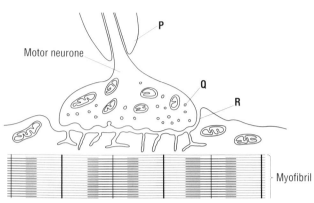

Figure 1.1

(a) Name structures **P** to **R**.
   P
   Q
   R [3]

(b) Using the information in Figure 1.1, describe and
explain how an impulse in the motor neurone stimulates
contraction of the myofibril. [7]
Nicotine acts at some neuromuscular junctions to
stimulate muscles to contract.

(c) Suggest a way in which nicotine may act at
neuromuscular junctions to have this effect. [2]

[Total: 12]
(OCR 2805 05 Jun03)

2 (a) Suckling in young mammals is an example of innate
behaviour.
Explain what is meant by *innate behaviour*. [2]

(b) The sympathetic nervous system responds rapidly to
danger. Resources are mobilised for strenuous activity.
These resources are diverted from the abdominal
organs, such as the stomach and intestines, to muscles.
This is often known as the 'fight or flight' response. The
sympathetic nervous system also stimulates the adrenal
glands to release adrenaline. One effect of adrenaline is
to stimulate the mobilisation of glucose from the liver.
State the actions of the sympathetic nervous system and
adrenaline in the 'fight or flight' response on
(i) the heart
(ii) the digestive system [4]

(c) Research was carried out on the physiological effects of
cold environmental temperatures on a species of small
mammal. The mammals were conditioned to press a
lever to increase their environmental temperature. The
conditioning lasted for three days. The researchers
measured changes in insulin and adrenaline in these
mammals for the three days of conditioning and for
seven days afterwards. These results are shown in
Figure 2.1.

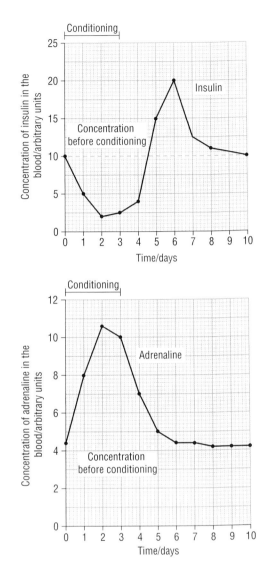

Figure 2.1

(i) Using the data in Figure 2.1, describe the effect of
the conditioning on the concentrations of insulin
and adrenaline over the 10 day period. [5]

(ii) Suggest an explanation for the changes you have
described in **(i)**. [4]

[Total: 15]
(OCR 2805 05 Jun03)

3 Plants must respond to changes in both their external and
internal environments. Communication in plants is achieved
by using a number of plant growth regulators.

(a) List **three** stimuli that plants respond to. [3]

(b) Describe how plant growth regulators are transported
within a plant. [2]

(c) In this question, one mark is available for the quality of
spelling, punctuation and grammar.
Students carried out an experiment to investigate the
role of plant growth regulators on the production of
α-amylase in germinating barley grains.

Answers to examination questions will be found on the Exam Café CD.

- Four sterile starch agar plates were prepared
- Barley grains that had been soaked for 24 hours were sterilised and then cut in half using a sterile razor blade, as shown in Figure 3.1.

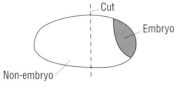

Figure 3.1

- Four non-embryo halves were placed cut side down onto each agar plate.
- 0.5 cm³ of the following solutions were added to the agar plates:
    - Plate 1 – distilled water (control)
    - Plate 2 – gibberellic acid (GA) solution
    - Plate 3 – abscisic acid (ABA) solution
    - Plate 4 – GA and ABA solution
- The plates were incubated at 20 °C for 96 hours.
- After incubation iodine solution was added to each plate.
- The appearance of a plate after adding iodine solution is shown in Figure 3.2.

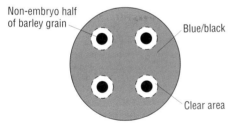

Figure 3.2

The maximum diameter of the clear area surrounding each non-embryo half was recorded.
The students' results are shown in Table 3.1

Table 3.1

Agar plate	Maximum diameter of clear area surrounding each of the four non-embryo halves after 96 hours/cm			
1 Distilled water	1.5	2.4	1.5	1.4
2 GA	2.0	1.3	2.5	2.2
3 ABA	0.9	0.8	1.5	1.2
4 GA and ABA	1.2	1.2	1.0	1.2

Using the results in Table 3.1, describe **and** explain the role of the plant growth regulators GA and ABA in seed germination. [8]
Quality of Written Communication [1]

[Total: 14]
(OCR 2804 Jun07)

4 In the potato plant, *Solanum tuberosum*, the stem develops from a bud on the parent tuber. Buds on the stem below ground level develop into underground stems during summer. The tips of these underground stems eventually swell into tubers, as shown in Figure 4.1.

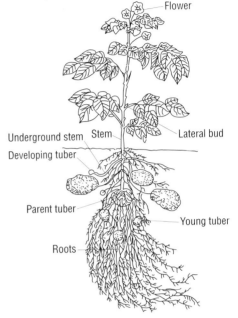

Figure 4.1

(a) Lateral buds on the stem above ground level remain inactive and growth only occurs at the apex of the stem. State the name of this phenomenon. [1]

(b) Figure 4.2 shows the effect on the development of lateral buds of applying auxin (IAA) and/or gibberellin (GA) to stems, each of which has had the bud at the apex removed. Lanolin is a chemical that has no effect on plant tissues.
In each case the lanolin was applied to the cut end of the stem as shown by the shaded regions in Figure 4.2.

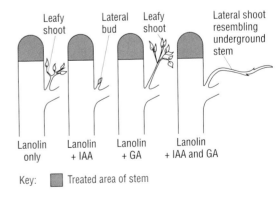

Figure 4.2

Describe the effect of IAA and GA on lateral bud development in potato plants. [4]

[Total: 5]
(OCR 2804/07 Jan07)

Answers

1.1.1 The need for communication

1 **(a)** and **(b)** Examples:
  Stimulus is feeling an insect bite – response is to move away – nervous coordination.
  Stimulus is seeing a large barking dog running towards you – response is to run away – nervous and hormonal coordination.
  Stimulus is smelling delicious food – response is to salivate – nervous coordination.

2 The thicker hair traps more air for insulation to prevent heat loss. The white colour gives it camouflage against the snow so its prey cannot see it.

3 A good communication system must:
  • be able to send messages or signals
  • be able to receive messages or signals
  • distinguish between different messages and signals
  • extend over the whole body.

1.1.2 Homeostasis and negative feedback

1 Negative feedback monitors a change and initiates a response that will reverse the change. It will maintain constant levels.
  Positive feedback monitors a change and initiates a process that will increase the change. It tends to be destabilising.

2 Thermostat – receptor.
  Boiler – effector.
  Electric cables – communication.

1.1.3 Maintaining body temperature – ectotherms

1 If temperature is too low, enzyme activity falls and the rate of metabolism drops. High temperatures may denature proteins such as channel proteins in membranes and possibly some enzymes.

2 Ectotherms: any examples of fish, reptiles, amphibians or invertebrates, e.g. lizard, frog, salmon, insect, worm.
  Endotherms: any examples of mammals or birds, e.g. human, dog, kangaroo, robin, eagle.

3 If the ectotherm is too cool it can absorb heat from the rock and directly from the sun. This will warm its skin and blood flowing through the skin. The extra heat is transported around the body to help warm up the muscles.

1.1.4 Maintaining body temperature – endotherms

1 A shrew is very small and has a large surface-area-to-volume ratio. It loses a lot of heat through its skin.

Therefore a lot of food must be used to replace the heat lost. An elephant has a small surface-area-to-volume ratio and therefore loses a much smaller proportion of its body heat.

2 **(a)** Australia is warm and the penguins do not need to be large to maintain their body temperature. Antarctica is very cold and larger penguins have a smaller surface-area-to-volume ratio – so they can maintain their body temperature more easily.

  (b) By huddling together the penguins provide insulation to prevent heat loss and gain warmth from each other. The huddle has a smaller surface-area-to-volume ratio than a solitary penguin.

3 Vasoconstriction reduces blood flow through the skin. The blood flows through vessels that are deeper in the body. Heat in the blood is less easily radiated out of the body.

1.1.5 Sensory receptors

1

Feature	Sensory neurone	Motor neurone
Position of cell body	Just outside the central nervous system (CNS)	Inside the CNS
Is it myelinated?	Yes	Yes
Length of dendrites or dendron	Long (often single dendron) – runs from the sensory receptor to the cell body just outside the CNS; has dendrites at end of axon	Many short dendrites
Length of axon	Short – runs from just outside the CNS into the CNS	Long – runs from inside CNS to effector
Direction of communication	Carries impulses from sensory receptor to CNS	Carries impulses from CNS to effector

2 Mitochondria produce ATP by aerobic respiration to maintain the resting potential and for exocytosis of neurotransmitter at synapses.

3 Charged ions cannot dissolve in a phospholipid bilayer membrane like that of the axon. Therefore they cannot

diffuse across the membrane. The ions need channel protein to produce a hydrophilic channel for diffusion or facilitated diffusion.

1.1.6 Resting potentials and action potentials

1 While the neurone is resting, or not conducting an impulse, the membrane is actively transporting sodium ions out of the cell and potassium ions into the cell.

2 The organic anions inside the neurone help to ensure that the inside of the cell remains negative compared to the outside.

3 The gated channels in the generator region have gates that are operated by energy coming from the stimulus. In a touch receptor the gates would be operated by movement. In the remainder of the neurone the gates are operated by potential difference changes. The gates would be opened by the movement of ions in the local circuit or by changes in potential difference as the membrane becomes depolarised.

4 A hyperpolarised membrane is much more difficult to depolarise. The cell membrane has a greater potential difference across it than the normal resting potential (–80 mV rather than –60 mV). This makes it more difficult to depolarise the membrane. As sodium ions enter, the threshold that causes voltage-gated sodium ion channels to open is not reached. Therefore an action potential is not reached. The sodium ions that moved into the cell to cause an action potential are still in the cell and would reduce influx of sodium ions if the sodium ion channels open. Also, some potassium ion gates are still open, allowing potassium ions out of the cell.

1.1.7 Transmission of action potentials

1 Each channel has a voltage-sensitive gate. A change in the potential around the gate causes it to move or change shape. This opens the channel.

2 The sodium/potassium ion pumps actively transport ions against their concentration gradients. Sodium ions are transported out of the neurone so that the concentration is low inside the neurone and high outside it. Potassium ions are transported into the neurone so that the potassium concentration is high inside the cell and low outside it.

3 The myelin sheath is a layer of fatty material tightly bound to the neurone. The sodium and potassium ions cannot move through the sheath. The sheath is incomplete and the neurone membrane is exposed at the nodes of Ranvier. The ions can diffuse through the membrane at these points. Therefore the action potential appears to jump from one node to the next.

1.1.8 Nerve junctions

1 An action potential in the presynaptic neurone can create another action potential in the postsynaptic neurone. This is achieved by releasing molecules of neurotransmitter (transmitter substance) into the cleft. The transmitter substance is released by the action of calcium ions that flood into the synaptic knob when the calcium ion channels are opened. The calcium ion channels have gates that respond to the voltage change caused by the original action potential. If the calcium ion channels responded to any other stimulus the synapse would release transmitter substance at the wrong time.

2 The synaptic cleft contains the enzyme acetylcholinesterase to break down the neurotransmitter, acetylcholine. If the acetylcholine is left in the cleft it will continue to stimulate the postsynaptic neurone and cause action potentials.

3 The presynaptic neurone ends in a synaptic knob as this provides a larger surface area. The larger surface area allows more vesicles of acetylcholine to be released and allows the presynaptic neurone to stimulate more than one postsynaptic neurone.

1.1.9 Signals and messages

1 Summation is the interaction of many small potentials (or EPSPs – excitatory postsynaptic potentials) to create one large potential that may pass the threshold and create an action potential.

2

Feature	Myelinated	Unmyelinated
Speed of conduction	Fast (100–120 m s^{-1})	Slow (0.5–10 m s^{-1})
Nodes of Ranvier	Yes	No
Myelin sheath	Yes – each neurone is wrapped in many Schwann cells	No – several neurones may be surrounded by one Schwann cell
Where found	In peripheral nervous system carrying impulses from sensory receptors to the CNS or from CNS to the effectors	In the CNS and in the nerves controlling the organs with no conscious control

1.1.10 The endocrine system

1 Exocrine glands contain cells that secrete into a duct that carries the secretion to where it is needed. Endocrine glands contain cells that secrete hormones directly into blood capillaries.

2 Target cells must have specific receptors to which the hormone can bind. This enables the hormone to activate a process inside the cell. The receptor must be specific so that the hormone can bind only to the correct cells that can respond to the hormone. Receptor and hormone have shapes complementary to each other.

3 Adrenaline has a wide range of effects all over the body. These include:
- Increasing heart rate and stroke volume – which increases cardiac output to pump more blood per minute around the body carrying more oxygen and glucose or fatty acids to the muscles (and removes more carbon dioxide and lactate).
- Dilation of bronchioles and increasing breathing rate – this ensures the blood is more fully oxygenated so that more oxygen is transported to the muscles.
- Constricting blood flow to the digestive system and other organs – raises blood pressure and leaves more blood to flow to the muscles.
- Conversion of glycogen to glucose – supplies more glucose to the muscles and blood so that the muscles can manufacture more ATP for contraction.
- Dilation of the pupils and increasing mental awareness – helps you to spot and respond to danger.

1.1.11 The regulation of blood glucose

1 Hepatocytes (liver cells) contain a store of glycogen. When there is excess glucose in the blood the cells absorb glucose and make more glycogen – therefore they need to respond to insulin. However, when there is insufficient glucose in the blood the cells break down glycogen and release glucose into the blood. Therefore they must also respond to glucagon.

2 Insulin and glucagon are first messengers. cAMP is a second messenger.

1.1.12 Regulation of insulin levels

1 Proteins are digested in the stomach and small intestine. If diabetics took insulin orally it would be digested to amino acids before being absorbed.

2 Hyperglycaemia means a blood glucose concentration that is too high ('hyper' = high or over, 'glyc' refers to glucose and 'aemia' refers to the blood). Hypoglycaemia means a blood glucose concentration that is too low ('hypo' = low or under).

3

Feature	Insulin secretion from β cells	Acetylcholine secretion in synapse
Stimulus	High concentration of glucose in blood	Action potential in axon membrane
Effect on membrane	Potassium ion channels close	Sodium ion channels open
Effect on ions	Potassium ions stop leaving the cell by diffusion	Sodium ions enter cell by diffusion
Effect on membrane potential	Potential difference across membrane is altered – it becomes less negative inside	Potential difference across membrane is altered – it becomes less negative inside
Effect on calcium ion channels	Opening of calcium ion channels	Opening of calcium ion channels
Effect on calcium ions	Calcium ions enter cell	Calcium ions enter cell
Effect on vesicles	Vesicles of insulin fuse to cell surface membrane	Vesicles of acetylcholine fuse to cell surface membrane
Mechanism of release	Exocytosis	Exocytosis

1.1.13 Control of heart rate in humans

1 Myogenic means that the heart muscle can initiate its own contraction.

2 Increased physical activity means that the muscles are using more energy. They need more ATP for contraction. The muscle cells need to respire more. Therefore they need more oxygen and more glucose or fatty acids for respiration. The heart rate rises to increase the cardiac output so that more blood is pumped per minute around the body. This increased blood flow supplies more oxygen and glucose or fatty acids to the muscles and removes the extra carbon dioxide or lactate produced.

1.2.1 Excretion

1 Excretion is the removal of metabolic waste from cell processes. Egestion is the removal of undigested food from the digestive system – it has never been in a body cell so it cannot be classed as excretion.

2 Respiratory acidosis is the result of high carbon dioxide levels in the blood. Carbon dioxide dissolves in the watery plasma and reacts to form the acid carbonic acid. The carbonic acid dissociates to release hydrogen ions, which make the blood acidic.

3 Ammonia is very soluble but is highly toxic. A fish can use lots of water to dilute the ammonia so that it is not toxic. This dilute ammonia solution can be released all the time so it does not accumulate in the body. Mammals do not live in water. They must conserve water, so they cannot release too much as urine. The urine must be concentrated and must be stored before being released. Therefore mammals must convert the ammonia to the less toxic compound urea which can be transported more safely in blood and stored more safely in the bladder.

1.2.2 The liver

1 Oxygenated blood comes from the lungs, via heart, aorta and hepatic artery, supplying the oxygen needed for aerobic respiration. The second supply comes from the digestive system. This blood carries the products of digestion which need to be removed from the blood so that excess amounts of amino acids/glucose/alcohol/ other toxic compounds do not get around the body. Excess glucose can be stored as glycogen and not wasted. Excess amino acids can be turned into other amino acids (transaminated) or deaminated.

2 Blood entering the liver flows into ever smaller vessels. The blood from the two sources mixes and flows along the sinusoids, which are lined by liver cells.

3 Mitochondria, ribosomes, endoplasmic reticulum.

1.2.3 Functions of the liver

1 Ammonia is highly soluble and very toxic; urea is less soluble and less toxic.

2 They contain valuable energy that can be converted to useable forms. Some amino acids can be converted into other amino acids.

3 The mitochondria provide ATP (metabolic energy) for the active or energy-requiring processes, e.g. protein synthesis, mitosis, active transport, endo- and exocytosis. The ribosomes manufacture the many enzymes that are needed in liver cells.

1.2.4 The kidney

1 To increase the length for greater surface area for absorption.

2 So that materials reabsorbed from the fluid in the tubule can re-enter the blood.

3 Some of the molecules in the nephron are waste and must be left in the fluid to be excreted. Other molecules are useful to the body and must be reabsorbed.

1.2.5 Formation of urine

1 Filtering on a molecular scale. Small molecules pass through the basement membrane, which acts as a filter, while larger molecules are held in the blood.

2 A large volume of very dilute urine would be produced and dehydration would occur.

3 Because the amino acids and glucose have been passed from blood plasma to the glomerular filtrate by ultrafiltration in the glomerulus.

1.2.6 Water reabsorption

1 This arrangement allows water to be reabsorbed from the collecting ducts back into the tissue fluid of the medulla. This concentrates the urine.

2 Terrestrial mammals gain water by eating and drinking. They lose water through sweat, exhaling, excretion and egestion. It is important not to lose more water than necessary, as it may not be readily available.

3 Beavers live beside or in water. Water is readily available and they do not need to conserve it so much.

1.2.7 Osmoregulation

1 Neurosecretory cells manufacture a hormone in their cell body; this is transferred down the axon. When it is released it goes straight into the blood rather than to another nerve cell.

2 Negative feedback occurs when a change in the internal conditions stimulates a reversal of that change – so the conditions are kept constant.

3 ADH must be broken down so that it is not continually acting on the walls of the collecting ducts.

1.2.8 Kidney failure

1 Water, glucose, salt and protein intake.

2 Haemodialysis fluid is kept sterile so that it does not cause infection to the recipient, as it is a good medium for microorganisms. It is kept at 37 °C so that it is at body temperature and does not cause either hypothermia or hyperthermia. (Heparin also added to prevent coagulation of blood in the dialysis tubing.)

3

Advantages	Disadvantages
replaces function of kidney	time-consuming
allows near-normal activity	risk of infection
	does not maintain blood composition constantly
	diet needs careful monitoring

4 In order to detect the presence of certain drugs there must be a marker to show at what point the drug appears on the chromatogram. This position can be compared with the urine to see if there is a deposit at the same point.

1.3.1 The importance of photosynthesis

1 Photosynthesis. Before photosynthesis there was no oxygen in the atmosphere. Photosynthesis released oxygen (from water) into the atmosphere and, over a very long time, the concentration of oxygen in the atmosphere became large enough to support aerobic respiration.

2

	Autotrophs	Heterotrophs
Do they carry out respiration?	yes	yes
Can they synthesise complex organic molecules from simple inorganic molecules?	yes	no
Do they use light energy?	yes	no
Can they hydrolyse complex organic molecules?	yes	yes
Examples	Plants, algae, some bacteria	Some bacteria, some protoctists, fungi, animals

1.3.2 How the structure of chloroplasts enables them to carry out their functions

1 Around 550 nm/530–570 nm/the green part of visible light.
2 Yes. They reflect light of around 530 nm/yellow/orange.
3 Consist of stacks of thylakoid membranes which give a large surface area for the photosystems/chlorophyll to trap light energy, the electron transport proteins and ATP-synthase enzymes. The space within the membrane stack allows the movement of protons for chemiosmosis and the generation of ATP. There are proteins, embedded in grana, which hold the photosystems in place.
4 Fluid-filled stroma has enzymes for Calvin cycle. It surrounds the grana so the products of the light-dependent reaction can easily pass into the stroma. DNA has the genetic code to make some of the proteins needed for photosynthesis and there are ribosomes for assembly of these proteins.

1.3.3 The light-dependent stage

1

	Cyclic	Non-cyclic
Photosystems involved	I only	I and II
Is photolysis involved?	No	Yes
Fate of electrons released from chlorophyll	Return to chlorophyll	Accepted by NADP
Products	ATP	ATP, reduced NADP, oxygen (by-product)

2 It is the source of hydrogen ions to be used in the light-independent stage to reduce GP to TP. It is an electron donor and replaces the electrons lost by PSII when light excites chlorophyll molecules. It keeps palisade cells turgid, so the full vacuoles push the chloroplasts to the outer edges of the cells where they can more readily trap sunlight and carbon dioxide can diffuse into them.
3 Light is trapped by photosystems (PSI) in chloroplasts in guard cells. ATP is made. This actively transports potassium ions into the guard cells. This lowers the water potential of the guard cells and water enters them, by osmosis, down the water potential gradient, from the surrounding epidermal cells. The guard cells swell and the tips, where cellulose in the walls is thinner and more flexible, bulge. This pushes guard cells apart from each other, opening the stoma between them.
4 A lack of iron may mean fewer electron carriers (proteins with haem/iron) in the thylakoid membranes and reduced light-dependent reactions. This could lead to less reduced NADP and fewer molecules of ATP, which would reduce the light-dependent reactions. This in turn reduces yield of amino acids, so fewer proteins for growth or for enzymes needed for the light-independent reactions.

1.3.4 The light-independent stage

1 Because it is continually being regenerated during the Calvin cycle.
2 It is the source of carbon. Once it is fixed it can be used to build larger, more complex organic molecules, all of which are carbon-based, such as hexose sugars, disaccharides, polysaccharides, amino acids, proteins, fatty acids and glycerol, fats, vitamins, chlorophyll and nucleic acids.

3 Lack of nitrates reduces production of amino acids, which means fewer proteins for growth. It also means fewer enzymes, such as rubisco, so a reduced rate of photosynthesis.

4 All except GP and sucrose.

1.3.5 Limiting factors

1 There is more kinetic energy so the rate of enzyme-catalysed reactions in the Calvin cycle increases; there will be an increased rate of collisions between enzyme and substrate molecules.

2 It increases the temperature, increasing the rate of enzyme-catalysed reactions, and produces carbon dioxide, a raw material for the light-independent stage.

3 Young trees are growing and absorb more carbon dioxide from the atmosphere – they act as carbon sinks. (And recycling paper produces dioxins that are harmful to the environment, as well as using a lot of water.)

4 A reduced light intensity reduces the levels of ATP and reduced NADP. GP will accumulate as it is not reduced to TP. Levels of RuBP will fall as it is not being regenerated.

1.3.6 Investigating the factors that affect the rate of photosynthesis – 1

1 and **2** Will depend on the results.

3 Gas collected is not pure oxygen. May contain some carbon dioxide so may not be a direct indication of rate of photosynthesis. Temperature may fluctuate, could affect photolysis as there is an enzyme involved. Errors with apparatus, timing, reading meniscus. Ambient light.

4 Use thermostatically controlled water bath to keep temperature constant. Use gas syringe/data logging, more precise measurement of volume/timing. Perform experiment in room with no ambient light.

1.3.7 Investigating the factors that affect the rate of photosynthesis – 2

1 As the leaf discs photosynthesise they produce oxygen. This collects in the air spaces in the spongy mesophyll layer and makes the leaf discs less dense.

2 To check reliability.

3 From narrow air spaces in palisade mesophyll and larger air spaces in the spongy mesophyll, out through open stomata.

4 (a) Light-dependent stage and light-independent stage (as no ATP or reduced NADP).

(b) Light-independent stage.

(c) Light-independent stage.

5 Could use seedlings or algae. Large numbers. All conditions except the one being measured kept constant. Sample at intervals, 10 individuals, from each condition. Dry to constant mass. Measure % increase in constant mass.

1.3.8 Limiting factors and the Calvin cycle

1 Electron carriers. Proteins that hold photosystems in place. High temperatures break hydrogen and ionic bonds and disrupt the 3D shape/tertiary structure of the proteins.

2 Less carbon dioxide gas enters.

3 Less acceptance of carbon dioxide by RuBP, which accumulates. Less GP formed, less TP as less GP to reduce to TP.

4 Less ATP and reduced NADP. GP cannot be reduced to TP so less TP to regenerate to RuBP.

1.4.1 Why do living organisms need to respire?

1 Chemical potential energy in food used to synthesise ATP, which contains chemical potential energy, which is changed to light energy.

2 RNA, as the sugar in adenosine is ribose sugar.

3 (a) and **(c)** anabolic; **(b)**, **(d)** and **(e)** catabolic.

4 It releases energy in small manageable amounts for powering chemical reactions in cells and it is found in all types of cells.

1.4.2 Coenzymes

1 Because these molecules are recycled/regenerated.

2 As the alcohol is metabolised, it undergoes dehydrogenation – removal of hydrogen atoms. These hydrogen atoms are combined with NAD, so there is less of it for respiration.

3 NAD contains ribose sugar (found in RNA), the nitrogenous base adenine and phosphate groups. (ATP also contains adenine, ribose and phosphate groups.)

1.4.3 Glycolysis

1 All the enzymes that catalysed the reactions of glycolysis and the ethanol fermentation pathway.

2 The coenzyme NAD accepts hydrogen atoms from the substrate molecules in stage 3 as triose phosphate is oxidised.

3 Because two molecules of ATP are used to activate hexose sugar at the beginning of the process. Four ATP are made (two during stage 3 and two during stage 4), so the net gain is $(4 - 2) = 2$.

4 During stage 3, hydrogen atoms are removed from triose phosphate (and combine with NAD). This is an oxidation reaction.

1.4.4 Structure and function of mitochondria

1 A mitochondrion from a skin cell would be smaller and have fewer cristae and shorter cristae in the inner membrane, because skin cells are not as metabolically active as heart muscle cells.

2 To provide a lot of ATP for synthesis of neurotransmitter molecules and to move the vesicles of neurotransmitter to the presynaptic membrane for exocytosis.

3 Chemiosmosis – the flow of hydrogen ions (protons) through channels of ATP-synthase enzymes, down the proton gradient, from intermembrane space to the mitochondrial matrix. Proton motive force – the force generated by this flow of protons. It changes the configuration of parts of the ATP-synthase enzyme and causes ADP and P_i to join and make ATP. Oxidoreductase enzyme – enzyme that catalyses a reduction reaction that is coupled with an oxidation reaction.

1.4.5 The link reaction and Krebs cycle

1 Because it is constantly being regenerated.

2 Each substrate has a different molecular configuration so each step needs an enzyme with a specifically-shaped active site that is complementary in shape to the substrate molecules.

3 It removes hydrogen atoms from pyruvate.

4 Once they have been changed to pyruvate, this is changed to acetate, which is carried onto Krebs cycle by CoA.

1.4.6 Oxidative phosphorylation and chemiosmosis

1 It combines with electrons at the end of the electron transport chain.

2 The protons have the potential to flow down their gradient into the matrix. This gradient allows chemiosmosis, which generates the proton motive force to drive the synthesis of ATP – a high-energy compound.

3 Oxygen will dissociate from haem in haemoglobin and diffuse into the plasma. As plasma in tissue fluid leaks out and bathes cells, oxygen diffuses across the plasma membrane into the cytoplasm and then across the mitochondrial membranes into the matrix.

4 If hydrogen atoms in the matrix combine with oxygen to form water, the concentration of hydrogen ions on the other side of the membrane/in the intermembrane space will become relatively higher.

1.4.7 Evaluating the evidence for chemiosmosis

1 (a) The lower pH in the intermembrane space indicates a high concentration of hydrogen ions (protons).

(b) This indicates an electrochemical gradient – more protons on the intermembrane space side of the membrane gives a positive potential, leaving the matrix side more negative, so there is a potential difference across the membrane.

(c) This shows that without the outer membrane, there is no intermembrane space where the proton concentration can build up.

(d) ATP synthase is removed – so no phosphorylation can take place.

(e) Oligomycin blocks the flow of protons through ATP synthase, therefore this flow of electrons is part of the process.

(f) This is how the proton gradient can be made.

1.4.8 Anaerobic respiration in mammals and yeast

1

	Yeast	Mammals
Hydrogen acceptor	Ethanol	Pyruvate
Is carbon dioxide produced?	Yes	No
Is ATP produced?	No	No
Is NAD reoxidised?	Yes	Yes
End products	Ethanol and carbon dioxide	Lactate
Enzymes involved	Pyruvate decarboxylase, ethanol dehydrogenase	Lactate dehydrogenase

2 2

3 They do not have the particular type of pyruvate decarboxylase that catalyses the change of pyruvate to ethanal.

4 The lowering of pH interferes with hydrogen bonds and disrupts the tertiary structure of enzymes involved in muscle contraction.

5 They have buffers in their blood. Non-diving mammals also have buffers (molecules that can accept or donate hydrogen ions and so resist changes in pH) but diving mammals will have more, or more efficient ones. Haemoglobin can act as a buffer. There are also

chemicals such as dipotassium hydrogenphosphate and potassium dihydrogenphosphate.

1.4.9 Respiratory substrates

1 Fat contains triglyceride, which contains fatty acids. These are long-chain hydrocarbons with many hydrogen atoms. This means there will be many protons for chemiosmosis, generating lots of ATP (high-energy molecules).

2 It is a fat so dissolves in the lipid bilayer of the mitochondrial membranes.

3 They will use muscle protein as a source of amino acids for respiration.

4 (a) As the fatty acids have many hydrogen atoms, many molecules of water will be produced.

  (b) Fats are changed to acetate and then enter Krebs cycle. They cannot enter glycolysis or any of the fermentation pathways.

2.1.1 How DNA codes for proteins

1 If the four nucleotide bases were arranged in pairs, the maximum number of different combinations would be 4^2, which is 16. This is not enough to code for the 20 amino acids involved in protein synthesis, plus some stop codes.

2

DNA coding strand	TTA	ATG	CGT	GGA	TAA
DNA template strand	AAT	TAC	GCA	CCT	ATT
mRNA	UUA	AUG	CGU	GGA	UAA
Amino acid	Leucine	Methionine	Arginine	Glycine	Stop

3 Each amino acid has more than one base triplet code. (This helps reduce the effects of point mutations.)

2.1.2 Translation

1 Prokaryotes, as the mRNA does not have to exit a nucleus and the DNA in prokaryotes is naked – there are no histone proteins.

2 21 – one each for the 20 amino acids and a stop.

2.1.3 Mutations – 1

1 C is substituted for T in the triplet TTA, which becomes TCA. It now codes for serine instead of leucine. The primary structure of the polypeptide is altered and this will probably change its tertiary structure.

2 A is substituted for T so TTA becomes TAA. This triplet now codes for stop and so the polypeptide will be truncated.

3 Although TTA has become TTG, this triplet still codes for leucine, so the polypeptide will be unaltered.

4 A deletion. T has been lost from TTA of the sixth triplet. This causes a frameshift and all amino acids after number 5 are changed.

5 All amino acids, except methionine, have more than one triplet of bases coding for them. Some have as many as six. So some point mutations will not change the amino acid that is added to the polypeptide at that position.

2.1.4 Mutations – 2

1 (a) Mutation 1 changes the base triplet at position 6 to a stop codon, so the polypeptide is truncated – nonsense.

  (b) Mutation 2 changes the amino acid at position 4 from valine to isoleucine. This will probably change the tertiary structure of the polypeptide – missense.

  (c) Mutation 3 changes the first triplet codon. All polypeptides start with methionine so this polypeptide will not start – it will not be produced – nonsense.

  (d) Mutation 4 is a deletion. The base T is deleted from the second triplet so there is a frameshift. All amino acids after the first one are changed. The polypeptide will have a very altered structure and probably not function at all.

2 (a) TAC AAA GGA CAA TTT ATG GTA GCG

  (b) AUG UUU CCU GUU AAA UAC CAU CGC

  (c) UAC AAA GGA CAA UUU AUG GUA GCG

  (d) Met Phe Pro Val Lys Tyr His Arg (methionine, phenylalanine, proline, valine, lysine, tyrosine, histidine, arginine)

2.1.5 The *lac* operon

1 A = III; B = I; C = IV; D = II

2 Binds to operator region and blocks promoter region so RNA polymerase can't bind and therefore genes Z and Y can't be transcribed.

3 Binds to repressor at another site. Causes its shape to change so it moves away from operator and leaves the promoter region free for RNA polymerase to bind. Z and Y genes are transcribed.

4 The bacteria only make enzymes to metabolise lactose when lactose is present in the culture medium/environment. The enzymes are inducible. Therefore resources are not wasted in making unnecessary enzymes.

2.1.6 Genes and body plans

1 It contains large amounts of vitamin A. Retinoic acid is a derivative of vitamin A and too much retinoic acid, particularly in the early stages of pregnancy when the body plan is being organised, can interfere with the normal expression of the homeobox genes and cause birth defects, including cranial (head) defects.

2 A morphogen is a substance that governs the pattern of tissue development by activating the homeobox genes.

3 Transcription factors are some of the polypeptides expressed by homeobox genes, which bind to other genes further along the DNA from homeobox genes, and switch them on. They regulate the expression of other genes and so influence the development of the embryo.

4 (a) They determine the embryo's polarity – which end is head (anterior) and which is tail (posterior).
  (b) They specify the polarity of each body segment.

2.1.7 Apoptosis

1 If the proton gradient cannot be built up across the inner membrane then there will be less chemiosmosis and less formation of ATP. The cell will not be able to carry out active transport or bulk transport and cannot maintain its metabolism – particularly anabolic reactions such as protein synthesis.

2 Apoptosis does not damage nearby cells because the hydrolytic enzymes are enclosed in vesicles and these are ingested by phagocytes.

3 Ineffective T lymphocytes would have no receptors on their cell surface membranes and could not recognise antigens. Harmful T lymphocytes would have receptors, on their cell surface membranes, with shapes complementary to self antigens, causing autoimmunity – the immune system attacking your own tissues. (Effective T lymphocytes have receptors that bind to foreign antigens.)

2.1.8 Meiosis

1 A homologous pair of chromosomes, each consisting of two sister chromatids, paired up for meiosis.

2 Mitosis.

3

	Mitosis	Meiosis
Number of divisions	One	Two
Products	Two genetically identical daughter cells	Four cells, genetically different from each other and from the parent cell
Chromosome number	Maintained	Halved
Do bivalents form?	No	Yes
Does crossing over occur?	No	Yes

2.1.9 The significance of meiosis

1 2^{23}

2 Because gamete production by meiosis and fertilisation increases genetic variation. During meiosis there is crossing over, independent assortment of chromosomes and chromatids. Fertilisation combines genetic material from two unrelated individuals.

2.1.10 Some terms you need to know

1 (a) $I^A I^B$ (b) $I^O I^O$
  (c) $I^A I^A$ or $I^A I^O$ (d) $I^B I^B$ or $I^B I^O$

2 (a) XX (b) XY

3 Because the production of the pigment at the extremities requires a cooler temperature.

4 At birth, they have been in the uterus and at an even temperature. Soon after birth they will be close to the mother and to each other and will be warm all over. After a few weeks they begin to explore and move away from the group so the extremities will be a little cooler.

2.1.11 Using the genetic diagrams – 1

1 Diagram should show male parent suffering from haemophilia, $X^h Y$, and female parent being a carrier, $X^h X^H$. One daughter $X^h X^h$ – haemophiliac; one daughter $X^h X^H$ – carrier; one son $X^h Y$ – haemophiliac; one son $X^H Y$ – normal.

2 The grandfather is the father of the boy's mother. The X chromosome does not pass from father to son.

3 Cross $X^r X^r$ with $X^R Y$. This produces $X^r Y$ males and $X^R X^r$ females.
  When these interbreed, they produce $X^R Y$ – male, red-eyed; $X^r Y$ – male, white-eyed; $X^R X^r$ – female, red-eyed; $X^r X^r$ – female white-eyed.

4 Mary is X^DX^d or X^DX^D. Astrid is X^DX^d. Jack is X^DY. Jane is X^DX^d.

5 1 in 4 (25%)

2.1.12 Using genetic diagrams – 2

1 The bull's phenotype is roan, genotype C^RC^W.

2 The alleles for colour are codominant. C^W = white feathers, C^B gives black feathers. C^WC^B gives grey colouring – a mixture of black and white feathers. To be sure of producing grey-feathered birds, cross white birds with black birds.

3 (a) A base substitution, which is a point mutation.

(b) When the red cells sickle, capillaries are blocked. This reduces oxygen supply to tissues.

2.1.13 Interactions between gene loci – 1

1 (a) **AaBb, AABb, AABB, AaBB** are purple

(b) **AAbb, Aabb** are pink

(c) **aaBB, aaBb, aabb** are white.

2 (a) **IIcc, Iicc, iicc, IICC, IiCC, IICc, IiCc** are all white

(b) **iiCC** and **iiCc** are coloured.

3 (a) **ccRR, ccRr, CCrr, Ccrr, ccrr**

(b) **CCRR, CcRr, CCRr, CcRR**

4 More than one gene are working together to influence the expression of one characteristic.

2.1.15 Interactions between gene loci – 2

1 AAbb, Aabb, aabb

2 AABB, AaBb, AaBB, AABb

3 aaBB, aaBb

4 Complementary gene action as the precursor substance has to be converted to black pigment and another reaction converts this to the banded pattern.

5 pprr × PpRr

6 This model would only give pea-, rose- and walnut-combed chickens. The F_2 contains single-combed chickens as well.

7 B is allele for brown; b is allele for black

C allele prevents pigment formation; c allele allows pigment formation

Parental phenotypes **black** × **white**
Parental genotypes **bbcc** × **BbCc**

Gametes: (bc) ×(BC) (Bc) (bC) (bc)

Offspring genotypes **BbCC Bbcc bbCc bbcc**
Offspring phenotypes **white brown white black**

Ratio of offspring phenotypes 50% white, 25% brown, 25% black

Parental phenotypes **brown** **black**
Parental genotypes **BBcc** **bbcc**

Gametes: (Bc) (bc)

Offspring genotypes **all Bbcc**
Offspring phenotypes **100% brown**

2.1.16 The chi-squared (χ^2) test

1 (a) yy

(b) 3 yellow: 1 grey

(c) Expected numbers of yellow and grey would be 156 yellow and 52 grey (divide 208 by 4, take ¾ and ¼ of them). Observed is 140 yellow and 68 grey. χ^2 is 6.56. With 1 degree of freedom, the critical value of χ^2 at 0.05 p level is 3.84.

As the value of χ^2 is greater than that, the difference is significant and not due to chance.

Therefore this is not a 3:1 ratio.

(d) As the ratio is nearer to 2 yellow: 1 grey, it is likely that the homozygous yellow, **YY**, mice do not survive. **YY** is a lethal combination of alleles. This allele must have another effect besides coat colours

(e) Yy × yy

Gametes: **Y, y** and **y**

Offspring genotypes: **Yy yy**

Offspring phenotypes: yellow : grey

Ratio: 1:1

2.1.17 Continuous and discontinuous variation

1 32 cm 16 cm 24 cm 24 cm 28 cm

2 Higher temperature provides more KE for enzyme-controlled reactions of photosynthesis (especially LI stage). This leads to increased yield so more carbohydrate and lipids to be stored in the seeds.

3 Different concentration of nitrates and other minerals – the availability of nitrates will limit amount of amino acid and therefore protein synthesis. Differences in pH may reduce enzyme activity and reduce synthesis of proteins and other chemicals.

2.1.18 Population genetics

1 Size of population, randomness of mating, mutation, competition and selection.

2 Because both alleles contribute to the phenotype, the genotypes of all phenotypes are known.

3 (a) The allele frequencies will not have dropped as the adult genotype frequencies will not change. It is only recently that people with cystic fibrosis are living long enough to reproduce, but many will be sterile due to thick mucus in the reproductive tracts, so many of the sufferers will not contribute to the gene pool in terms of reproduction.

(b) For some conditions, such as thalassaemia, where many pregnancies are terminated, this may alter the allele frequencies in a population. The assumption that mating is random is not necessarily the case. Many parents who know they are carriers will have IVF and pre-implantation screening to reduce the chance of producing a baby with a genetic defect.

2.1.19 The roles of genes and environment in evolution

1 Biotic: predators, disease/pathogens, food availability, mate availability. Abiotic: temperature, water availability, soil type, habitat availability, shelter.

2 The characteristics selected for as useful to humans may make the animal less adapted to its natural environment.

3 If there is a genetic defect in the family's gene pool, the chance of two individuals, both carriers, mating and producing an affected individual are increased.

4 This is known as the founder effect. If one person in the original small group of settlers carried a recessive allele, it would be passed to his offspring and grandchildren. Cousins may marry so two carriers may mate. The frequency of this allele will increase within the population.

2.1.20 What is a species?

1 **(a)** Because bacteria do not reproduce sexually, the biological concept is not appropriate. Their biochemistry, morphology, ecological niche can all be studied and analysed.

(b) The phylogenetic (cladistic) system, as we can't deduce anything about their sexual reproduction and ability to produce fertile offspring, but we can deduce information about morphology, genes (DNA can be obtained from old bones) and ecological niche.

2.1.21 Natural and artificial selection

1

	Natural selection	Artificial selection
Agent of selection	Environment	Humans
Effect on allele frequencies	Alters it	Alters it
Effect on evolution	Both contribute to evolution of the species concerned	
Speed	Slow	Quick

2 The cladistic concept. Using the biological concept they would be classified as being in the same species (different varieties/races/cultivars) as they can interbreed and produce fertile hybrid offspring.

3 **(a)** It is carbon dioxide produced from respiration (probably aerobic and anaerobic) of yeast.

(b) More gluten/sticky rubbery mass is needed to hold the gas bubbles and give the light, bread texture so bread flour needs higher protein content.

2.2.1 Clones in nature

1 Asexual reproduction does not require the development of sex organs or specialised haploid gametes. The organism does not need to release male gametes, nor do cells need to be transferred to another individual.

2 Sexual reproduction leads to offspring showing genetic variety. This has advantages in long-term survival of the species in terms of allowing adaptation to changes in the environment. Being able to reproduce asexually too allows for the production of offspring even if there are no other individuals of the same species sufficiently close by, or if the production of seeds fails.

3 Asexual reproduction in plants uses vegetative structures, i.e. structures that are a normal part of the growing organism, often associated with overwintering, not separate 'sex organ' structures.

4 Select the plants that are most resistant to Dutch elm disease. Subject them to stress such as drought or severe damage (or use an auxin inhibitor) to make them produce suckers from the root buds. Separate the plantlets produced by suckers. They will all be resistant as they are all clones of the resistant plants.

2.2.2 Artificial clones and agriculture

1 Meristem cells are plant cells that retain the ability to divide. Most plant cells cannot divide, mainly because the cell wall is too rigid.

2 Rootstock varieties that are particularly hardy, or suited to the surrounding conditions, can be chosen. It may be that the complete fruit tree would not grow well in these conditions.

3 A limit to total area covered in one location limits the amount of crop that can be lost if a new pathogenic organism arrives. The distance between areas of the same crop is important in that direct transfer of the new pathogenic organism is not possible from location to location, and farmers/growers can put in place interventions to help limit the spread of the pathogenic organism – such as halting any transportation between sites.

2.2.3 Cloning animals

1

Splitting of embryos	Nuclear transfer
The nucleus comes from an egg fertilised *in vitro* and allowed to divide to form an embryo.	The nucleus comes from an adult, differentiated cell taken from the animal to be cloned.
The embryos produced are clones of the original zygote.	The embryos produced are clones of the donor adult organism.
The zygote is a product of a fertilised egg. Since this is formed from fusion of sperm and egg cells which are themselves a product of meiosis, it is impossible to know exactly what characteristics the cloned organism will possess.	The adult cell is a product of an adult with known characteristics. It is therefore known what characteristics the clones will possess.
All cellular components are derived from the original zygote. The mitochondrial DNA component will be identical in each clone.	The cellular components are derived from the egg cell used therefore the mitochondrial DNA component will be different from that of the original adult cell.

2 Heart transplant surgery carries with it risks of infection associated with any major surgery. There are also risks associated with use of general anaesthetic. The transplanted heart is foreign tissue – as such it may be rejected by the host's immune system. Cloned cells are derived from the host and it may be possible to get them to the right location using relatively minor surgical methods.

2.2.4 Biotechnology basics

1 Microorganisms can make use of a variety of human waste products as a source of nutrients. Waste water treatments use a mixed population of microorganisms that grow on the organic matter in the waste and so purify the water. The microorganisms grown can then be used in other processes – for example incinerated to generate heat energy.

2 A number of organisms produce chemicals that are toxic to other organisms. Fungi such as *Penicillium* spp. produce antibiotic chemicals and secrete them into their surroundings, killing off many other organisms (especially bacteria). This has the advantage of preserving the food source that the fungus is growing on.

3 (a) (i) Pharmacogenomics is developing drug treatments tailored to individuals, so that the patient does not suffer side effects. Each of us has a different genome, and because genes and molecules of the drugs interact, our reactions to a specific drug may differ.

   (ii) Nutrigenomics is developing diets that suit different individuals. Each of us has a different genome, and because genes and food molecules interact, we may metabolise foods in different ways.

   (b) The Human Genome Project may make it possible for each of us to know our genetic profile. The effects of certain molecules on us can then be predicted.

2.2.5 The growth curve

1 (a) Addition of nutrients in the lag phase would be unlikely to have any effect. During this phase organisms are becoming accustomed to the surroundings and synthesising the enzymes they need, and it is likely that nutrient supply is not limiting during this period.

   (b) Addition of nutrients would most likely lead to an increase in growth. The additional nutrient would be used by the organisms in order to support further growth because in the stationary phase, nutrient supply is likely to be limiting. The carrying capacity has been reached. However, additional nutrient would have no effect if the reason for entering stationary phase was an accumulation of waste products or lack of space, but with a plentiful supply of nutrients remaining.

2 Primary metabolites are all of the proteins and other products associated with growth of the organism itself, therefore any growth in overall numbers of primary metabolites is because of an increase in the overall level of microorganisms.

3 During the lag and exponential phases, the nutrient supply is not limiting and the microorganisms' metabolic processes are dedicated to growth. As the main growth phase ends, microorganisms produce secondary metabolites, for example antibiotics. These molecules tend to protect the dwindling nutrient supply from being taken by competitor microorganisms.

2.2.6 Commercial applications of biotechnology

1 The fermentation products differ because the processes of aerobic and anaerobic respiration differ. In yeasts, anaerobic respiration results in the production of ethanol, whereas aerobic respiration yields carbon dioxide and water.

2 An obligate anaerobe is an organism that cannot survive in the presence of oxygen (these organisms are sometimes referred to as fastidious anaerobes). Likewise, an obligate aerobe requires oxygen to survive.

3 (a) In a batch culture, contamination will only lead to loss of one batch of the culture product. In continuous culture, contamination may lead to large-scale loss of culture product. It may be some time before contamination is detected and, when it is, all product must be discarded. It is also necessary to shut down the fermenter and sterilise all equipment.

  (b) Contamination of a bench or starter culture is contained in very small amounts of culture, whereas large-scale fermentation tanks can contain hundreds of litres. Contamination necessitates the loss of all product and culture from the fermentation tank.

2.2.7 Industrial enzymes

1

Adsorption	Covalent bonding	Entrapment	Membrane separation
Non-covalent bonds hold enzymes to support	Covalent bonds hold enzymes to support	No bonding of enzymes	No bonding of enzymes
Leakage likely	Leakage very unlikely	Leakage unlikely	Leakage unlikely
Enzymes are not physically separated from reaction mixture		Enzymes are physically separated from reaction mixture	
Immobilisation can reduce reaction rate because bonding to support can affect active site. However, close contact with substrate can increase reaction rate		Immobilisation does not reduce reaction rate in terms of its effect on active site – active site is not affected. However, separation from reaction mixture can reduce reaction rate as substrate must penetrate the separating barrier	
Enzyme immobilisation means that product is easily separated from enzyme in biotechnological process			

2 Covalent bonding to the support might alter the tertiary structure of the enzyme, and so may alter the shape of the active site. If the active site is changed significantly, then the enzyme can no longer catalyse the reaction. Even if the shape of the enzyme is unchanged, bonding to a support could be close enough to the active site to slow down or stop the substrate entering.

3 The most likely targets on bacterial cells for antibiotics that would be of use to humans would be any feature of the cell that differs from human cells. This would include the cell wall components – penicillin for example interferes with cell wall manufacture in some types of bacteria – or bacterial ribosomes, which are smaller than eukaryotic ribosomes – aminoglycoside antibiotics inhibit prokaryotic ribosome function and so inhibit bacterial protein synthesis.

2.2.8 Studying whole genomes

1 It was originally thought that much of the non-coding DNA had no function, in that it was left in the genome following evolution of more developed characteristics and was simply there taking up space. We now know that some of this DNA acts as regulatory genes, that is, it influences the expression of other genes.

2 Mutations of DNA can lead to the production of a non-functioning protein. In haploid organisms, the affected organism will not produce the working protein and so will exhibit any condition/disease associated with lacking that protein. In diploid organisms, the presence of the (usually harmless) protein product of the mutation does not affect production of a functioning protein when the functioning allele on the second (homologous) chromosome is transcribed. Such organisms are described as heterozygous and do not exhibit the condition associated with the mutation.

3 (a) There are a number of possibilities; for example, since all organisms are based on DNA as the molecule of inheritance, a gene that codes for a protein enzyme such as DNA polymerase.

  (b) Mammals have fur and feed live young on milk from mammary glands. Genes associated with lactation would be good examples.

  (c) All photosynthetic organisms use some type of chlorophyll pigment in order to trap light energy. The gene for a chlorophyll molecule would be a good example.

2.2.9 DNA manipulation – separating and probing

1 Hydrogen bonding.

2 Electrophoresis is very like chromatography in that materials are separated by size. The separating mechanism involves the balance between 'retardation' or slowing down in the stationary phase (paper or gel in

chromatography, gel in electrophoresis) against solubility in the moving phase (solvent in chromatography, electric charge across buffer solution in electrophoresis).

3 The single-stranded probes anneal to sequences that are complementary. A long probe may partially anneal to a variety of sequences that show some complementary bases. Such a probe may bind in several places and would be of no use in determining the presence of the very specific sequence required.

2.2.10 Sequencing and copying DNA – 1

1 By multiplying the minute fragment several times, it is amplified enough to be used in gel electrophoresis and analysed.

2.2.11 Sequencing and copying DNA – 2

1 Because it works well at higher temperatures, the process does not have to be cooled in between breaking H bonds and copying the strands. This speeds up the process and is therefore more cost effective.

2 This fragment will be the smallest fragment passing though the machine. It will pass through first as small fragments move through the electrophoresis gel faster than large fragments.

2.2.12 An introduction to genetic engineering

1 Restriction enzymes, like all enzymes, have a specific active site, complementary in shape and charge distribution to their substrate. The DNA backbone is built from A, T, C and G nucleotides. The specific nucleotides at the restriction site give that site a very specific shape and charge distribution because, although similar, A, T, C and G are not identical. This means that a large range of restriction enzymes, each with its own unique restriction site, is possible.

2 Viruses which infect bacteria do so by inserting their DNA (or RNA) into the bacterial cell. The inserted DNA then goes on to cause the bacterial cell to manufacture new virus particles, according to the instructions in this DNA, eventually destroying the cell. If suitable restriction enzymes are present in the cytoplasm, then inserted DNA from a virus is cut into pieces and can no longer infect the bacterial cell.

3 If the bacterial cell's chromosome contained restriction sites that were complementary to the restriction enzymes in the cell, then these enzymes would cut up the bacterial chromosome and it would no longer be able to function properly.

4 Restriction enzymes are all named by using an identifier for the bacterial species from which they were isolated, so those from *E. coli* are called *Eco*, plus a list number for the order of identification – so R1 stands for first restriction enzyme isolated.

2.2.13 Genetic engineering and bacteria

1 The S-strain bacteria and the R-strain bacteria both contain the bacterial chromosome; however, the S-strain bacteria also contain a plasmid. The gene for production of the pneumococcal toxin is present on this plasmid.

2 Mutation.

2.2.14 Engineering case studies – 1: human insulin

1 Insulin mRNA is only produced in those cells which transcribe the insulin gene in order to lead to the production of insulin – these cells are the β cells of the islets of Langerhans in the pancreas. Other cells of the body do not generate insulin mRNA because the gene is turned off.

2 All eukaryotic genes contain introns. These are the parts of the gene which are not transcribed into mRNA. Since cDNA is copied back from mRNA it will not be identical to the original gene because it lacks the introns present in the original gene.

3 Lack of a functioning allele to produce the functioning insulin receptors in cell surface membranes of cells of insulin target tissue. As with all hormones, the effect of the hormone on target tissues depends on the target tissue cells displaying, on their cell surface membranes, a functioning receptor to the hormone. In type II diabetes mellitus, it is the insulin receptor molecule that does not function. This is because either the target cells don't have insulin receptors or the receptors are the wrong shape and are not complementary to the insulin molecule. The presence of insulin therefore has no effect on such target cells. However, type 2 diabetes can become insulin-dependent as β cells may eventually cease to produce insulin.

2.2.15 Engineering case studies – 2: Golden Rice™

1 (a) Rice plants produce beta-carotene in their leaves: it is a photosynthetic pigment. Since all rice cells contain the same set of the organism's DNA, then all cells in the plant contain the genes that code for the production of beta-carotene. However, part of the differentiation of endosperm tissue is the switching off of these genes in endosperm cells. The endosperm cells do not require beta-carotene as they do not photosynthesise.

(b) The production of beta-carotene is a complicated metabolic sequence. As with all metabolic sequences, intermediate products are produced that serve other functions within organisms. Some of the enzymes of the beta-carotene pathway catalyse reactions that are not only required as part of the beta-carotene pathway, but are also involved with the production of other metabolites that are required in endosperm tissue.

2 Vitamin A is a fat-soluble vitamin which is stored in fat deposits within the liver. In those people who are underweight, there will be little stored vitamin A. If these people consume a diet deficient in vitamin A, the stores will quickly run out and the symptoms of deficiency will show much earlier than in people who have greater stored reserves.

3 Your answers must be based on facts, not 'gut feelings'.

2.2.16 Gene therapy

1 From an individual who neither has the condition nor carries the allele for the condition (i.e. a homozygous individual without the condition).

2 Gene therapy relies on getting functioning copies of an allele into target cells which need that allele in order to perform their functions. It is extremely difficult to get one gene into a cell so that it is expressed. The added complication of trying to insert a number of genes into a cell means that the delivery system (usually a liposome) would have to be big enough to carry the extra DNA, and the chances of a large number of genes being inserted into a cell and all functioning correctly is minimal.

2.2.17 The rights and wrongs of genetic manipulation

1 Identical twins have the same genetic make up and so the immune system of one will not recognise as foreign any cellular component of the other. Siblings are the result of sexual reproduction and so carry a variety of different alleles from each parent. Whilst similar, the immune system of one sibling will recognise some cellular components of other siblings as foreign. This is the basis of graft rejection.

2 A gene for pest resistance can be engineered into a crop plant. On producing pollen, the gene would be passed into pollen grains which are dispersed from the plant. Some of this pollen will land on wild plants in the local area. It is possible, especially where the wild plants are closely related to the crop plant, that the pollen could fertilise the wild plant and so the seeds produced carry a combination of genes from the crop and the wild plant. If wild plants and crop plants express pest resistance genes, then many of the pest organisms will be killed, even in areas where crops are not grown. This could lead

to starvation of the organisms that rely on the pest as a food source. These predator organisms may die out, or, if they use other organisms as a food source, compete with other organisms for these other food sources. Overall, the balance of food chains and diversity could be seriously disturbed.

3 Natural events – examples include: volcanic eruptions, floods, drought, disease epidemics, tsunami, earthquakes. Human activities – examples include: vaccination, surgery, selective breeding of crop plants and livestock, use of prophylactic drugs such as anti-malarial tablets, use of insecticides to reduce yield loss of crops, use of fertilisers to increase crop yields.

2.3.1 Ecosystems

1 (a) Parasitism, competition.
  (b) Water pH, light intensity.
2 (a) A producer is autotrophic, converting light energy to chemical energy, while a consumer gains energy and materials by eating other living things.
  (b) A habitat is the place where an organism lives, whereas its niche is its role in its ecosystem.
3 At least three descriptions of how a rabbit interacts with biotic and abiotic factors are acceptable. Examples could include the following: eats grass in the field; provides waste for decomposers to feed on; hence provides minerals for grass; reproduces with other rabbits; digs burrows; provides food source for foxes and birds of prey.
4 If they occupy the same niche, they will compete. One will be better at competing than the other, and will survive and reproduce, whereas the other will fail to reproduce and die out.
5 Bacteria; chemical energy.

2.3.2 Understanding energy transfer

1 (a) To show energy flow through a whole ecosystem rather than just through one food chain.
  (b) Because a pyramid of energy provides a more accurate picture of energy at each trophic level. This is because dry mass in different organisms may release different amounts of energy.
2 (a) Because energy leaves the food chain at each trophic level, there is less energy available at each successive trophic level, meaning fewer and fewer organisms can survive at higher and higher trophic levels.
  (b) Because energy leaves the food chain at each trophic level, eventually there is not enough energy to sustain any individuals at the highest trophic levels.

3 Because there are higher temperatures and greater light intensity near the equator, which increases the rate of photosynthesis. As a result, vegetation is more extensive and more incident light from the Sun is absorbed by plants.

2.3.3 Manipulating energy transfer

1 (a) Selective breeding can be used to develop plants that grow well in response to high fertiliser levels. It can produce drought-resistant crop plants, or disease- and pest-resistant plants to prevent loss of yield. Likewise it can be used to produce animals that are fast growing, or have high egg or milk production levels.

(b) Steroid hormones given to animals can cause them to invest more energy into growth, producing faster growth rates.

(c) Genetic modification has been used with plants to provide resistance to pests or fungal infection.

(d) Growing plants in a greenhouse increases the temperature – cold temperatures can slow down chemical reactions in plants. Using light banks in the greenhouse can also increase photosynthetic rates.

(e) Fertiliser given to plants ensures that low mineral levels do not limit rates of photosynthesis.

(f) Pesticides kill pests such as insects, nematodes or caterpillars. Without pesticides these would remove biomass and stored energy from the food chain, reducing eventual crop yield.

2 Ectotherms would be most efficient, as they would not use energy to maintain body temperature as endotherms need to. Hence, more energy would be allocated to growth, and contribute to eventual yield.

3 Across the world, a lot of land is allocated to producing meat, through rearing cows, sheep, pigs or other farm animals. That land could be allocated to arable crop production. Because arable crop production involves shorter food chains (e.g. wheat → humans), less energy is wasted rather than in a longer food chain producing farm animals (e.g. grass → cows → humans). If fewer people ate meat, more land would be used for arable crop production, and less rainforest would need to be felled.

2.3.4 Succession

1 (a) A directional change in a community of organisms over time, beginning from bare ground.

(b) The living organisms which first begin to colonise bare ground.

(c) The stable community that emerges at the end of a process of succession.

2 Any sensible similarities and differences are appropriate. Possible responses include:

Bare rock	Sand dune
Pioneer community lichens and algae	Pioneer community comprises higher plants
Climax community woodland	Climax community woodland
No windblown sand	Windblown sand provides shelter
Dead and decaying organisms provide minerals for successors	Dead and decaying organisms provide minerals for successors

3 Because the sea deposits sand on the beach, the sand nearest to the sea is deposited more recently than the sand further away. Sand just above the high water mark is at the start of the process of succession, whereas the sand much further away already hosts its climax community. By walking up the beach and through the dunes, it is possible to see each stage in the process of succession.

2.3.5 Studying ecosystems

1 Because it is often impossible to count or measure all the individuals in a population.

2 (a) No bias influences the individuals chosen, and hence we avoid skewing the eventual results.

(b) Every part of the habitat is sampled to the same extent.

3 (a) Use a belt transect.

(b) Use a quadrat to take samples from the two areas.

4 A continuous belt transect gives more detailed information on abundance of each species, rather than just presence or absence provided by a line transect. The amount of work involved in a belt transect is much greater than in a line transect.

2.3.6 Decomposers and recycling

1 Because they secrete enzymes onto dead and waste material. The enzymes digest the material into small molecules, which are then absorbed into the organism's body.

2 *Rhizobium* – fixes nitrogen. *Nitrobacter* – converts ammonium to nitrite. *Nitrosomonas* – converts nitrite to nitrate.

3 Nitrification – conversion of ammonium to nitrate; nitrogen fixation – conversion of inert nitrogen gas into nitrogen compounds like nitrate or ammonium compounds.

4 Aerobic.

2.3.7 What affects population size?

1 Any sensible suggestions. Could include biotic – predation, competition, parasitism, nesting sites, shelter; abiotic – temperature, water, light, oxygen.
2 (a) Almost equal, but reproduction rate slightly higher.
  (b) Reproduction rate is higher.
  (c) Equal.
3 In the wild, predators often eat more than one type of prey, and there are several other limiting factors, which a simple controlled experiment does not include.
4 The carrying capacity is the maximum population size that can be maintained over a period of time in a particular habitat. Limiting factors are those factors that prevent the carrying capacity being any bigger than it is.

2.3.8 Competition

1 Intraspecific competition is competition between members of the same species. Interspecific competition is competition between members of different species.
2 (a) As factors (such as food supplies) become limiting, individuals compete. Those individuals best adapted will survive and reproduce, while those not so well adapted fail to reproduce or die. This slows down population growth and the population enters the stationary phase. If the population size drops, competition reduces, and the population size increases. If the population size increases, competition increases, and the population size drops.
  (b) If individuals of two species compete together, one species tends to out-compete the other. Its population size stays high (because the individuals are able to survive and reproduce), while the other species' population size declines, or it may even die out altogether.
3 There is a wide range of other variables involved that influence the population size of each species. The values of these variables may also vary over time.

2.3.9 Sustainable management

1 Management techniques ensure that biodiversity is maintained, whilst providing a continuous and regular supply of wood.
2 Differences – different-sized timbers produced; coppicing vs felling; habitat destruction happens more in large-scale production; large-scale felling can reduce soil quality – coppicing does not; coppicing does not involve planting new trees, whereas large-scale production does. Similarities – selective cutting and growing standards are similar strategies to gain high-quality large timbers.

2.3.10 Conservation

1 (a) Ethical reasons, economic reasons and social reasons.
  (b) Habitat destruction and fragmentation, over-exploitation of resources, introduction of invasive species into ecosystems.
2 A species has direct financial value if it can be harvested or otherwise exploited directly for money. A species has indirect value if its activities provide a function that would otherwise cost money.
3 (a) Because every ecosystem will go through a process of succession.
  (b) Understanding which species were part of the original community is not always clear, and succession is likely to take a long time before it allows such a community to survive again.

2.3.11 Humans and the Galapagos

1 Habitat disturbance due to increased population size, over-exploitation of resources, effects of invasive species.
2 Goats feed on Galapagos rock purslane, a species unique to the islands, and trample and feed upon giant tortoises' food supply.
3 Because its management takes into account the views of all stakeholders. It is managed collaboratively by all stakeholder organisations, including the National Park Service, the Charles Darwin Research Station, and representatives of local fishermen, the tourist industry and naturalist guides.
4 Any sensible suggestions are acceptable. These could include the following. Do prevent invasive species entering the islands. Do try to eradicate invasive species already in the islands. Do plan conservation to take into account all stakeholders' views. Do control fishing levels. Do limit human immigration and tourist numbers in the islands. Don't clear forest for agricultural land. Don't allow immigrant species to invade habitats in the islands. Do try to reclaim habitats. Don't impose conservation solutions without proper discussion with interested stakeholders.

2.4.1 Why plants respond to the environment

1 They are chemical messengers that can be transported away from their site of manufacture to act in other parts (**target cells** or **tissues**) of the plant. When they reach their target cells, they bind to receptors on the plasma membrane.
2 Specific hormone molecules have specific shapes which can only bind to specific receptors with complementary shapes on the membranes of particular cells.

3 Because the flowers release ethene in gaseous form, which causes the bananas to ripen (an ageing process), and become over-ripe.

2.4.2 How plants respond to the environment

1 Cause cell elongation. Auxin increases the stretchiness of the cell wall by promoting the active transport of H^+, by an ATPase enzyme on the plasma membrane, into the cell wall. The resulting low pH provides optimum conditions for wall-loosening enzymes (expansins) to work. These break bonds within the cellulose (at the same time the increased hydrogen ions also disrupt hydrogen bonds within cellulose) so the walls become less rigid and can expand as the cell takes in water.

2 (a) Auxins stimulate shoot growth by causing cell elongation. The extent to which cells elongate and the shoot elongates is proportional to the concentration of auxins.

(b) Usually auxin inhibits abscission by acting on cells in the abscission zone. However, leaf senescence causes auxin production at the tip of the leaf to drop. This makes cells in the abscission zone more sensitive to ethene, and causes an increase in ethene production. This increases production of the enzyme cellulase, which digests the walls of the cells in the abscission zone, eventually separating the petiole from the stem.

3 Plants whose shoots grow towards the light would have an advantage over other plants, as they can photosynthesise more. Plants whose roots are negatively phototropic would have an advantage as they would grow down and absorb more water. Even to a small extent, these features would make them more likely to survive longer and reproduce. After many generations, the whole population of plants would respond to light through phototropisms.

2.4.3 Controlling plant growth

1 The two variables may have no effect on each other, but could coincidentally both be affected by a third variable.

2

Auxins	Gibberellins
Cause stem elongation	Cause stem elongation
Acidic	Acidic
Cause cell elongation	Cause cell elongation
Don't cause cell division	Cause cell division
Loosen cell walls	Loosen cell walls
Cause growth behind shoot tip	Cause growth in internodes

3 Abscisic acid stops gibberellins promoting internode elongation. When there is little oxygen in deep water, this stimulates ethene production, which reduces abscisic acid levels. This means that deep in the water, gibberellins cause internode elongation, whilst near to the surface (where there is more oxygen), gibberellin action is inhibited, so the upper parts of the plant grow more slowly. Together, these effects make sure the plant grows in response to the depth of the water.

2.4.4 Commercial uses of plant hormones

1 (a) Gibberellins.
(b) Ethene.
(c) Auxins.

2 Cytokinins.

3 Any plausible answers are acceptable. Hormones: Costly and labour intensive; versatile – option to produce seedless or seeded fruit still remains. Parthenocarpy: Triploid plants can't reproduce; reduction in genetic diversity as most banana plants are cloned, hence increased susceptibility to disease; no hormones needed so less labour intensive.

4 Silver salts from the coin inhibit ethene synthesis.

2.4.5 The brain

1 Damage to the cerebellum results in a general lack of muscular coordination. The ability to walk properly is also affected as impulses from the balance organs in the inner ear are not effectively coordinated with muscular movements associated with balance. People with such damage appear to be intoxicated by alcohol in terms of how they move.

2 This is because the motor centres at one side of the brain are responsible for stimulating movement in the muscles in the opposite side of the body.

3 The diagram shows the areas associated with speech; these are present on the left cerebral hemisphere in almost all people.

2.4.6 Organising the nervous system

1 A reflex action involves sensory and motor neurones passing through the CNS. These actions involve a stimulus and an appropriate and immediate response to that stimulus. Pupil diameter is controlled by autonomic neurones. Where too much light enters the eye, a reflex action involving these neurones leads to constriction of the pupil by contraction of the circular muscles of the iris.

2	During the exercise period	Increased blood flow to muscles	Increased activity of the sympathetic nervous system is responsible for the changes that occur during exercise. The cardiac, vasomotor and respiratory centres in the brain receive input from pH, temperature and stretch receptors in the carotid body, aortic body and medulla which leads to the increased sympathetic output from these centres.
		Increased heart and ventilation rate	
		Decreased blood flow to gut	
		Vasodilation in skin to accommodate sweating and increase blood flow to surface	
	After the exercise period	All of the above are reversed over the period of recovery	Increased activity of the parasympathetic nervous system is responsible for the recovery back to resting state.

2.4.7 Coordinated movement

1 Fine control such as that achieved by muscles that control the motor movements of the fingers requires many motor units, each with relatively few muscle cells. This allows the greatest gradation of response.

2 The movement of the joint operates in one plane only, much as the opening of a door around a hinge.

3

Synapses	Neuromuscular junction
Neurone to neurone	Neurone to sarcomere (muscle cell)
Postsynaptic stimulation leads to action potential in postsynaptic neurone	Postsynaptic stimulation leads to depolarisation of the sarcolemma, muscle contraction
Synaptic knob tends to be smooth and rounded	End plate has appearance of microvilli ('brush border') and is flattened up to muscle fibre
Neurotransmitter located in vesicles in presynaptic cytoplasm	
Vesicles release neurotransmitter into cleft on stimulation (arrival of an action potential)	
Neurotransmitter diffuses across the gap and binds to postsynaptic membrane receptor	
Binding of neurotransmitter results in (opening of sodium ion channels) and depolarisation of postsynaptic membrane	
Enzymes present to degrade neurotransmitter to avoid continual stimulation of postsynaptic membrane	

2.4.8 Three types of muscle

1 Circular muscles are needed in order to constrict blood vessels; however, the dilation of the blood vessels is achieved by relaxation of the circular muscles only, because the pressure of blood in the vessel dilates the vessel – effectively the force for dilation is derived from the pumping of blood by the heart.

2

Involuntary	Voluntary	Cardiac
Short spindle-shaped cells, each with a single nucleus	Cells form long multinucleate fibres	Cells form branched fibres with intercalated discs joining cells at their ends
Contracts slowly and fatigues slowly	Contracts quickly and fatigues quickly	Contracts quickly but does not fatigue
Contraction is under autonomic nervous system control	Contraction is under voluntary nervous system control	Contraction is myogenic. Autonomic nervous system controls the rate of contraction
Involved in movement of materials along internal tubes such as gut. Involved in autonomic reflexes such as pupil dilation/constriction	Involved in voluntary movements of the bones of the skeleton about the joints	Involved in pumping blood around the body
Unstriated appearance under microscope	Striated appearance under microscope	Striated appearance under microscope

3 The structure of the heart is such that the blood exits the ventricles from vessels towards the top of the chamber. It is essential that the ventricles contract smoothly from the bottom (apex of the heart) upwards. If the electrical activity passed from atrial walls to ventricular walls, they would begin contracting from the top downwards. It is also essential to allow full contraction of atrial walls, pushing blood into the ventricles, before ventricular contraction begins. The impulse only being able to pass via the AV node introduces a short delay which allows full atrial contraction before ventricular contraction begins.

2.4.9 The sliding filament model

1 Contraction of sarcomeres in the power stroke; protein synthesis; regeneration of creatine phosphate when the muscle is at rest.

2 Anaerobic exercise leads to the production of lactate during explosive exercise. On completion of exercise, most of the lactate is converted back to pyruvate. Anaerobic exercise uses anaerobic respiration, which can only use glucose as respiratory substrate. To respire fat/fatty acids, aerobic respiration, therefore aerobic exercise, has to be used.

3 **(a)** Aerobic respiration.

(b) Creatine phosphate regeneration of ATP.

2.4.10 Pulling together – muscles, nerves and hormones

1

Physiological changes	Advantage in preparedness for action
Pupils dilate	Allows more light to enter eye so more detailed visual information can be collected
Heart rate and blood pressure increase	Increases the flow of blood around the body and directs blood to the muscles which may be called upon for explosive and sustained action
Arterioles to the digestive system and skin are constricted whilst those to the muscles and liver are dilated	
Blood glucose levels increase	To supply respiratory substrate to muscles for ATP generation
Metabolic rate increases	Heightened levels of activity require increased metabolic rate to mobilise all of the nutrients required, e.g. the mobilisation of lipid from stores in case the level of heightened activity needs to be sustained
Erector pili muscles in the skin contract – making hairs stand up	In many animals hair standing on end makes the animal look bigger and more fierce. This is intended to frighten off the 'attacker'
Ventilation rate and depth increase	Increased removal of carbon dioxide and supply of oxygen to allow for heightened activity
Endorphins (natural painkillers) are released in the brain	Painkillers may be required in a situation where damage is possible
Sweat production increases	Increased metabolism generates heat which needs to be removed from the body

2.4.11 Innate behaviour

1 If the innate responses lead to enhanced capacity to survive then they will be passed on to the next generation.

2 Innate responses, being carried in the DNA, are used from the beginning of the organism's life to enhance survival. The short lifespan of invertebrates gives them little time to learn survival skills. Such skills are generally learned from parent organisms. The short lifespan means the parent is likely to be dead by the time the offspring are born.

3 Male sees egg-laden (gravid) female; he responds by swimming towards his nest.
Female sees the red belly of the male; she responds to this stimulus by following the male.
Male sees female at nest; he responds by swimming behind her to push her in.
Once in the nest, male prods the base of female's tail with trembling motion.
The prodding to tail stimulates female to release eggs.
Release of eggs stimulates male to push female out of nest.
Male enters nest containing eggs and releases sperm.

2.4.12 Behaviour can be learned

1 **(a)** Operant (or trial and error) learning.

(b) A number of other species of moth larvae carry similar markings and, although they don't taste bad, birds will associate their colours with the taste of the cinnabar moth larvae and avoid these other moths. These survive and pass the colouring to their offspring. This is known as mimicry.

2 It is thought that the young chimpanzees had learned something about building structures in a 'latent' manner through playing with boxes on previous occasions. This play was without particular purpose or reward. It is thought that the extended period of development in

primates is particularly useful in allowing for play, which is particularly important in learning the skills required of an adult of that species.

2.4.13 Primate behaviour

1 Gorilla groups are very small, often around 10–15 individuals. The presence of mature males would destabilise the group in terms of conflict with the silverback male. The movement away from the group of mature offspring is also important in avoiding inbreeding. The mature females are the silverback's offspring and mating with them could lead to increased risk of genetic defects or inbreeding depression. The mature male and female offspring are all those of the silverback, so again staying with the group would cause problems if one of the males became the lead silverback male following conflict with the original silverback.

2 Innate social behaviours in insects for example: allow reproduction to take place by innate recognition of mate; allow food location to be communicated to others as in bees.

Learned social behaviour in primates for example: allow learning what food sources are good or to be avoided; allow learning how to use simple tools and techniques for gaining particular foods.

2.4.14 Human behaviour, dopamine and DNA

1 An allele is an alternative form of a gene that gives a slightly different protein product. The variants of DRD4 each result in the production of a dopamine receptor number 4; however, in many cases it is difficult to know whether the final protein product is any different or functions differently. Some of the variants may be differences in intron sequences. It would not necessarily be incorrect to call the variants which generate slightly different forms of the receptor protein alleles, but it is also at present unknown how and whether each variant affects the protein product.

2 ADHD is a term used to describe a range of human behavioural traits. Not all of these traits are present in every individual diagnosed as having ADHD. Also, the intensity of the behaviours varies considerably between individuals. Such diverse effects are most likely due to a combination of a range of environmental factors acting on a range of genes and their regulatory functions. Cystic fibrosis is a condition that is entirely due to the absence of functional chloride ion channel proteins (CFTR) which leads to the production of thick, drier mucus. This mucus cannot perform its usual function and clumps together to block tubular structures within the body such as the pancreatic duct, the bronchioles and in males the seminiferous tubules. In the lungs the cilia are not fully hydrated and can't waft. Only individuals who are homozygous for the faulty CFTR allele have the condition.

3 It had been thought that identical twins, because they have the same genome, would express genes in an identical way. The differences seen in identical twins were thought to be due to environmental influences on learning and differences caused perhaps by different nutrient levels and different environmental factors such as toxin exposure and so on. Differential gene expression in identical twins suggests that environmental conditions can in some way have an effect on the level of expression of some genes in ways that we do not fully understand. This added dimension to nature and nurture complicates the picture on genetic and environmental influences on phenotype and behaviour.

Stretch and challenge answers

1.1.2 Homeostasis and negative feedback

A As temperature increases the rate of enzyme action increases. An increase of 10 °C will double the rate of reaction. Above a certain temperature (around 50 °C) the enzymes will begin to denature and the reaction rate falls quickly.

B The enzymes begin to denature and the enzyme-controlled reactions in the cells slow down. (Other proteins, e.g. membrane carriers/channels/pumps, electron carriers, are also denatured.)

C Adrenaline causes the 'flight or fight' reflex. This may have evolved because any response that allows the organism to run away more quickly or to fight more aggressively will help it to survive. If the organism survives it can pass on the alleles that enabled it to survive. A mother who runs away or fights may lose her offspring or may be injured too badly to look after her offspring. Oxytocin causes a 'tend or befriend' response. This may have evolved because the mother who pacifies a threatening predator survives to pass on those alleles that helped her survive and look after her offspring. She also looks after her offspring, who carry her alleles.

1.1.3 Maintaining body temperature – ectotherms

A Bees in the centre of the swarm will be warmer than those on the outside. Warmer bees move towards the outer parts of the swarm while colder bees move toward the centre. This transfers heat from the centre to the outer parts of the swarm.

In hot weather the bees create more passages for air flow; the passages are also wider. Thus more air can pass through the swarm and carry heat away. In cooler weather there are fewer air passages and they are narrower.

1.1.4 Maintaining body temperature – endotherms

A If the climber is unable to find shelter, the low temperature could reduce the body temperature to the point where enzyme activity is severely reduced. Vasodilation caused by the alcohol in the brandy will increase the rate of heat loss from the body, because more blood carries heat from the body's core to the surface where it can be lost. Therefore hypothermia and death will happen sooner in a person who has drunk alcohol.

1.1.7 Transmission of action potentials

A Positive feedback: a small change in one direction (making the p.d. less negative) causes a larger change in that direction.

1.1.8 Nerve junctions

A The smooth endoplasmic reticulum is associated with the metabolism of the neurotransmitter and with packaging it into vesicles. The vesicles contain the neurotransmitter ready for exocytosis into the cleft. The mitochondria supply the ATP needed to recycle the acetylcholine, make the vesicles and move them towards the presynaptic membrane.

1.1.9 Signals and messages

A Cytokines are similar to hormones. They carry signals to certain cells in the immune system. They stimulate the action of phagocytes, macrophages, B lymphocytes and T killer cells. If too much cytokine is released it may overstimulate the action of phagocytic cells and T killer cells which could cause damage to many body cells and produce inflammation.

1.1.11 The regulation of blood glucose

A Cells that are specialised to manufacture proteins such as insulin and glucagon will have a lot of ribosomes and rough endoplasmic reticulum. This is because the ribosomes are the sites of protein synthesis. These cells will also have a lot of Golgi apparatus, as substances such as hormones are prepared and packaged into vesicles in the Golgi apparatus. There will also be many secretory vesicles in these cells as these vesicles are used to transport the hormone to the cell surface (plasma) membrane for secretion by exocytosis. These cells are also likely to have many mitochondria to supply much ATP from aerobic respiration as an energy source for the active processes in the cell.

1.2.3 Functions of the liver

A The P450s are proteins; these are manufactured by the ribosomes that are attached to the endoplasmic reticulum – from here they can be packaged in vesicles and transported to where they are needed.

B Each person may have slightly different enzymes. These may break the drugs down in slightly different ways producing different by-products.

C Genetic variation means that different people have different alleles – these will produce slightly different enzymes.

1.2.5 Formation of urine

A Proteins are normally filtered by the basement membrane. If this has been damaged by high blood pressure then proteins can enter Bowman's capsule from the blood and pass into the urine.

1.2.6 Water reabsorption

A A higher salt concentration in the medulla means that a greater water potential gradient can be achieved between the urine in the collecting ducts and the medulla. This means that more water can be reabsorbed from the collecting ducts and then pass into blood capillaries, and the urine is made more concentrated. There will be less urine produced and less water lost.

B The blood vessels running down into the medulla are looped in a similar fashion to the loop of Henle. This allows the exchange of solutes between the ascending vessel and the descending vessel.

1.2.7 Osmoregulation

A Alcohol inhibits the release of ADH. Therefore the collecting ducts are not very permeable and less water is reabsorbed. This means that more water is lost in urine and dehydration occurs. The ethanal produced from the metabolism of ethanol also contributes to the headache.

B (i) Diuretic drugs can be used to relieve water retention, which can cause swelling and high blood pressure.

　(ii) Antidiuretic drugs can be used to relieve diabetes insipidus (a form of diabetes caused by a lack of ADH, resulting in very large amounts of watery urine) and bed wetting.

1.3.1 The importance of photosynthesis

A The chloroplasts would photosynthesise and some of the organic compounds produced may be used as respiratory substrates/building blocks by the cells of *E. viridis* and *E. chlorotica*. These cells may be able to use oxygen produced by the chloroplasts for aerobic respiration. The green coloration may provide camouflage.

1.3.3 The light-dependent stage

A It accepts excited electrons released from chlorophyll molecules that have been oxidised. It may also accept hydrogen ions from the photolysis of water.

B Ice-cold temperature to prevent/reduce enzyme activity. 2% sucrose to prevent water entering or leaving the chloroplasts by osmosis.

1.3.4 The light-independent stage

A When chloroplasts are illuminated, protons (hydrogen ions) are pumped from the stroma across the thylakoid membranes into thylakoid spaces. Hydrogen ions reduce pH so removing them raises the pH of the stroma to pH 8 – the optimum pH for rubisco. Rubisco is present in the stroma, where the light-independent reaction takes place.

B When photosynthesis first evolved, there was very little free oxygen in the atmosphere, not enough to compete with carbon dioxide for the active site of rubisco. Since the concentration of oxygen has increased, perhaps the average temperatures have not been high enough to make this competition significant enough to select rubisco enzymes that have reduced oxygenase activity. Perhaps an altered rubisco with reduced oxygenase activity would also have reduced carboxylase activity and that would confer a selective disadvantage.

C If changed rubisco was still an efficient carbon dioxide acceptor but less susceptible to photorespiration (which may become more of a problem with climate change/global warming) there would be less wastage of ATP generated by photophosphorylation and a greater yield as more ATP for the light-independent reaction.

D Increased temperatures increase photorespiration. This wastes ATP generated in light-dependent reaction and so will reduce the rate of photosynthesis. Increased temperature may also lead to closure of stomata, so less carbon dioxide enters leaves.

E To break down toxic hydrogen peroxide, generated during photorespiration, to water and oxygen.

F Because it is essential for photosynthesis and nearly all organisms depend upon photosynthesis for food and for oxygen.

1.3.7 Investigating the factors that affect the rate of photosynthesis – 2

A The plant will absorb carbon dioxide (which could react with the water to become acidic) and so this raises the pH of the solution.

B The *Elodea* cannot photosynthesise in the dark, but it continues to respire, consuming oxygen and producing

carbon dioxide. The carbon dioxide dissolves in the indicator solution, lowering the pH.

C Equal volumes/mass/sized pieces of *Spirogyra*. Equal volumes and concentrations of hydrogencarbonate indicator solution. Same temperature. Vary light intensity/wavelength (but not both). Take absorbance reading of hydrogencarbonate indicator solution as blank. Use same filter and measure absorbance of hydrogencarbonate indicator solution after 15 minutes of *Spirogyra* photosynthesising. Could express rate as change in absorbance × 1000 (makes numbers more friendly) divided by time.

1.3.8 Limiting factors and the Calvin cycle

A Reduced chlorophyll production so less chlorophyll/fewer photosystems and less light energy trapped, less ATP and reduced NADP, less conversion of GP to TP.

B Reduced water uptake. Palisade cells won't be turgid so chloroplasts not near edges of cells. Stomata may close and then less diffusion of carbon dioxide into leaves.

C Blocking the flow of electrons will reduce the number of hydrogen ions pumped across the thylakoid membrane into the thylakoid space and reduce chemiosmosis and therefore production of ATP and reduced NADP. This, in turn, reduces the light-independent reactions (GP to TP and TP to RuBP). Plant fails to grow.

1.4.2 Coenzymes

A The Mexican Indians soaked their corn in lime water. This is alkaline and so after this treatment the nicotinamide in the corn can be absorbed from the intestine. The corn eaten in the Southern United States was not exposed to alkali and so much less of the nicotinamide was absorbed from the intestine. Pellagra is caused by a deficiency of nicotinamide.

1.4.3 Glycolysis

A Stage 1, when glucose-6-P is changed to fructose-6-P.

B This pathway evolved a long time ago in early life forms. It has remained in all cell types as new species have arisen.

1.4.4 Structure and function of mitochondria

A Size – most are 1–5 μm long – similar to the size of many bacteria. They have 70s ribosomes – the same type of ribosomes as are found in bacteria. They have circular DNA, similar to that of bacteria.

1.4.5 The link reaction and Krebs cycle

A They do not have any mitochondria. Both of these reactions take place in the mitochondrial matrix.

B Because this reduced NAD is in the matrix and does not have to cross any mitochondrial membranes. It offloads the hydrogen atoms (protons and electrons) on the matrix side of the inner membrane.

C Krebs cycle will occur in the cytosol (cytoplasm) and oxidative phosphorylation on the cell membrane, which is likely to be invaginated (folded inwards) into structures called mesosomes.

1.4.6 Oxidative phosphorylation and chemiosmosis

A There may be protein channels in the membranes that allow malate and aspartate to pass through the membrane. There are no channels that allow oxaloacetate and reduced NAD through the membrane.

1.4.8 Anaerobic respiration in mammals and yeast

A Because lactate can be recycled back to pyruvate (which can be changed to acetate and enter Krebs cycle) or changed into glycogen and stored for later use. Ethanol is not recycled and is not respired by yeast.

1.4.9 Respiratory substrates

A 7 reduced NAD from beta-oxidation gives $7 \times 2.6 = 18.2$ ATP

8 turns of Krebs gives 3×8 reduced NAD,
so this gives $24 \times 2.6 = 62.4$ ATP plus 8 ATP
Net gain is $(18.2 + 62.4 + 8) - 16 = 72.6$ ATP

2.1.2 Translation

A cAMP will inhibit glycogen synthase, by binding to an allosteric site and changing the enzyme's shape so that the active site becomes buried/hidden.

B Glucose 6-P activates glycogen synthase by binding to an allosteric site, changing its shape and exposing the active site.

2.1.4 Mutations – 2

A Scientists may be able to change differentiated cells into stem cells and then direct their development into any kind of cell, for example to repair certain tissues, e.g. nerve tissues and heart muscle; or the cells could be used for drug testing to reduce the need for tests on animals.

2.1.7 Apoptosis

A Pros: It will greatly reduce cervical cancer (which could spread to other parts of the body) and so save lives and save money on treatment.

Cons: The vaccine needs to be given before sexual maturity is reached as it is ineffective if given after infection. Because some girls reach sexual maturity earlier than others and not all girls wait until the legal age of consent before having sex, it needs to be given to all girls at age around 12. Some people fear that it will encourage promiscuity among girls, although this is debatable: many people do not realise that cervical cancer is a sexually transmitted disease and so it is not a factor that delays having sex. Some people do not want to be told to have a vaccine as they do not want their civil liberties infringed. However, no vaccine in the UK is compulsory. Because a vaccine is available does not mean everyone has to be vaccinated.

2.1.9 The significance of meiosis

A Some gametes will have more chromosomes than the haploid number and some will have fewer.

B It will cause them to pair up during meiosis and independently assort.

2.1.12 Using genetic diagrams – 2

A The fetal haemoglobin has two alpha chains and two gamma chains. The mutation that leads to sickle-cell disease is on the gene that codes for the beta chains.

B The T cells transfused into the patient reject the patient's tissues.

C Red blood cells live for about 3 months.

D Normal Hb migrates further along the gel towards the positive electrode than does abnormal Hb because normal Hb has less positive charge.

E Red blood cells do not have a nucleus and so do not have any DNA to analyse.

2.1.14 What determines sex?

A Males are homogametic and so a male with white spot could be homozygous or heterozygous. Females with black heads must have the genotype X^bY.

B The female moths are black – a rarity amongst pale brindled moths. Their genotype is X^BY. Normal males are genotype X^bX^b. When crossed, all males are X^BX^b – black. All females are X^bY – pale/normal.

C Starting with crossing black winged male, X^BX^B and cinnamon female X^bY. This produces X^BY black females and X^BX^b – black males.

Cross these black males with a cinnamon female and this gives half the males cinnamon, X^bX^b, half the males black X^BX^b, half females black, X^BY, and half females cinnamon, X^bY.

Now just cross cinnamon males with cinnamon females.

Parental genotypes:	X^BX^B	X^bY
Gametes	$\textcircled{X^B}$	$\textcircled{X^b}$ $\textcircled{Y}$
Offspring	X^BX^b	X^BY

Second cross:	X^BX^b		X^bY	
Gametes	X^B	X^b	X^b	Y
Offspring	X^BX^b	X^bX^b	X^BY	X^bY

Finally X^bX^b to be crossed with X^bY

2.1.15 Interactions between gene loci – 2

A Gametes in both cases are: **DE, De, dE, de**
Genotypes of offspring and their phenotypes:

DDEE	DDEe	DdEE	DdEe
dark brown	med brown	med brown	light brown

DDEe	DDee	DdEe	Ddee
med brown	light brown	light brown	dark blue

DdEE	DdEe	ddEE	ddEe
med brown	light brown	light brown	dark blue

DdEe	Ddee	ddEe	ddee
light brown	dark blue	dark blue	light blue

Phenotype ratio: 1 dark brown: 4 med brown: 6 light brown: 4 dark blue: 1 light blue

B Yes, two parents with dark blue eyes can have a child with pale brown eyes.
Parents' genotypes may be Ddee and ddEe and they can have children of genotype DdEe.

C That it is more complex than described here, with interactions between other genes.

2.2.1 Clones in nature

A Plant can reproduce even if it is one isolated individual, and can eventually colonise a whole area. No reliance on agents of pollination such as wind or insects. Reproduction is faster. No genetic variation so all individuals are equally susceptible to disease. Tubers and bulbs may be edible by humans. Growers can easily propagate many natural clones of individual plants with desirable characteristics.

2.2.3 Cloning animals

A Only the mother's.

B Such polypeptides would be involved in respiratory processes – for example electron transport chain proteins or enzymes associated with Krebs cycle. (However, some can affect other systems; there is a type of blindness caused by a mutation to mitochondrial DNA.)

2.2.5 The growth curve

A The newly arrived species would have to be able to cope with the conditions in the location – this includes biotic factors such as the capacity to avoid predation, compete for food sources and find mates in order to reproduce, plus the abiotic factors such as surviving the temperature and fluctuation of temperature, availability/quality of water, light, nesting space and so on.

B The competitive exclusion principle states that no two organisms can occupy the same niche in one location. If the new arrival shares a niche with another organism, either the newly arrived species or the original species will eventually disappear completely.
The newly arrived species could be a predator of other species present, so reducing their numbers, or could be a food source for other species present and increase their numbers.

C Eventually, a newly arrived species, if it can survive and reproduce, will reach the carrying capacity for that species. This number might be fairly static over time if the environment does not change, or it might vary, for example as the seasons change. As long as the environment can support a number of organisms, the population will not decline as it would in a closed system, because nutrients are continually recycled through the system as energy flows though it.

E The term *biological control* is used to describe the placing of organisms of a species where they will reduce the population size of a pest or unwanted organism. A simple example is keeping a cat on board a ship to reduce the rat population.

2.2.10 Sequencing and copying DNA – 1

A The antisense strand can anneal to the mRNA strand produced by transcription because it is complementary to that strand. This forms double-stranded RNA which cannot bind to, or be read by, a ribosome. Double-stranded RNA like this also triggers destructive enzymes which chop up the double-stranded RNA.

B Transfer RNA and ribosomal RNA.

C Cancerous cells divide continually and without differentiating. It is hoped that it may be possible either to stimulate the natural RNA interference mechanisms which have been lost in these cells, or artificially interfere with the genes which are switched on and cause the continual cell division process. If this can be done, the cancerous cells would not be able to divide to form tumours.

D Cells produced by mitosis in a multicellular organism go on to differentiate into one of the organism's adult cells. Differentiated adult cells have specific roles and functions and so require specific genes to be switched on and others switched off in order to fulfil the particular role of the cell. It is thought that a significant step in differentiation is the use of natural RNA interference mechanisms to switch off (silence) the genes that are not required in the differentiated cell.

2.3.7 What affects population size?

A Secondary succession.

B Because they usually have a very fast rate of reproduction, which is usually much faster than *k*-strategists.

C *k*-strategists. Because their population size is determined by the carrying capacity, it stabilises at the carrying capacity. A climax community is a stable population.

2.3.8 Competition

A Because allelopathic chemicals stop a plant's neighbours from using the resources in a habitat.

2.3.9 Sustainable management

A The woods would go through a process of succession, reaching a climax community in which biodiversity would be much lower than is currently the case.

2.4.1 Why plants respond to the environment

A Any plausible and relevant ideas are acceptable. For example, the response of sensitive plants may make it more difficult for herbivores to eat their leaves.

B Variation in extent of response between individuals. Those with most pronounced response are less likely to be eaten, as the leaves droop more quickly, and hence are more likely to survive longer and reproduce. After many generations, eventually the whole population of plants would respond more dramatically.

C Impose an artificial day length using artificial lights.

D Finding out which day length promotes flowering enables growers to manipulate the day length to encourage sustained flowering. Fertilisation of flowers leads to fruit and seed production. Hence more flowering gives higher yield.

2.4.2 How plants respond to the environment

A Any relevant and plausible points are valid, but you should show an awareness of the limitations of a definition of growth as 'getting larger', and some appreciation of growth being irreversible, with an increase in dry mass and including cell division.

B Roots grow in response to low levels of auxin, but high levels of auxin inhibit growth. Auxin migrates away from the light side of the root, to the dark side of the root. This inhibits growth on the dark side (because of the high concentration of auxin), but promotes growth/cell enlargement on the light side (because of the low concentration of auxin). As a result, the root bends away from the light, and into the soil where it is dark.

2.4.5 The brain

A (i) Broca's area is a motor area – damage results in the inability to make motor movements associated with speaking.

(ii) Wernicke's area is an association area – damage results not in the loss of sensory function but the ability to make sense of the sensory input received.

B (i) Visual sensory, visual association, Wernicke's area, Broca's area, hearing sensory area.

(ii) Hearing sensory area, hearing association area, Wernicke's area, Broca's area.

2.4.10 Pulling together – muscles, nerves and hormones

A An imagined threat, generated in the cerebrum, sends impulses to other parts of the brain. The other parts of the brain respond to the perception of threat generated by the cerebrum. These other areas do not differentiate real or imagined threat; the responses that follow are the same and are related to the cerebral perception or imagination of threat.

B In order to create something new or original there has to be a capacity to consider how the muscular operations required to carry out the creation will work. This involves being able to consider materials and operations that might never have been used before. This is the basis of imagination, without which the intentional creation of a structure or piece of music for example would not be possible.

C As with question **A** above, the response to perceived threat is the same as to imagined threat and to the remembering of previous threat. In PTSD it is thought that previous events are held in the memory and become accessed uncontrollably, producing flashback – a sense of reliving the event. This access stimulates the fight or flight response as the threat seems to be there again and leads to uncontrolled stress responses in individuals.

D Prolonged and repeated activation of the fight or flight response can cause digestive disorders because the sympathetic nervous system slows down the digestive processes. The blood supply to the gut is reduced as more blood is diverted to muscles. This leads to problems with the movement of food along the gut and with the release of enzymes onto the food. The food stays in the gut for longer, where toxins can be generated by the action of gut microbes. Food also becomes compacted in the gut. Disorders such as colon cancers are also significantly more likely in individuals suffering from prolonged stress.

E Elevated blood pressure is a common feature of prolonged stress. This can lead to many conditions, including kidney damage and circulatory problems such as aneurysms which can lead to cerebrovascular accident (stroke). Long-term stress is also implicated in heart conditions such as angina, myocardial infarction and an enlarged heart because sympathetic stimulation leads to increased heart rate. Continued prolonged stress can also result in mental health problems, the features of which include inappropriate responses such as uncontrolled crying without cause, or anger management problems.

2.4.13 Primate behaviour

A Human milk is that of a 'carry animal'. Its nutrient value is balanced to support brain development in the relatively underdeveloped human infant. It also contains specific human proteins and fatty acids and antibodies to any infectious organism that the mother has encountered. The delivery of milk has evolved to be in a pattern of very regular small feeds so that the relatively underdeveloped gut is not overwhelmed.

Cow's milk is that of a 'follow animal'. Its nutrient value is balanced to support the physical development of the relatively developed calf. It contains specifically bovine proteins and fatty acids, some of which have been shown to provoke allergic responses in some humans. The delivery of milk has evolved to be in a less regular pattern of larger feeds because the gut is fairly well developed in calves at birth.

B (i) Primate birth canals are not large enough to allow birth of a fully developed infant. Relatively short gestation means relatively small skull, which can pass down the birth canal.

(ii) Once the infant is born there is a relatively long period of development. During this period there is time for development of the brain and so development of all of the learned behaviours required of an organism of that species.

C Temperature sensors in skin of mother send impulses to mother's brain. This input is compared in mother's brain against normal set point of mother (37 °C). If temperature detected at sensors is below set point, mother's autonomic response is to raise core temperature (probably through increased respiration in liver and brown fat tissue). Vasodilation to skin in contact with baby is also stimulated. Raised temperature of baby detected in skin receptors would trigger the reverse of the above.

D (i) Scientific research into the effects of caring for children in a particular way cannot be carried out (ethically) by subjecting children to anything that it is anticipated might be of harm to them. For example, it would be unethical in a study into whether carrying offspring is of benefit to development of balance to deprive some children of being carried, since the expectation would be that this would hinder their development. However, where human groups have particular behaviour patterns, then it is ethical to study these and attempt to draw conclusions about the value of such behaviour patterns. These conclusions may then be of benefit to other social groups. For example the research into human social groups has shown that there are some disease states (in particular allergy) that do not occur, or very rarely occur, in breastfed infants and in mothers who breastfeed. Such societies also have an almost nil occurrence of unexplained 'cot death', perhaps because close contact with parents 're-stimulates' breathing when it occasionally stops in infants. The psychological research is continuing into theories about the detachment effects of mothers going to work when infants are very young.

(ii) All disease and psychological states are to some extent affected by the genetic make-up of the individual, for example in the particular development of their immune system. The differences also apply to groups of humans. All groupings of humans, especially where these groups have been separate from other groups for a long period of time, have particular sub-sets of alleles of some genes. It is the difference in allelic make-up that gives different groups of humans their particular and recognisable features. These differences do not only impact on physical features, they occur in metabolic pathways within the individual groups. A clear example of this is seen in Inuit populations. Individuals can live on a diet which is almost completely of animal products – such a diet would cause severe health problems for individuals in most other human populations. This means that findings from the study of one particular group may not apply to all groups.

Practice answers

1.1 Communication and homeostasis

1 Change in temperature; change in water availability; change in light intensity.

2 A stimulus is a change in the environment that causes a response. A response is the change in an organism caused by a stimulus.

3 Cell signalling occurs whenever one cell releases a molecule that is detected by another cell.

4 Temperature, pH, water potential, blood glucose concentration.

5

Response to feeling warm	Explanation – how does this response help to cool the body?
The skin releases more sweat	**The water in sweat evaporates from the surface of the skin. Evaporation requires energy (latent heat of vaporisation) which is absorbed from the skin, blood and tissues beneath the skin**
The blood vessels in the skin dilate	**Dilated blood vessels allow more blood to flow close to the skin and more heat can be lost by radiation to the surroundings**
Body hairs lie flat	Less air is trapped close to the skin so that the body is less well insulated. More heat can be lost by convection and radiation
Sit in the shade	**Less direct sunlight and heat reach the body – so less heat is absorbed from the Sun**

6 Transducers absorb energy and re-emit it in a different form. The light receptors in the retina of the eye absorb light energy and convert it to electrical energy in the nerve impulse.

7 A stimulus (a change in the environment) causes gated sodium ion channels in the membrane of the receptor to open. A small stimulus will open a few gates and allow a few sodium ions into the cell. This creates a small change in the potential difference across the membrane – this is the generator potential.

8 (a) Sodium ion channels open and sodium ions diffuse across membrane into cell. This depolarises the membrane and, when it reaches the threshold potential, more sodium ion channels (voltage-gated sodium ion channels) open and lots of sodium ions diffuse in (positive feedback). When the membrane potential difference reaches +40 mV (action potential), the sodium ion channels close, Potassium ion channels open and potassium ions diffuse out of the cell, repolarising the membrane. The membrane becomes temporarily hyperpolarised (p.d. about –90 mV). The sodium–potassium pump restores the resting potential.

(b) When sodium ions diffuse into the cell they increase the local concentration of sodium ions in that region of the cell. The ions diffuse away from the local high concentration producing a local current. This causes sodium ion channels further along the membrane to open.

9 (a) Myelinated neurones have a myelin sheath. This is a layer of Schwann cells closely bound to the neurone. Non-myelinated neurones do not have a myelin sheath. There may be several neurones loosely wrapped inside one Schwann cell.

(b) The myelin sheath insulates the neurone and prevents the ions moving across the cell membrane. Gaps in the sheath allow ionic movements at 1–3 mm intervals along the neurone (at the nodes of Ranvier). Therefore the action potential jumps along the neurone. (This is described as *saltatory.*) The action potential is transmitted much more quickly.

10 (a) Synapses join neurones together; allow some neurones to connect to many others; create nervous pathways; ensure impulses travel in the correct direction; allow acclimatisation or accommodation to occur; filter out unwanted impulses; amplify small signals by summation.

(b) When an action potential reaches the synaptic knob it causes calcium ion channels to open. Calcium ions diffuse into the synaptic knob and cause vesicles of acetylcholine to move towards and fuse with the presynaptic membrane. The transmitter substance acetylcholine is released into the synaptic cleft by exocytosis.

(c) Postsynaptic membranes contain special sodium ion channels. These have acetylcholine receptors. When the acetylcholine binds to the receptors the sodium ion gates open.

11 Pituitary, pancreas, thyroid, adrenal glands, testes, ovaries, stomach, duodenum and parathyroids.

12 Each hormone molecule has a particular shape which is complementary to the shape of the receptor. The receptor is found only on the target cell membranes.

When the hormone binds to the receptor it causes that cell to activate a particular series of enzyme-controlled reactions.

13 α and β cells in islets of Langerhans monitor the concentration of glucose in the blood. If the concentration rises too high this is detected by the β cells. In response the β cells secrete insulin. The insulin is transported around the body in the blood. The liver cells and muscle cells have insulin receptors on their cell surface membranes. When insulin binds to the receptors the cells absorb more glucose from the blood. The cells use more glucose in respiration, or convert it to storage compounds such as glycogen or fats. If the blood glucose concentration gets too low this is detected by the α cells. The α cells secrete glucagon. Liver cells have glucagon receptors. When glucagon binds to the receptors the cells convert glycogen to glucose. They also convert amino acids and fats to glucose. They release more glucose into the blood.

14 Type I diabetes is juvenile onset. It is due to an inability of the cells in the pancreas to make enough insulin. Type II diabetes is due to a lack of or damage to the receptors on the liver cells so that the liver cells do not respond to insulin properly. It may be compounded by lower levels of insulin production.

15 When we exercise, the muscles need more energy, in the form of ATP, for contraction. The muscles need to respire at a higher rate to supply more ATP. Therefore the muscles need more oxygen and glucose (or fatty acids). The heart must beat more frequently (increase in heart rate) to increase the cardiac output. This will pump more blood per minute to the lungs and to the muscles to increase the supply of oxygen. As the muscles respire more they produce more carbon dioxide. This reacts with the water in blood plasma and makes the plasma more acidic (lowers the pH). The change in pH is detected by chemoreceptors in the carotid artery and in the brain. These send nervous impulses to the cardiovascular centre in the medulla oblongata of the brain, to increase heart rate. Exercise will also cause adrenaline to be secreted into the blood. The adrenaline acts directly on the heart muscle to increase heart rate.

1.2 Excretion

1 The removal of waste products from cell metabolism.
2 (a) Carbon dioxide and urea.
  (b) The lungs and the kidneys.
3 (a) The hepatic artery and the hepatic portal vein.
  (b) A sinusoid is a channel between the liver cells; they are found in the liver lobules.

4 A liver cell.
5 (a) The removal of the amino group from an amino acid forming ammonia and leaving a ketose residue.
  (b) The conversion of ammonia to urea by the addition of carbon dioxide.
6 (a) Cortex, medulla and pelvis.
  (b) Nephrons.
  (c) Ureter.
7 Any two from: numerous glycogen granules for storage; many mitochondria to produce lots of ATP for protein synthesis and endo/exocytosis; many ribosomes/a lot of endoplasmic reticulum to assemble the many enzymes needed for all the liver's metabolic activities.
8 (a) Carbon dioxide combines with water to form carbonic acid. This is buffered by proteins including haemoglobin. This reduces the oxygen-carrying capacity of the haemoglobin. It also combines directly with haemoglobin to form carbaminohaemoglobin. Excess carbon dioxide can also cause respiratory acidosis in which the blood pH drops too low.
  (b) Ammonia is too soluble and very toxic – it needs to be diluted with a lot of water. Land and sea-dwelling mammals (many of which don't drink sea water but obtain water from their food) need to conserve water.
9 (a) In the cell body of neurosecretory cells in the hypothalamus.
  (b) The posterior pituitary gland.
  (c) The cells lining the collecting ducts and distal convoluted tubules in the kidneys.
10 (a) Bowman's capsule.
  (b) Proximal convoluted tubule.
  (c) Collecting ducts.
11 (a) To provide a large surface area for reabsorption. Increased cell membrane therefore more space for protein carriers or channels.
  (b) To provide a lot of ATP to actively reabsorb glucose and amino acids.

12

Substance	Concentration relative to blood plasma	
	In urine	In Bowman's capsule
Glucose	Less	Same
Amino acids	Less	Same
Urea	More	Same
Protein	Less	Much less/none

13 (a) The hypothalamus.
   (b) When the water potential of the blood drops too low (too negative) ADH is released from the posterior pituitary gland. This acts on the walls of collecting ducts to make them more permeable to water by increasing the numbers of aquaporins in the cell membranes and more water is reabsorbed into the blood. This increases the water potential of the blood (makes it less negative). If the water potential rises too high less ADH is released, the collecting duct walls become less permeable to water and less water is reabsorbed.

14 (a) Alcohol is oxidised to ethanal. This step is catalysed by ethanol dehydrogenase. Ethanal is oxidised to acetate. This is catalysed by ethanal dehydrogenase. Both steps require the removal of hydrogen atoms, which are accepted by NAD. The acetate is combined with coenzyme A. (Acetyl coenzyme A can enter Krebs cycle.)
   (b) Metabolising alcohol uses NAD. The NAD is therefore not available to help oxidise fatty acids. The fatty acids are changed into fats, which accumulate in the liver cells.

15 (a) A tube that turns back on itself so that the fluid in the tube runs close to another part of the tube. It increases the efficiency of exchange of materials from one part of the tube to another.
   (b) The loop of Henle turns back on itself. Sodium ions are actively transported out of the ascending limb into the medulla. This lowers the water potential of the tissue fluid in the medulla and causes water to leave the collecting duct by osmosis. This conserves water.

16 The cells lining the proximal convoluted tubule have microvilli that increase surface area for protein channels. Sodium ions are actively removed from these cells into the tissue fluid. This allows sodium ions to diffuse into the cells from the tubule fluid. They diffuse in through special co-transporter proteins in the cell membrane (this is facilitated diffusion). These proteins allow sodium ions to enter in association with glucose molecules. The glucose molecules can then diffuse out of the cells into the tissue fluid (this may be enhanced by active transport). From here they diffuse into the blood.

17 Any two from: freedom from dialysis and freedom from the time commitment it takes; diet is less limited; feeling better physically; a better quality of life, e.g. able to go on holiday; no longer seeing oneself as chronically ill.

1.3 Photosynthesis

1 (a) Organisms that can produce complex organic molecules; from simple inorganic molecules and energy, either from light or chemicals.
   (b) Organisms that cannot make complex organic molecules from inorganic ones, but have to eat food to obtain energy and organic molecules.

2 Most plants and animals rely on aerobic respiration; this requires oxygen; oxygen is a by-product from photosynthesis; until photosynthesis evolved there was no free oxygen in the Earth's atmosphere.

3 (a) Thylakoid membranes/grana.
   (b) Stroma.
   (c) Chloroplast ribosomes.

4 An organic molecule; contains magnesium/porphyrin ring; absorbs light energy; associated with light of a particular wavelength/range of wavelengths; found in chlorophyll/includes carotene/xanthophyll/chlorophyll a and b.

5 Oxygen; ATP; reduced NADP.

6 (a) Water is an electron donor; it gives an electron to replace one lost by chlorophyll a in photosystem II; is also a source of hydrogen ions/reduces NADP.
   (b) A 5-carbon compound present in the stroma of chloroplasts; is a carbon dioxide acceptor; combines with carbon dioxide to give unstable intermediate that changes into GP.
   (c) A hydrogen acceptor; electron acceptor; carries hydrogen atoms/ions from light-dependent stage to Calvin cycle/light-independent stage; helps to maintain proton gradient across thylakoid membranes by removing hydrogen ions.
   (d) Energy currency/high-energy intermediate compound; makes light energy available to light-independent stage as chemical energy; used to change GP to TP and to change TP to RuBP.
   (e) Source of carbon; for the organic molecules being synthesised; in Calvin cycle/light-independent stage.
   (f) Enzyme; catalyses acceptance of carbon dioxide; by RuBP.

7 (a) The pigments reflect light; of wavelength 650–700 nm; this stimulates cells in the retina and the brain interprets what we see as red.
   (b) It absorbs light of 550 nm (green); which leaves not in shade will have reflected; it avoids/reduces competition.

8 Will absorb more light energy; maximises photosynthesis; absorbs more heat; may increase rate of enzyme-controlled reactions in leaf; e.g. light-independent stage of photosynthesis (and respiration).

9 (a) Photosynthesis depends on it; this process converts light energy into chemical energy; for consumers; is at the beginning of nearly all food chains.

(b) It contains more than one polypeptide chain; there are eight large subunits and eight small subunits.

(c) The enzyme still works when the small subunits are removed.

(d) The shape of the CABP molecule is very similar to the shape of a 6-carbon sugar; CABP is a competitive inhibitor of this enzyme and so it can occupy its active site; when RuBP and carbon dioxide occupy the active site it is likely that they react together to give a very short-lived 6 C compound.

10 (a) Many stacks of thylakoid membranes; gives large surface area; for photosystems; for ATP-synthase enzymes; for electron carriers; liquid stroma surrounds thylakoids; short distance for products of light-dependent stage to reach stroma; enzymes for Calvin cycle in the stroma.

(b) Allows them to capture more of the limited light energy; to reduce limiting effect of light intensity; to maximise rate of photosynthesis.

1.4 Respiration

1 Active transport; exocytosis; endocytosis; protein synthesis; movement of organelles using microtubule motors; DNA replication, mitosis and meiosis.

2 (a) cytoplasm/cytosol;

(b) matrix of mitochondria;

(c) matrix of mitochondria;

(d) cristae of mitochondria.

3 4

4 Active transport.

5 (a) An activated nucleotide; contains adenine; ribose sugar; 3 phosphate/phosphoryl groups.

(b) (i) ATP: activates glucose at beginning of glycolysis.

(ii) NAD: a coenzyme; accepts hydrogen atoms; from respiratory substrates; carries hydrogen to inner mitochondrial membranes; where hydrogen atoms split into electrons and protons; these are used to generate ATP during chemiosmosis.

(iii) Electron carriers: pass the electrons (from hydrogen atoms) along the chain; energy released from the electrons is used to generate the proton gradient across the inner mitochondrial membranes.

(iv) Cristae: provide large surface area; for electron carriers; and ATP-synthase enzymes.

(v) Acetyl coenzyme A: carries acetate onto Krebs cycle.

(vi) Oxygen: final electron acceptor; accepts electrons and hydrogen ions from cristae/end of electron transport chain; to form water; this helps maintain the proton gradient across the cristae.

6 Inner membrane with many folds/cristae; provides large surface area; for electron carriers; ATP-synthase channels/enzymes; liquid matrix contains enzymes for Krebs cycle; DNA and ribosomes for synthesis of some respiratory proteins/enzymes; short distance for reduced NAD from Krebs to travel to cristae.

7 Proton gradient established across inner mitochondrial membrane; using energy released from electron transfer; to pump protons from reduced NAD from matrix to intermembrane space; electrochemical/proton gradient is a source of potential energy; hydrogen ions cannot pass through lipid bilayer of inner membrane; pass through protein channels; associated with ATP-synthase enzymes; force of proton flow; down electrochemical gradient; causes ADP and P_i to be joined to give ATP.

8 (a) Organic molecules that can be respired to release energy used to make ATP; e.g. glucose, lipid and amino acids.

(b) Lipid/fatty acid molecules contain many more hydrogen atoms; these are used to form the proton gradient; and to flow through ATP-synthase and make ATP.

9 In yeast, pyruvate is decarboxylated to release carbon dioxide and ethanal; ethanal acts as hydrogen acceptor to reoxidise NAD; and is changed to ethanol; in mammalian muscle cells pyruvate is the hydrogen acceptor; pyruvate is reduced to lactate as NAD is reoxidised.

10 (a) They can be transported in the blood – blood plasma is largely water.

(b) As ketone bodies are changed to acetate; and enter Krebs cycle; they can only be respired aerobically.

(c) Hydrolysis.

(d) One molecule of glycerol; and three of fatty acids.

(e) Beta-oxidation.

(f) There will be very little glucose in the blood; so energy stores need to be used.

2.1 Cellular control

1 The code is DNA; instructions for assembling polypeptides; the sequence of base triplets determines the sequence of amino acids/primary structure, in the polypeptides.

2 (a) Many amino acids have more than one base triplet code.

(b) It reduces the effects of substitution mutations; as some of these may alter one base but the triplet still codes for the same amino acid.

3

	DNA replication	DNA transcription
Enzyme	DNA polymerase	RNA polymerase
Does DNA unwind?	Yes	Yes
Amount of DNA involved	Whole molecule/ chromosome	A gene/length of DNA which is part of a molecule
When in cell cycle does it occur?	During interphase in synthesis (S) phase	During interphase in growth (G₁) phase
Where in the cell does it occur?	Nucleus	Nucleus/ nucleolus
What are the products?	Two molecules of DNA, both identical to each other and to the parent molecule	mRNA – a copy of the coding strand of DNA (tRNA molecules – copies of parts of the template strands of DNA)

4 Mutation during DNA replication; crossing over at prophase I; independent assortment of chromosomes at metaphase I; independent assortment of chromatids at metaphase II (haploid gametes, each of which can combine with another gamete).

5 It increases biodiversity, variation within a population; greater variety of alleles in the gene pool; whatever the change in the environment, some individuals will be able to survive; these will be able to reach reproductive age and pass on the favourable alleles to their offspring; over many generations the allele frequency will change; in domesticated/cultivated organisms, breeds with desired characteristics will be selected; in the wild, speciation may occur.

6 Discontinuous variation – individuals have or do not have the characteristic; controlled by one gene/ monogenic; less influenced by environment. Continuous variation – shows a range, e.g. height in humans; controlled by many genes; more influenced by environment.

7 Fastest-running horses (mares and stallions) selected; by humans/breeders; bred together; fastest-running offspring selected; continued over many generations; pedigree records kept; one fast stallion may be used to impregnate many mares; slower horses not used for breeding. (Artificial insemination not used for horse breeding but in theory it could be, as could embryo cloning and use of surrogate mares).

8 (a) It is on the X chromosome.

(b) No RNA-binding protein will be made.

(c) An expansion of the CGG repeat; from between 52 to any number up to 230.

(d) Some body cells/tissues will have more CGG repeats than others.

(e) The number of CGG repeats increases; during meiosis; to produce gametes; and is passed on to the offspring.

9 (a) The gene for the polypeptide is on one of chromosomes 1–22, and not on X or Y. The abnormal allele codes for and still produces (is transcribed to give) an abnormal protein even if the normal allele is present.

(b) Sons inherit the Y chromosome from their father and X chromosome from their mother.

(c) Something else needed from the environment (male cells/tissues); such as (higher levels of) testosterone.

10 (a) Mutation in gene gives allele for lactase production that stays switched on in adult; (mutation may be in a regulatory gene). In Europe (more temperate climes) dairy produce was an important constituent of the diet; lactose-intolerant individuals died before reproducing/were selected against (as lack of dairy produce meant lack of vitamin D and so increased risk of rickets – women with resulting small pelvises would die in childbirth); over time, this leads to an increase in the frequency of the allele that stays switched on for lactase production in the population.

ACKWORTH BRANCH
COLLEGE
LEARNING RESOURCES

(b) The lactose in milk has been digested to glucose and galactose by the culture bacteria.

(c) The gene coding for lactase is switched off, so they are lactose-intolerant and cannot digest the lactose sugar in milk.

(d) Heated (or treated with enzyme) to hydrolyse lactose to glucose and galactose.

2.2 Biotechnology and gene technologies

1 (a) shoot tip/root tip/apical bud/cambium/nodes between areas of growth;

(b) cells are, totipotent/not differentiated/can form all cell types; throughout plant, cloned cells are genetically identical; easier to modify one cell, which then divides by mitosis; meristem cells have large nucleus and little cytoplasm.

2 (a) DNA from two different organisms; usually from different species; joined/combined;

(b) bind at specific sequences of DNA; cut DNA leaving sticky ends; exposed nucleotide bases; use same enzyme to cut plasmid/vector and remove gene from donor; vector and donor DNA can bind as sticky ends complementary; hydrogen bonds form between complementary bases;

(c) e.g. Golden Rice™; gene for beta-carotene expressed in rice grains;

(d) people obtain more beta-carotene and can make more vitamin A/avoid blindness; possibility of genes form cultivated rice passing in pollen to other strains

3 They could bind to mRNA; making double stranded RNA; by hydrogen bonds between complementary nucleotide bases; this cannot be translated at ribosomes; so no protein product of the gene made

4 (a) enzyme, attached to/enclosed in, an insoluble material;

(b) encapsulation in alginate beads; adsorption onto collagen/resin/glass; covalent chemical bonding to cellulose/collagen fibres; trapped inside gel;

(c) urine can be processed; can recycle it into pure drinking water; can reuse same enzymes; product not contaminated with enzyme; enzyme more stable/heat resistant

5 (a) as source of oxygen; for aerobic respiration; as growth or organism faster if respiration aerobic; as more ATP made per mole glucose; the metabolite produced when organisms respires aerobically; needs to be sterile so no contaminants/bacteria/ viruses introduced;

(b) (i) carbon needed to form all organic molecules; such as carbohydrate (or named example), protein/amino acids/named example, and fats/ triglycerides/fatty acids/nucleic acids (named example); to be used for structure of cell; or as respiratory substrate/source of energy; nitrogen needed for amino acids/protein for growth/protein synthesis; and for nucleic acids/coenzymes such as NAD; and for ATP;

(ii) in the nutrient medium; as sugars; or proteins or chemicals such as urea;

(c) microorganisms produce heat; waste product form their respiration; small SA/V ratio of large fermenter so heat cannot be easily dissipated; water jacket carries away heat/heat exchange; prevents temp rising too much which would denature enzymes and other proteins; so prevent metabolism and production of needed metabolites;

(d) waste products of microorganisms metabolism/ carbon dioxide/lactate; would alter/lower pH; would interfere with H bonds in enzyme molecules and disrupt 3D/active site shape; and slow metabolism;

(e) can grow microorganisms in controlled conditions; growth not dependent on climate; so can grow anywhere in world; easy to harvest wanted chemical; can make human versions of the proteins by using GMOs; can produce large amounts cheaply

6 (a) (for any of first three) allows field of crops to be sprayed with herbicide; reduces competition from weeds; for light/water/minerals; increases yield; (for potato) removes need to use insecticide; so do not kill useful insects; reduces insect damage to potato and increases yield; no danger to agricultural workers as they do not have to spray insecticide;

(b) low risk in terms of survival of transgenic plants; as rape, maize and sugar beet extinct after 4 years; potatoes in one habitat survived longer; possibility of cross pollination to non GM counterparts; may cause allergies/toxicity; genes may transfer to certified organic crops; may interfere with gene expression in another organism; may have unforeseen effect; these results may not be valid for other habitats; however most of the food we eat has not been tested and some will cause allergies in some people; increases yield could provide food in areas where food is short; need to weigh up benefits and possible hazards; social change very slow and hard to implement; if more land taken for growing biofuels crops, need to increase yield in other plants; other valid points

2.3 Ecosystems and sustainability

1 (a) (i) The place where an organism lives.

(ii) The role of an organism in its ecosystem.

(iii) All the populations of different species who live in the same place at the same time, and who can interact with each other.

(iv) All of the organisms of one species, who live in the same place at the same time, and who can breed together.

(b) (i) feed on waste material or dead organisms; recycle nutrients;

(ii) supply chemical energy to all other organisms; usually plants which convert light energy in photosynthesis

2 (a) to represent the amount of energy stored at each trophic level.

(b) (i) If different species are different sizes, pyramids of number don't accurately represent how much tissue (and stored energy) is actually found at each trophic level.

(ii) The same mass of tissue in different species may store different amounts of energy.

(iii) They only take a snapshot of an ecosystem at one moment in time. Because population sizes can fluctuate over time, this may provide a distorted idea of the efficiency of energy transfer.

(c) The gross primary productivity is the rate at which plants convert light energy into chemical energy. However, because energy is lost when the plant respires, less energy is available to the primary consumer. This remaining energy is called the net primary productivity (NPP).

3 More light: grow plants under light banks; sow/plant crops early. More water: irrigate crops; use drought-resistant strains. Warmer temperature: grow plants in greenhouses; sow/plant crops early. Minerals: use crop rotation with nitrogen-fixing crop; use strains responsive to high levels of fertiliser. Competition: use herbicides to kill weeds. Pests: use fungicides or insecticides to kill pests; use resistant strains of crops.

4 Energy, nutrients, nitrogen, pioneer, climax, biotic, interspecific, population.

5 (a) A quadrat is used to sample the plants in an ecosystem. Place the quadrat either randomly (using random number generator), or at evenly distributed points across the habitat. Record simply presence or absence of each species (distribution), or count the number of individuals (abundance) of each species. For some plants, like grass and moss, it is difficult to count individuals, so estimate percentage cover instead.

(b) Lower the frame into the quadrat and record any plants touching the needles. If the frame has 10 needles, and you lower it 10 times in each quadrat, you will have 100 readings. Every time an individual touches a needle, this represents 1% cover.

(c) Stretch out a tape measure through a habitat, and then take samples at regular intervals along the tape, making a note of which species is touching the tape.

(d) Stretch out a tape measure through a habitat, and then take samples at regular intervals along the tape, by placing a quadrat next to the line (interrupted belt transect) and looking at the species within it. Alternatively, place a quadrat next to the line, moving it along the line after looking at each quadrat (continuous belt transect).

6 (a) Ammonium ions are released by bacteria involved in putrefaction of proteins in dead or waste organic matter.

(b) Nitrogen-fixing bacteria, such as *Rhizobium*, live inside root nodules of leguminous (bean family) plants. In anaerobic conditions, they use nitrogen reductase to reduce nitrogen gas to ammonium ions. Nitrogen-fixing bacteria are also found living freely in soil.

(c) Nitrification happens when chemoautotrophic bacteria in the soil absorb ammonium ions. They oxidise ammonium to nitrite (*Nitrosomonas* bacteria), or nitrite to nitrate (*Nitrobacter* bacteria).

(d) When denitrifying bacteria grow under anaerobic conditions (without oxygen), such as in water-logged soils, they use nitrate as a source of oxygen for aerobic respiration and produce nitrogen gas (N_2) and nitrous oxide (N_2O).

7 (a) (i) Where the rate of a natural process is affected by a number of factors, the limiting factor is the one whose magnitude limits/slows down the rate of the process.

(ii) The maximum population size that can be maintained over a period of time in a particular habitat.

(b)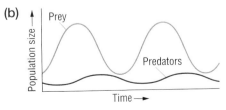

- When the predator population gets bigger, more prey are eaten.
- The prey population then gets smaller, leaving less food for the predators.
- With less food, fewer predators can survive interspecific competition and their population size reduces.
- With fewer predators, fewer prey are eaten, and their population size increases.
- With more prey, the predator population gets bigger, and the cycle starts again.

8 (a) Competition happens when resources (like food or water) are not present in adequate amounts to satisfy the needs of all the individuals who depend on those resources.

(b) Sometimes, interspecific competition could result in one population being much smaller than the other, with both population sizes remaining relatively constant. In other circumstances, one population would be out-competed and would die out

9 (a) Cut the stem of a deciduous tree close to the ground; once cut, several new shoots grow from the cut trunk, and eventually mature into stems of quite narrow diameter; after cutting, new shoots start to grow again, and the coppice cycle continues.

(b) Selective cutting – remove only the largest, most valuable trees; leave each section of the forest to mature for a long time before felling (50–100 years); control pests and pathogens; only plant particular tree species where they will grow well; position trees an optimal distance apart.

10 (a) Maintenance of biodiversity, including diversity between species, genetic diversity within species, and maintenance of a variety of habitats and ecosystems.

(b) *Max one from each. Ethics:* Every living thing has a right to survive; humans have a responsibility to look after living things; *Economic:* any example of direct economic value; any example of indirect economic value; *Social:* recreation.

(c) *Max two from each. Management:* Raise carrying capacity by providing extra food; move individuals to enlarge populations/ encourage natural dispersion of individuals between fragmented habitats by developing dispersal corridors of appropriate habitat; restrict dispersal of individuals by fencing; control predators and poachers; vaccinate individuals against disease; preserve habitats by preventing pollution or disruption; intervene to

restrict the progress of succession. *Reclamation:* Clean up pollution; remove unwanted species; recolonise with original species.

11 (a) Forests of *Scalesia* trees and shrubs have been almost eradicated on Santa Cruz and San Cristóbal to make way for agricultural land.

(b) The red quinine spreads rapidly, changing the highland ecosystem from low scrub and grassland to closed forest canopy. As a result, *Cacaotillo* shrub almost eradicated and Galapagos petrel nesting sites have been lost. Red quinine also out-competes native *Scalesia* trees.
The goat out-competes the giant tortoise for grazing, and changes the habitat to reduce the number of tortoise nesting sites.

(c) Depletion of shark populations; depletion of sea cucumber populations, which has had drastic effect on underwater ecology; giant tortoises over-harvested for food.

2.4 Responding to the environment

1 (a) needs no conscious thought;
(b) carried out in the same way by all individuals in a species;
(c) response/reflex, modified following exposure to new stimulus; response to stimulus associated with normal stimulus

2 (a) neurotransmitter; released from synaptic knob/ presynaptic membrane; attaches to receptors on postsynaptic membrane; causes sodium ion channels to open so action potential initiated in postsynaptic neurone;

(b) enzyme; in synapse; breaks down acetylcholine (into choline and ethanoate); so they can diffuse back into synaptic knob to be resynthesised

3 CNS consists of brain; and spinal cord; peripheral NS consists of somatic NS; and autonomic NS; somatic NS consist of sensory neurones; and motor neurones; autonomic NS consists of parasympathetic NS; and sympathetic NS

4 (a) depolarisation of sarcolemma; spreads down T tubules; calcium channels in sarcoplasmic reticulum open and calcium ions diffuse out; calcium ions bind to troponin; troponin moves tropomyosin; exposes myosin binding sites on actin filaments; myosin heads bind/cross bridges form; power stroke; release of ADP and P_i; actin filament move along so more overlap of actin and myosin filaments; cross bridges break; ATP hydrolysed for release of myosin heads

(b) glycogen hydrolysed to glucose; to be respired/named stage of respiration; aerobically or anaerobically; produce ATP; glycogen enters muscle cells form blood, under influence of insulin; can be stored as insoluble and does not upset osmotic balance of cell; branched structure enables much to be stored in small space; ends stick out so easy for enzyme to break off a few molecules of glucose as needed.

5 (a) (i) growth response of plants to a stimulus;

(ii) plant (shoots) grow towards, light/direction of stimulus;

(iii) plant (shoots) grow away from, pull of gravity/direction of stimulus;

(b) (i) roots are positively geotropic, so grow downwards in soil, which is where water is/anchorage;

(ii) shoots positively phototropic, so grow towards light source, obtain more light for photosynthesis

(c) plant hormones made in several places/types of plant tissue; animal hormones made in endocrine glands; animal hormones act on different tissue; plant hormones may act on different tissue or in cells where made; animal hormones travel in blood; plant hormones travel by diffusion/active transport/mass flow in xylem or phloem; both affect target cells/tissue; by fitting onto receptors, in cell membranes, that have complementary shape to the hormones

6 (a) promoted cell elongation; as makes cellulose cell walls more flexible so calls can take up more water; inhibits growth of side shoots;

(b) auxin moves away from illuminated side of shoot; active transport or facilitated diffusion involved; makes cells on side away from light source elongate, so shoot bends towards light;

(c) drop in production of cytokinins; fewer nutrients reach leaves and senescence begins; leads to drop in auxin production at tip of leaf; cells in abscission zone become more sensitive to ethene; drop in auxin concentration also leads to increase in ethene production; positive feedback effect; ethene causes increase in production of cellulase enzyme; digests walls of cells in abscission zone so leaf petiole falls away from stem

7

Area of brain	Functin
Cerebellum	Coordination of posture
Medulla oblongata	Control of heart rate
Hypothalamus	Control of temperature regulation
Cerebral hemisphere	Control of speech and voluntary movements
Visual association area in cerebrum	Interpreting a photograph

8 (a) (i) habituation;

(ii) no waste of energy responding to a stimulus that poses no threat; less stress;

(b) (i) at first not sure if stimulus poses a threat so run away; learn by association that stimulus is neutral so ignore it;

(ii) associate stimulus of sound of van with food; conditioned response/reflex; food is reward/reinforcer; use association centre in brain; involves learning

Glossary

α cells   Cells in the islets of Langerhans that release glucagon in response to low blood glucose levels.

Acetylcholine   A neurotransmitter (transmitter substance) found in cholinergic synapses.

Acetylcholinesterase   An enzyme in the synaptic cleft that breaks down the transmitter substance acetylcholine.

Actin   A protein found in muscle cells. It is the main component of the thin filaments.

Action potential   A brief reversal of the resting potential across the cell surface membrane of a neurone. All action potentials have a value of +40 mV.

Adenyl cyclase   The enzyme found inside cells, associated with hormone receptors, that converts ATP to cAMP.

Afferent   Incoming or leading towards.

Allele   An alternative version of a gene.

All or nothing   Refers to the fact that a neurone either conducts an action potential or it does not.

Allotransplantation   Transplantation of organs between individuals of the same species, for example transplantation of a human heart into another human.

Amplification (DNA)   The making of multiple copies of the same short section of DNA. The process of PCR is used in automatic amplification of DNA sections.

Anabolic steroids   Drugs that mimic the action of steroid hormones and increase muscle growth.

Anabolism   Type of metabolism: biochemical reactions that synthesise large molecules from smaller molecules. This requires energy/ATP.

Antagonistic   Working against each other in a pair.

Annealing   The term used to describe hydrogen-bond formation between complementary base pairs when sections of single-stranded DNA or RNA join together. Annealing is seen when complementary sticky ends join and where DNA probes attach to a complementary DNA section.

Antidiuretic hormone (ADH)   The hormone made in the hypothalamus and released from the pituitary gland that acts on the collecting ducts in the kidneys to increase the reabsorption of water into the blood.

Apical dominance   The growing apical bud at the tip of the shoot inhibits growth of lateral buds further down the shoot.

Apoptosis   Programmed cell death. An orderly process by which cells die after they have undergone the maximum number of divisions.

Ascending limb   The limb of the loop of Henle that carries fluid from the medulla towards the cortex of the kidney.

Asepsis   Literally means without contamination. In biotechnology, this refers to lack of contamination by foreign, unwanted microorganisms.

Aseptic techniques   Any techniques/manipulations of equipment or materials that are designed to prevent contamination by foreign and unwanted microorganisms.

Association area   A region of the cerebral cortex where the information in the form of impulses from sensory areas is made sense of by comparison with previous experience.

ATP   Molecule (nucleotide derivative) found in all living cells and involved in energy transfer. When it is hydrolysed energy is released.

ATP synthase   Enzyme associated with stalked particles in mitochondria and chloroplasts. It catalyses the joining of ADP and inorganic phosphate to make ATP.

Audus microburette   *See* photosynthometer.

Autonomic nervous system   The system of motor neurones that controls the non-conscious actions of the body. The autonomic system controls the actions of involuntary muscles and glands.

Autotroph   Organism that makes its own food using simple inorganic molecules, such as carbon dioxide and water, and energy. Photoautotrophs (plants, some protoctists and some bacteria) use light as the source of energy. Chemoautotrophs (some bacteria) use chemical energy.

β cells   Cells in the islets of Langerhans that release insulin in response to high blood glucose levels.

Basement membrane   A layer of connective tissue – mostly collagen – that holds an epithelium in place.

Batch culture   A culture of microorganisms that takes place in a single fermentation. Products are separated from the mixture at the end of the fermentation process.

Bilirubin   One of the waste products produced from breaking down haemoglobin.

Biodiversity   The number and variety of living things to be found in the world, an ecosystem or habitat.

Biofortified   Any food substance in which a particularly valuable nutrient is in higher than usual levels. Golden rice™ is biofortified with the accumulation of vitamin A.

Bioremediation   Use of microorganisms to remove waste products from a location or substance. The most important example is waste water (sewage) treatment.

Biotechnology   Use of microorganisms or biochemical reactions to generate useful products.

Bivalent   Pair of synapsed (joined) homologous chromosomes during prophase and metaphase of meiosis I.

Bowman's capsule   The cup-shaped end of a nephron tubule.

Callus   A mass of undifferentiated plant cells formed by meristem tissue extracted from the plant and grown in tissue culture.

Cardiovascular centre   Region in the medulla oblongata of the brain that controls heart rate.

Carrying capacity   The maximum population size that can be maintained over a period of time in a particular habitat.

Catabolism   Type of metabolism: biochemical reactions that produce small molecules by hydrolysis of larger molecules.

Cell metabolism   The result of all the chemical reactions taking place in the cell cytoplasm.

Central nervous system   The brain and spinal cord. It has overall control over the coordination of the nervous system.

Chemiosmosis   The flow of hydrogen ions (protons) through ATP synthase enzymes. The force of this flow allows the production of ATP. Occurs across the thylakoid membranes during the light-dependent stage of photosynthesis. Also occurs across the inner mitochondrial membrane during oxidative phosphorylation (in respiration).

Chemoautotrophs   *see autotrophs*.

Chiasmata (sing. chiasma)   The points where non-sister chromatids within a bivalent join, where they cross over.

Chi-squared (χ^2) test   Statistical test that can be carried out on data that are in categories. It enables the investigator to determine how closely an observed set of data corresponds to the expected data.

Chloroplasts   Organelles, in plant and some protoctist cells, where photosynthesis occurs.

Cholinergic synapse   A junction between neurones that uses acetylcholine as the neurotransmitter.

Chorionic gonadotrophin   A hormone released by the cells of an embryo.

Chromatogram   A chart produced when substances are separated by movement of a solvent along a permeable material such as paper or gel.

Chromosome mutation   Random change to the structure of a chromosome. There are different types: inversion (a section of chromosome turns through 180°); deletion (a part is lost); translocation (a piece of one chromosome becomes attached to another); non-disjunction (homologous chromosomes fail to separate properly at meiosis 1 or chromatids fail to separate at meiosis 2; if this happens to a whole set of chromosomes, polyploidy results). The shuffling of alleles in prophase 1 is **not** an example of mutation.

Clade   A monophyletic taxonomic group; that is, a single ancestor and all its descendants,

Cladistics   A method of classifying living organisms based on their evolutionary ancestry.

Classical conditioning   A form of learning in which two unrelated stimuli are applied to an animal, one a 'normal response' (for example salivation in the presence of food) another unrelated (for example the ringing of a bell). After repeated exposure to both stimuli together the animal will eventually respond with the normal response to the unrelated stimulus.

Closed culture   A culture of microorganisms set up in a reaction vessel and then allowed to grow without the addition of nutrients or the removal of products or wastes.

Codominant   A characteristic where both alleles contribute to the phenotype.

Coenzyme A   A coenzyme that carries acetate from the link reaction of respiration to Krebs cycle.

Coenzymes   Molecules that help enzymes carry out oxidation or reduction reactions. They work like shuttles, carrying atoms or molecules from one enzyme-controlled reaction to another. Many important coenzymes are involved in respiration and photosynthesis. In respiration, many coenzymes are concerned with removing hydrogen atoms from substrates.

Community   All the populations of different species that live in the same place at the same time, and who can interact with each other.

Comparative genome mapping   The comparison of DNA sequences coding for the production of proteins/polypeptides and regulatory sequences in the genomes of different organisms of different species. Comparisons include the search for sequences that make some organisms pathogenic whilst related organisms are not.

Competition   A struggle between individuals for resources (like food or water) that are not present in amounts adequate to satisfy the needs of all the individuals who depend on those resources.

Complementary genes   Genes that interact together to govern the expression of a single characteristic.

Conjugation (in bacteria)   Bacterial cells can join together and pass plasmid DNA from one bacterial cell to another. This process can take place between bacteria of different species and is of concern in terms of passing plasmid-located genes for antibiotic resistance.

Conservation   Maintenance of biodiversity, including diversity between species, genetic diversity within species, and maintenance of a variety of habitats and ecosystems.

Consumers   Living organisms that feed on other living organisms.

Continuous culture   A culture of microorganisms set up in a reaction vessel to which substrates are added and from which products are removed as the fermentation process continues.

Continuous variation   Genetic variation, also called quantitative variation, where there is a wide range of phenotypic variation within the population. There are no distinct categories. It is controlled by many genes. Examples include height in humans.

Coppicing   Cutting a tree trunk close to the ground to encourage new growth.

Co-transporter proteins   Proteins in the cell surface membrane that allow the facilitated diffusion of simple ions to be accompanied by transport of a larger molecule such as glucose.

Cotyledons   Food store in seeds of dicotyledonous plants. In some plants, these appear above the soil after germination and act as the first leaves.

Cross-bridge   In voluntary muscle, the joining of a myosin head group to an actin thin filament in the presence of calcium ions.

Crossing over   Where non-sister chromatids exchange alleles during prophase I of meiosis.

Cytokines   Cell-signalling molecules.

Deamination   The removal of the amine group from an amino acid to produce ammonia.

Decomposers   Organisms that feed on dead organic matter, releasing molecules, minerals and energy that then become available to other living organisms in that ecosystem.

Dehydrogenation   Removal of hydrogen atoms from a substrate molecule.

Depolarisation   The loss of polarisation across a membrane – when the membrane loses its resting potential.

Descending limb   The limb of the loop of Henle that carries fluid from the cortex towards the medulla of the kidney.

Detoxification   Conversion of toxic substances, such as alcohol, to less toxic substances.

Diabetes mellitus   A condition in which the patient is unable to control blood glucose levels.

Dialysis   Treatment for patients with kidney failure, in which metabolic wastes and excess salts and water are removed from the blood.

Dialysis fluid   The fluid used in dialysis; it consists of a complex solution that matches the composition of body fluids.

Dialysis membrane   A partially permeable membrane that separates the dialysis fluid from the patient's blood in a dialysis machine.

Diffusion   Movement of molecules down their concentration gradient. It may be through a partially permeable membrane.

Digest   Hydrolyse a large molecule to smaller molecules.

Diploid   Having two sets of chromosomes (eukaryotic cell or organism). Denoted by $2n$.

Discontinuous variation   Also called qualitative variation. Genetic variation where there are distinct phenotypic categories. Usually controlled by one gene. Examples include cystic fibrosis, shape of earlobes in humans and height in pea plants.

Distal convoluted tubule   The coiled portion of the nephron between the loop of Henle and the collecting duct.

DNA ligase   An enzyme capable of catalysing a condensation reaction between the phosphate group of one nucleotide and the sugar group of another. This results in DNA backbone molecules being joined together and is an essential part of recombinant DNA procedures.

DNA mutation   A change to the DNA structure. May be substitution of one base pair for another; inversion of a base triplet; deletion of a base pair or triplet of bases (on both strands); addition of a base pair or triplet of bases (on both strands); or a triple nucleotide repeat – a stutter.

Dominant   Characteristic in which the allele responsible is expressed in the phenotype even in those with heterozygous genotypes.

DRD4   Gene that codes for a dopamine receptor molecule.

Ecosystem   All the living organisms and all the non-living components in a specific habitat, and their interactions.

Ectotherms   Organisms that rely on external sources of heat and behavioural activities to regulate their body temperature.

Efferent   Outgoing or leading away from.

Electron acceptors   Chemicals that accept electrons from another compound. They are reduced while acting as oxidising agents.

Electron carriers   Molecules that transfer electrons.

Electrophoresis   A method used to separate molecules in a mixture based on their size. The method relies on the substances within the mixture having a charge. When a current is applied, charged molecules are attracted to the oppositely charged electrode. The smallest molecules travel fastest through the stationary phase (a gel-based medium) and in a fixed period of time will travel furthest, so the molecules separate out by size. The method is particularly important in separating DNA fragments of different sizes in DNA sequencing and profiling (fingerprinting) procedures.

Endocrine gland   A gland that secretes hormones directly into blood capillaries.

Endocytosis   The transport of large molecules or fluids into the cytoplasm of the cell, by the invagination (folding inwards) of the cell surface membrane to form a vesicle.

Endothelium   The tissue which lines the inside of a blood vessel or nephron.

Endotherms   Organisms that can control production and loss of heat to maintain their body temperature.

Energy   The ability to do work. From the Greek *energos*, meaning active work.

Envelope   Double membrane. Double lipid bilayer.

Environmental resistance   The combined action of biotic and abiotic factors that limits the growth of a population.

Epistasis   The interaction of genes concerned with the expression of one characteristic. One gene may mask the expression of another gene.

Epithelium   The tissue that covers the outside of a structure.

Eukaryotes   Organisms with eukaryotic cells – protoctists, fungi, plants and animals.

Evolution   The process of gradual change in the inherited traits passed from one generation to the next within a population. It results in the formation of new species.

Excretion   The removal of metabolic waste (waste from the reactions inside cells) from the body.

Exergonic   Chemical or biochemical reaction that releases heat energy.

Exocrine gland   A gland that secretes substances into a duct.

Exocytosis   A mechanism of secretion from a cell involving vesicles that fuse to the cell surface membrane and release their contents to the outside. It uses ATP.

Explant   A piece of tissue taken from a particular plant (which includes meristematic tissue) then sterilised in order to grow a callus in tissue culture micropropagation.

Facilitated diffusion   Diffusion that is enhanced by the action of protein channels or carriers in the cell membrane.

Fermentation   (1) The process of anaerobic respiration in microorganisms, used to yield specific products. For example, the anaerobic respiration of yeast is used in the fermentation of grapes to produce wine. (2) The process of culturing any microorganism in order to generate a specific product, either aerobically or anaerobically. All industrial biotechnological processes using whole microorganisms are referred to as fermentation.

Fertilisation   Fusion of male and female gamete nuclei.

Fight or flight response   The set of responses in an animal that accompany the perception of threat. The response is driven by the sympathetic nervous system and sets the body at a higher level of capacity to respond to the threat; for example increased respiration rate in muscles and increased blood flow to muscles to prepare for explosive muscle action necessary to fight or run away.

First messenger   A hormone that acts as a message in the bloodstream.

Gametes   Specialised sex cells. In many organisms the gametes are haploid and are produced by meiosis.

Gas chromatography   A technique used to separate substances in a gaseous state.

Gene   A length of DNA that codes for one (or more) polypeptides/proteins. Some genes code for RNA and regulate other genes.

Gene pool   Total genetic information possessed by the reproductive members within a population of organisms.

Gene therapy   In humans, any therapeutic technique where the functioning allele of a particular gene is placed in the cells of an individual lacking functioning alleles of that particular gene. Can be used to treat some recessive conditions but not dominant conditions such as Huntington disease.

Generator potential   A small depolarisation of the membrane in a receptor cell.

Genetic drift   Also called allelic drift. The change in allele frequency in a population, as some alleles pass to the next generation and some disappear. This causes some phenotypic traits to become rarer or more common.

Genetic engineering   The branch of biotechnology characterised by the obtaining of a particular gene, either by removal from a donor organism's genome using restriction enzymes or by manufacture, usually from mRNA transcript using reverse transcriptase enzyme. Once obtained, the gene is inserted into the genome of a recipient organism – often of a different species from the donor organism. The inserted gene is transcribed into protein, so giving the recipient organism a characteristic/capacity that it did not have previously. Such organisms are referred to as being transgenic or genetically modified.

Genetic fingerprinting (genetic profiling)   The use of DNA fragmentation and electrophoresis gives banding patterns that are unique to each individual. Samples of DNA, for example from crime scenes, are fragmented using a range of restriction enzymes, and, because each individual's DNA has differences, the number and size of fragments produced is slightly different. Electrophoresis and staining of the DNA gives a banded pattern that can be compared with other samples of DNA treated with the same set of restriction enzymes.

Genetic markers   Antibiotic resistance genes held on bacterial plasmids are used as genetic markers to identify the bacteria that have taken up the required gene. The gene is inserted into a plasmid that carries a resistance to a particular antibiotic. If a bacterium can grow on the particular antibiotic, then the plasmid, and so the required gene, is present in the bacterium.

Genetic variation   Variation of genetic information in a gene pool.

Genome   All the genetic information within an organism/cell.

Genome sequencing   The technique used to give the base sequence of DNA of a particular organism. The sequencing reaction can only identify up to around 1000 base pairs of sequence in a fragment. In order to sequence the whole genome, overlapping fragments are sequenced, then reassembled by computer software in order to generate the original sequence detail.

Genomics   The study of the whole set of genetic information in the form of the DNA base sequences that occur in the cells of organisms of a particular species.

Genotype   Alleles present within cells of an individual, for a particular trait/characteristic.

Germ line gene therapy   This involves placing the gene into embryonic cells. This technique is not currently legal and is deemed unethical.

Glomerulus   A small network of capillaries found inside the Bowman's capsule.

Glucagon   A hormone released by the α cells in the islets of Langerhans in the pancreas – it causes the blood glucose level to rise by converting glycogen in liver cells to glucose.

Glycerate-3-phosphate (GP)   Intermediate compound produced during the Calvin cycle in the light-independent stage of photosynthesis.

Glycolysis   Metabolic pathway. The first stage of respiration. It is anaerobic and occurs in the cytosol (cytoplasm). Although anaerobic, it involves oxidation as substrate molecules are dehydrogenated.

Golden rice™   A variety of rice that is genetically engineered to carry large amounts of the vitamin A precursor beta-carotene. The rice appears golden brown, unlike its non-engineered relative, which is white in colour.

Grana (sing. granum)   Stacks of thylakoid membranes, found in a chloroplast.

Habitat   The place where an organism or population of organisms lives.

Haemodialysis   A form of treatment for kidney patients in which blood is taken from a vein and passed through a dialysis machine so that exchange can occur across an artificial partially permeable membrane.

Hairpin countercurrent multiplier   An arrangement of a tubule or blood vessel involving a 180° bend so that the fluid in one end of the tubule flows back past the fluid at the other end. This arrangement facilitates the exchange of materials by ensuring that there is a concentration gradient all along the tubule.

Half-life   The time taken for the concentration of a substance to drop to half its original value.

Haploid   Eukaryotic cell or organism having only one set of chromosomes. Denoted by n.

Hardy–Weinberg principle   The concept that both genotype frequencies and gene frequencies will stay constant from generation to generation, within a large interbreeding population where mating is random, there is no mutation and no selection or migration.

Hayflick constant   The number of times that a normal body cell divides before undergoing apoptosis. The number of divisions is about 50.

Hemizygous   Cell or individual having only one allele for a particular gene.

Hepatic portal vein   An unusual blood vessel that has capillaries at both ends – it carries blood from the digestive system to the liver.

Hepatocytes   Liver cells.

Heterotroph   Organism that gains its nutrients from complex organic molecules. It digests them to simpler, soluble molecules and then respires some of them to obtain energy, or uses the products of digestion to synthesise the organic molecules it needs. Heterotrophs are consumers in food chains. Parasites and saprotrophs are also heterotrophs. Animals, some bacteria and some protoctists are heterotrophs.

Heterozygous   Eukaryotic cell or organism that has two different alleles for a specific gene.

Hierarchy (social)   Within a group individuals have a place in the order of importance within the group. This is often shown by individuals higher up in the hierarchy receiving more food or having rights of access to mate with other individuals.

Homeobox genes   Genes that control the development of the body plan of an organism.

Homeostasis   The maintenance of a constant internal environment despite external changes.

Homeotic selector genes   These direct the development of individual body segments. They are master genes that control other regulatory genes.

Homozygous   Eukaryotic cell or organism that has two identical alleles for a specific gene.

Hormone   A molecule released into the blood that acts as a chemical messenger.

Hox clusters   Groups of homeobox genes. More complex organisms have more Hox clusters. This is probably due to a mutation that duplicated the Hox clusters.

Human chorionic gonadotrophin (hCG) A hormone released by the human embryo. Its presence in a pregnant woman's urine can be detected to confirm pregnancy.

Hydrolysis Splitting of large molecules into smaller molecules with addition of water.

Hyperglycaemia A high blood glucose concentration.

Hyperpolarised The condition of a membrane that is more highly polarised than the usual resting state. The resting potential is lower than usual.

Hypertension A condition in which the resting blood pressure (particularly the diastolic pressure) is raised for prolonged periods.

Hypoglycaemia A low blood glucose concentration.

Hypostasis Where two alleles interact to control the expression of one characteristic one is epistatic and one is hypostatic. Where a homozygous recessive allele at the first locus (place on a chromosome) prevents the expression of another allele at a second locus, the alleles at the first locus are epistatic and the alleles at the second locus are hypostatic.

Hypothalamus A portion of the brain that contains various receptors that monitor the blood. Also involved in controlling the autonomic nervous system.

Innate behaviour A behaviour that an animal is capable of from birth without any learning or practice. Such behaviours appear to be very inflexible in their operation although they may often be slightly modified in individuals by some elements of learning.

Insulin A hormone released by the β cells in the islets of Langerhans in the pancreas – it causes the blood glucose level to fall.

Islets of Langerhans Patches of endocrine tissue in the pancreas – they consist of α and β cells.

Isolating mechanism Mechanism that divides populations of organisms into subgroups.

Kangaroo mother care The term used to describe a method of human infant care which involves extended skin-to-skin contact and breastfeeding on demand.

Krebs cycle Third stage of respiration. It is aerobic and in eukaryotes it occurs in the matrix of the mitochondria.

Kupffer cells Specialised macrophages that move around in the sinusoids and are involved in the breakdown and recycling of old red blood cells.

Lamellae A pair of membranes that contain chlorophyll. Intergranal lamellae in the chloroplasts link the thylakoids of one granum with the thylakoids of another granum.

Light-dependent stage First stage of photosynthesis.

Occurs in the thylakoid membranes of the chloroplasts. It involves using light energy to make ATP. Other products are reduced NADP and oxygen.

Light-independent stage Second stage of photosynthesis. Occurs in the stroma of the chloroplasts. Involves using ATP, reduced NADP and carbon dioxide to make organic molecules.

Light intensity A measure of the amount of energy associated with light. The relative light intensity of a source can be calculated using the formula $I = 1/d^2$ where d is the distance between source and object receiving the light.

Limiting factor A variable that limits the rate of a particular process. If the factor is increased then the process will take place at a faster rate. Where the rate of a natural process is affected by a number of factors, the limiting factor is the one whose magnitude limits the rate of the process.

Linkage Genes for different characteristics that are present at different loci on the same chromosome are linked.

Link reaction Stage of aerobic respiration that links glycolysis with the Krebs cycle. In eukaryote cells it occurs in the mitochondrial matrix.

Local currents Movements of ions along a neurone close to the cell surface membrane, caused by influx or efflux of ions through the membrane.

Locus Specific position on a chromosome, occupied by a specific gene.

Maternal chromosome Member of a homologous pair of chromosomes that originally came from the female gamete.

Medulla oblongata A portion of the brain (the brain stem) that contains centres for the control of various unconscious bodily functions and via the autonomic nervous system.

Meiosis Type of nuclear division. A reduction division. The chromosome number is halved. It involves two divisions. It produces cells that are genetically different from each other and from the parent cell.

Meristem Growth points in a plant where immature cells are still capable of dividing.

Metabolic waste Waste substances that may be toxic or are produced in excess by the chemical reactions inside cells.

Micropropagation A form of artificial vegetative propagation using sterile explant tissue grown to form a callus culture from which many new plants are grown by separation and growth of small parts of the callus. Particularly useful in generating vast numbers of genetically identical plants following the genetic engineering of a particular gene into the callus.

Microvilli Microscopic folds of the cell surface membrane that increase the surface area of the cell.

Monoclonal antibodies   Antibodies that are identical because they have been produced by cells that are clones of one original cell.

Monogenic   Characteristic coded for by one gene.

Monophyletic   A monophyletic group is one that includes an ancestral organism and all its descendent species.

Morphogen   A substance that controls the pattern of tissue development. It is produced in a particular region of a developing organism. It diffuses to other cells, which then enter a specific developmental pathway.

Motor area   An area of the cerebral cortex within which the neurones are responsible for driving motor functions.

Motor unit   Refers to the innervation of a cluster of muscle fibres by a single motor neurone. The number of muscle fibres within a motor unit is governed by the level of manipulation required in the muscle. Muscles responsible for very fine motor functions have as few as 3 muscle fibres in a motor unit. Muscles requiring less fine motor control may have over 200 muscle fibres in a motor unit.

Mutation   Structural change to genetic material – either to a gene or to a chromosome.

Mutualism   A relationship between two organisms from which both benefit.

Myelin   A fatty sheath around a neurone that consists of many layers of the plasma membranes of Schwann cells.

Myogenic   Contraction of the muscle is generated from within the muscle itself. The term is used to describe the contraction of the heart, which is controlled by the action of the sinoatrial node.

Myosin   The protein that forms the thick filament in muscle cells. This protein has head groups that form the cross-bridges associated with muscular contraction.

NAD   Coenzyme involved in respiration. It removes hydrogen atoms from substrates. It becomes reduced NAD, which carries hydrogen atoms (protons and electrons).

NADP   Coenzyme involved in photosynthesis. It accepts hydrogen atoms from photolysis of water during the light-dependent stage and carries them to the light-independent stage.

Natural selection   Mechanism for evolution. Organisms that are well adapted to their environment are more likely to survive and reproduce, passing on the alleles for the favourable characteristics.

Necrosis   Disorderly, often accidental cell death.

Negative feedback   A process in which any change in a parameter brings about the reversal of that change so the parameter is kept fairly constant.

Nephrons   Tubules in the kidney that are used to produce urine.

Neurosecretory cells   Cells in the hypothalamus that are similar to neurones but release a hormone into the blood instead of a transmitter substance into a synapse.

Neurotransmitter   A transmitter substance – a chemical that is released from the presynaptic membrane of one neurone to pass a signal to another neurone.

Niche   The role that a species plays in an ecosystem.

Nitrogen fixation   Conversion of nitrogen gas into a form which is usable by plants, such as nitrate or ammonium ions.

Non-disjunction   Failure of members of a homologous pair of chromosomes, or of a pair of chromatids, to separate during nuclear division.

Non-reproductive cloning   Also known as therapeutic cloning. The use of stem cells in order to generate replacement cells, tissue or organs, which may be used to treat particular diseases or conditions of humans. For example, the use of stem cells to generate replacement heart cells in patients suffering from myocardial infarction (heart attack).

Operant conditioning   Also known as trial-and-error learning. The term is used to describe learning that takes place in animals given punishment or reward to reinforce the performance of a particular operation. Most famously, this type of learning is seen in rats and pigeons in a 'Skinner box', where operation of a lever rewards the animal with a food pellet.

Operon   A unit consisting of genes that work together under the control of an operator gene. An example is the *lac* operon, which consists of two structural genes and an operator gene. Operons were first discovered in prokaryotes but later found in eukaryotes.

Organelles   Structures within cells. Each carries out a specific function.

Ornithine cycle   A process that occurs inside liver cells to convert ammonia to urea.

Osmoreceptors   Receptor cells that monitor the water potential of the blood and detect any changes.

Osmoregulation   The control and regulation of the water potential of the blood and body fluids.

Oxidation   Chemical reaction involving loss of electrons, gain of oxygen or loss of hydrogen atoms.

Oxidative phosphorylation   The formation of ATP, in the presence of oxygen, by chemiosmosis.

Oxytocin   A hormone released by the posterior pituitary gland to facilitate birth and breastfeeding.

Pancreas   A small organ in the abdomen that secretes digestive fluids and hormones.

Pancreatic duct   A duct leading from the pancreas to carry digestive juices to the small intestine.

Paraphyletic group   A classification group for living organisms that includes the most recent ancestor but not all of the descendants.

Partially permeable membrane   A membrane that is permeable to certain substances, such as water, but is not permeable to other substances.

Paternal chromosome   Member of a pair of homologous chromosomes that originally came from the male gamete.

Peripheral nervous system   The sensory and motor neurones connecting the central nervous system to the sensors and effectors around the body.

Peritoneal dialysis   A form of treatment for kidney patients in which dialysis fluid is pumped into the body cavity so that exchange can occur across the peritoneal membrane.

Phagocytosis   Endocytosis of large solid molecules into a cell.

Phenotype   Observable characteristics of an organism.

Photoautotroph   *see* Autotroph.

Photolysis   Enzyme-catalysed reaction where water molecules are split, using light energy. It occurs in photosystem II, during the light-dependent stage of photosynthesis.

Photophosphorylation   Formation of ATP in the presence of light energy. It takes place in thylakoid membranes of chloroplasts, during the light-dependent stage of photosynthesis. There are two types: cyclic photophosphorylation and non-cyclic photophosphorylation.

Photosynthetic pigments   Chemicals that absorb light energy. Found in thylakoid membranes, in photosystems. Each pigment absorbs energy associated with light of a specific wavelength. Examples include chlorophyll a, chlorophyll b, carotenoids and xanthophylls.

Photosynthometer   Apparatus to measure rate of photosynthesis by collecting and measuring the volume of oxygen produced in a certain time.

Photosystems   Group of photosynthetic pigments in the thylakoid membrane. Consists of a primary reaction centre and accessory pigments.

Phylogenetic group   Group of organisms that share evolutionary ancestry.

Podocyte   A specialised cell that makes up the lining (endothelium) of the Bowman's capsule. Podocytes have finger-like processes. They aid ultrafilitration as fluid entering the renal capsule from the blood goes through the gaps in these processes, making ultrafiltration more efficient as the podocytes do not provide a barrier to filtration.

Polarised   Membrane with a potential difference across it.

Polygenic   Characteristic coded for by many genes. Examples include height and intelligence in humans. Polygenic characteristics are more influenced by environmental factors than are monogenic characteristics.

Polypeptide   Large polymer molecule made of many amino acids joined by peptide bonds.

Polyploid   Eukaryotic organisms or cell with more than two sets of chromosomes.

Population   All of the organisms of one species, who live in the same place at the same time, and who can breed together.

Population genetics   The study of the gene pools and the allele and genotype frequencies of populations of organisms.

Positive feedback   A process in which any change in a parameter brings about an increase in that change.

Posterior pituitary gland   The hind part of the pituitary gland, which releases ADH.

Power stroke   The term describes the action of the myosin head in muscular contraction. The head group attached to the actin filament bends, pulling the thick filament to overlap further with the thin filament. Energy from ATP is used up in the power stroke.

Precursor   Literally means 'coming before'. In biology, a precursor molecule is one which is used in order to form another more useful molecule. For example, beta-carotene is the precursor molecule for vitamin A.

Primary metabolite   Any metabolite which is formed as part of the normal growth of a microorganism. During growth the lipids, proteins, carbohydrates and waste products generated by the microorganism in order to grow in numbers are described as primary metabolites.

Primary pigment reaction centre   The primary photosynthetic pigments in a photosystem act as reaction centres. In PSI the primary pigment reaction centre is a molecule of chlorophyll a that has a peak absorption of 680 nm. This means that its greatest absorption is of light with a wavelength of 680 nm. In PSII the primary pigment reaction centre is a molecule of another type of chlorophyll a that has an absorption peak of 700 nm.

Primer (DNA)   Short single-stranded sequences of DNA, around 10 bases in length. They are needed, in sequencing reactions and polymerase chain reactions, to bind to a section of DNA because the DNA polymerase enzymes cannot bind directly to single-stranded DNA fragments.

Producers   Autotrophic organisms (plants, some protoctists and some bacteria) that convert light energy to chemical energy, which they then supply to consumers.

Productivity   Primary productivity is the rate of production of new biomass by producers. It is the energy captured by their chlorophyll and used to synthesise organic molecules. This minus the energy released via their respiration is the net primary productivity – the energy available to heterotrophs through consumption of producers' biomass.

Programmed cell death *see* apoptosis.

Protein A macromolecule. A polymer of many amino acids joined by peptide bonds. May also be called a polypeptide.

Protoctist Eukaryotic organism classified as belonging to the kingdom Protoctista. This kingdom includes organisms that do not fit into/cannot be classified as belonging to the other four kingdoms. It includes algae, protozoa and slime moulds. Some members of this phylum are photosynthetic. Some have undulipodia and some have cilia.

Proton motive force Force produced as hydrogen ions flow, through ATP synthase channels, down their concentration gradient. The force causes ADP and P_i to combine and form ATP.

Proto-oncogene Gene that can undergo mutations to become an oncogene, which induces tumour formation (cancer).

Quadrat A square frame used for sampling in fieldwork.

Recessive Characteristic in which the allele responsible is only expressed in the phenotype if there is no dominant allele present.

Recombinant DNA A section of DNA, often in the form of a plasmid, which is formed by joining DNA sections from two different sources.

Reduction Chemical reaction involving the gain of electrons, gain of hydrogen atoms or loss of oxygen atoms.

Refractory period The short period of time after firing during which it is more difficult to stimulate a neurone.

Replica plating The process of growing bacteria on an agar plate, then transferring a replica of that growth to other plates using a sterile velvet pad. The replica plates usually contain different antibiotics. Analysis of growth patterns on the replica plates gives information about the genetic properties of the growing bacteria.

Response The reaction to a stimulus.

Resting potential The potential difference or voltage across the neurone cell membrane while the neurone is at rest.

Restriction enzyme An enzyme originally derived from bacteria, in which it has a role in defence against infection by viruses. The enzymes catalyse a hydrolysis reaction that breaks the phosphate–sugar backbone of the DNA double helix. The two backbones are usually broken at slightly different points on the restriction site, leaving a staggered cut known as a sticky end. The restriction site for each restriction enzyme is unique.

Restriction site The specific location on a stretch of DNA which is the target site of a restriction enzyme. Restriction sites are around eight bases long.

Reverse transcriptase An enzyme originally derived from retroviruses. The enzyme catalyses the construction of a DNA strand using an mRNA strand as a template. Effectively the reverse of transcription.

Ribulose bisphosphate (RuBP) 5-carbon compound, present in small amounts in stroma of chloroplasts. It is a carbon dioxide acceptor. It is regenerated from triose phosphate.

Ribulose bisphosphate carboxylase (rubisco) Enzyme that catalyses the carboxylation (addition of carbon dioxide) to ribulose bisphosphate.

RNA interference Describes the process in which short fragments of single-stranded RNA bind to complementary regions on mRNA molecules and in doing so form sections of double-stranded mRNA which cannot be translated. The binding of the interfering RNA section often triggers cellular destruction of the mRNA. The term gene silencing is used to mean that RNA interference can prevent the formation of the product of a particular gene within a cell/organism. RNA interference has been shown to operate in natural regulation of gene expression as well as in the targeted silencing of genes in the laboratory.

Saltatory Refers to way in which the action potential appears to jump from node to node.

Saprotrophs Organisms (like bacteria and fungi) that feed by secreting enzymes onto food, and absorbing digested nutrients across their outer walls.

Sarcomere In voluntary muscle, the span between one Z-line and the next Z-line. The Z-line is the central part of the I band, which alternates with the A band. The sarcomere is the smallest unit of contraction of voluntary muscle, consisting of the thick and thin filaments responsible for muscular contraction.

Second messenger A chemical inside the cell released in response to a hormone binding to the cell surface membrane, e.g. cAMP.

Secondary metabolite A metabolite produced by a microorganism, usually in the latter stages of growth as the culture ages. Secondary metabolites are not specifically required for the organism to grow. They usually have antibiotic properties.

Segmentation genes Genes that control the development of polarity (which end is head and which end is tail) in organisms.

Selection pressure Environmental factor that confers greater chances of surviving and reproducing on some members of the population than on others.

Selective reabsorption   The absorption of certain selected molecules back into the blood from the fluid in the nephron tubule.

Sensory area   An area of the cerebral cortex within which the neurones associated with receiving sensory information from the receptors are found. These neurones often pass information to association areas in order to make sense of the incoming information.

Sex linkage   Gene with its locus on one of the sex chromosomes, X or Y. As there are few genes on the Y chromosome, in humans, most sex-linked genes are on the X chromosome. However, there are some genes on the Y chromosome, notably the gene SRy that stimulates development of the testes and subsequent development of the embryo into a male. There is also a Y STR (short tandem repeat on the Y chromosome) used in genealogy DNA testing. (Note that in some organisms it is not the presence of a Y chromosome that controls development into a male. For example in *Drosophila* it is the number of X chromosomes, 1 for male and 2 for female. In turtles, sex is determined by incubation temperature of the eggs.) In birds, butterflies and moths, males are XX (or ZZ) and females are XY (ZW). In grasshoppers and crickets females are XX and males are XO (just one X chromosome). In bees and wasps, diploid individuals are female and haploid individuals are male. Earthworms and some snails are hermaphrodite (have both male and female anatomies). Some organisms, such as oysters and some fish, can change sex during their life cycle.

Sexual reproduction   Production of new organisms involving fusion of nuclei from male and female gametes, usually from unrelated individuals. Increases genetic variation in the population.

Sinoatrial node (SAN) or sinus node   The region of the heart right atrial muscle wall, about 3 mm wide, 15 mm long and 1 mm deep, which consists of specialised muscle fibres (cells) that have no contractile filaments but connect directly to atrial muscle fibres. Any excitation (electrical activity) starting in the SAN spreads immediately to the rest of the atrial wall. Hence the pacemaker controls the synchronised rate of beating of the whole heart. (Artificial pacemakers are battery-powered devices, usually inserted under the skin and connected via wires in the subclavian vein and vena cava to the heart muscle.)

Social behaviour   Behaviour of organisms of a particular species living together in groups with relatively defined roles for each member of the group.

Sodium–potassium pumps   Protein carriers embedded in the membranes of some cells, which use energy from ATP to move sodium ions and potassium ions in opposite directions simultaneously, against their concentration gradients. They are chemically gated ion channels.

Somatic cell gene therapy   Involves the placing of the gene in adult differentiated cells. Examples include the placing of CFTR genes into the respiratory system cells of individuals with cystic fibrosis.

Somatic cell nuclear transfer   The technique of cloning organisms involving the removal of the nucleus from an adult, differentiated cell, which is then placed into the enucleated egg cell taken from a donor organism. The cell formed is placed into a surrogate mother in order to develop. The resulting organism is a clone of the organism which provided the adult, differentiated cell nucleus.

Species   The **biological species** concept is a group of similar organisms that can interbreed and produce fertile offspring. The **phylogenetic species** concept is a group of organisms with similar morphology, physiology, embryology and behaviour, and that all occupy the same ecological niche.

Stabilising selection   A type of natural selection in which the allele and genotype frequency within populations stays the same because the organisms are already well adapted to their environment.

Stem cells   Undifferentiated cells. Embryonic stem cells are totipotent and are able to differentiate into any type of specialised cell found in organisms of that species. Umbilical stem cells and adult stem cells may become specialised into a more limited range of cell types.

Stimulus   Any change in the environment of an organism that causes a response.

Stomata (sing. stoma)   Pores between guard cells in the epidermis of leaves.

Stroma   Fluid-filled matrix of chloroplasts. This is where the light-independent stage of photosynthesis takes place.

Substrate-level phosphorylation   Formation of ATP from ADP and P_i during glycolysis and the Krebs cycle.

Succession   A directional change in a community of organisms over time.

Summation   The way that several small potential changes can combine to produce one larger change in potential difference across a neurone membrane.

Synaptic knob   The swelling at the end of a neurone where it forms a junction (synapse) with another neurone.

Synovial joint   A type of joint in the skeleton characterised by the presence of a synovial membrane that produces synovial fluid to lubricate the joint. Synovial joints are found where a large movement range is required, such as the elbow and the hip.

Target cells   Cells that have receptors embedded in the plasma membrane that are complementary in shape to specific hormone molecules. Only these cells will respond to that specific hormone.

Taxon (pl. taxa) Group of organisms used in a hierarchical classification. Examples are Kingdom, Phylum, Class, Order, Family.

Threshold potential A potential difference (usually −50 mV) across the membrane. If the depolarisation of the membrane does not reach the threshold potential then no action potential is created. If the depolarisation reaches the threshold potential then more sodium ion channels open and an action potential is created.

Thylakoid Inner membrane in chloroplast. Site of photosystems and ATP synthase.

Tissue A group of cells, with a common origin and similar structures, which performs a particular function; for example, blood, bone, epithelium, muscle, nervous tissue, xylem and phloem.

Tissue culture Also called micropropagation. The cloning of isolated cells or small pieces of plant tissue in special culture solutions, under controlled aseptic conditions.

Totipotent stem cells Stem cells that can differentiate into any type of specialised cells found in organisms of that species.

Transcription The formation of an RNA molecule, using a length of DNA as a template. Complementary base pairing is used. The enzyme RNA polymerase catalyses the reaction.

Transect A line taken through a habitat, which helps with systematic sampling of changes across a habitat.

Transformation Bacteria that take up DNA from their surroundings (e.g. from dead bacteria) are transformed.

Translation Stage of protein/polypeptide synthesis in which the amino acids are assembled at ribosomes. The order in which the amino acids are joined together, by peptide bonds, is determined by the sequence of codons on the mRNA, which is itself determined by the sequence of nucleotide triplets on the coding strand of a length of DNA (gene). The genetic code is translated.

Triose phosphate (TP) 3-carbon compound formed when a molecule of glycerate phosphate is reduced, during the Calvin cycle in the light-independent stage of photosynthesis.

Trophic level The level at which an organism feeds in a food chain.

Tropism A directional growth response in which the direction of the response is determined by the direction of the external stimulus.

Ultrafiltration Filtration at the molecular level in the glomerulus of kidneys. Some molecules are filtered out of the blood of the glomerulus into the renal capsule. Molecules with relative molecular masses above 69 000 are retained in the blood capillaries.

Urea An excretory product formed from the breakdown of excess amino acids.

Vector Carrier. In DNA technology, refers to the agent that carries a piece of DNA from one cell into another, e.g. a bacterial plasmid.

Vegetative propagation Asexual reproduction in plants making use of specialised vegetative structures that grow to form new and separate individual organisms.

Voltage-gated channels Channels in plasma membranes that allow the passage of ions. They respond to changes in potential difference (voltage) across a membrane and, as a result, open or close.

Xenotransplantation The transplantation of cells or organs from one species into the body of an organism of another species.

Zygote Cell formed, during sexual reproduction, from the fusion of two gametes.

Index

Your Exam Café CD-ROM

In the back of this book you will find an Exam Café CD-ROM. This CD contains advice on study skills, interactive questions to test your learning, a link to our unique partnership with New Scientist, and many more useful features. Load it onto your computer to take a closer look.

Amongst the files on the CD are PDF files, for which you will need the Adobe Reader program, and editable Microsoft Word documents for you to alter and print off if you wish.

Minimum system requirements:
- Windows 2000, XP Pro or Vista
- Internet Explorer 6 or Firefox 2.0
- Flash Player 8 or higher plug-in
- Pentium III 900 MHz with 256 Mb RAM

To run your Exam Café CD, insert it into the CD drive of your computer. It should start automatically; if not, please go to My Computer (Computer on Vista), click on the CD drive and double-click on 'start.html'.

If you have difficulties running the CD, or if your copy is not there, please contact the helpdesk number given below.

Software support
For further software support between the hours of 8.30–5.00 (Mon-Fri), please contact:
Tel: 01865 888108
Fax: 01865 314091
Email: software.enquiries@pearson.com

ASKHAM BRYAN
COLLEGE
LEARNING RESOURCES